Foundations of
Corporate Law

Interdisciplinary Readers In Law
ROBERTA ROMANO, *General Editor*

Foundations of Corporate Law
ROBERTA ROMANO

Foundations of Tort Law
SAUL LEVMORE

Foundations of Adminstrative Law
PETER SCHUCK

Foundations of Contract Law
RICHARD CRASWELL AND ALAN SCHWARTZ

Foundations of Corporate Law

ROBERTA ROMANO

New York   Oxford
Oxford University Press
1993

Oxford University Press

Oxford   New York   Toronto
Delhi   Bombay   Calcutta   Madras   Karachi
Kuala Lumpur   Singapore   Hong Kong   Tokyo
Nairobi   Dar es Salaam   Cape Town
Melbourne   Auckland   Madrid

and associated companies in
Berlin   Ibadan

Library of Congress Cataloging-in-Publication Data
Foundations of corporate law / [edited by] Roberta Romano
p. cm.—(Interdisciplinary readers in law)
Includes bibliographical references.
ISBN 0-19-507412-2
ISBN 0-19-507413-0 (pbk)
1. Corporation law—United States.   2. Corporate governance—
United States.   I. Romano, Roberta.   II. Series.
KF1414.F68   1993   346.73'066—dc20
[347.30666]   92-34550

9 8 7 6 5 4 3 2 1

Printed in the United States of America
on acid-free paper

Preface

Corporate law underwent a revolution over the past decade. In the midst of an extraordinary period of innovation in business organization and acquisitive activity, legal scholarship was transformed by the use of the new analytical apparatus of the economics of organization and modern corporate finance. This learning has already had, and will increasingly have, a profound impact on corporate practice and, accordingly, on the teaching of corporate law. This book of readings seeks to provide an accessible introduction to the enduring policy debates in corporate law as well as the intuition for the fundamental economic concepts of the new learning that informs the debates. In addition, a concerted effort has been made to provide a realistic sense of the institutional landscape, which is foreign to many students, by extensive referencing of the burgeoning empirical research on corporate governance.

The key feature of the public corporation is Adolph Berle and Gardiner Means's insight concerning the separation of ownership and control: managers of the firm, who run the business, are not the owners. This separation creates a host of organizational problems, because managers' incentives are not always aligned with the owners' interest; such problems are generically referred to as agency problems. Much of corporate law is directed at mitigating agency problems, as selections in the reader illustrate. The readings also indicate how the economic theory of organization as well as corporate finance clarify different facets of the agency problem and suggest ways of mobilizing the legal system to address this master problem.

A word on the format is in order. I have used materials in this reader to supplement casebooks in my courses in corporate law and corporate finance. The reader was crafted with the intention that it be used as a springboard for class discussion in a corporations course, but there are comprehensive notes and questions to ensure that it is sufficiently self-contained for independent, self-directed use. All of the selections have been extensively edited to facilitate accessibility. Mathematics appearing in original works has been suppressed, although simple numerical examples have been retained or included in the notes to illustrate concepts. A danger with such an approach is that complexities of the literature can easily be lost, and this may convey the misimpression that there is no ambiguity to policymaking. I have sought to temper this risk by juxtaposing sharply differing positions in the selections or accompanying notes. In addition, references and most footnotes have been omitted from excerpts. Precision and bibliographic convenience have been sacrificed for the pedagogic benefit of greater readability. Readers who are sufficiently intrigued by an excerpt can follow up on arguments and references by recourse to the original source.

This book would not have been completed without the superb assistance of Cathy Briganti and the unflagging encouragement and support of Albert Romano. I cannot begin to thank them; I can only end by acknowledging that fact.

New Haven   R.R.
October 1992

Contents

Foundations of
Corporate Law

Theory of the Firm and Capital Markets

The readings in this chapter provide a framework for understanding corporate law. The selections in part A on the theory of the firm offer explanations of why individuals organize their economic activity into firms and why certain institutional arrangements are so prevalent. They also suggest that the markets in which firms operate affect their organizational structures. A critical market for public corporations is the capital market. This is because one of the explanations for organizing a business as a corporation is its greater accessibility to capital, which, as discussed in Chapter II, is facilitated by corporate characteristics of free transferability of shares and limited liability. The selections in part B introduce the building-block concepts of modern corporate finance that are prerequisites for understanding the operation of capital markets.

In neoclassical economics, the firm is a black box, represented by a production function. Although firms have an objective, profit maximization, the neoclassical approach focuses on aggregate firm behavior (markets) rather than the individual firm. The readings in this chapter on the theory of the firm, however, take a more microanalytical approach, delving inside the firm and differentiating the players and their interests.

Michael Jensen and William Meckling's analysis of the firm proceeds from the key organizational problem of the modern corporation identified by Adolph Berle and Gardiner Means over sixty years ago, the separation of ownership and control: ownership rights (stock) are not held by the individuals who manage the corporation. Berle and Means, *The Modern Corporation*

3

and Private Property (New York: Macmillan, 1932). This separation creates a potential for the divergence of shareholder and manager interests, which is referred to as a principal-agent problem. Firms are, therefore, not necessarily profit maximizers. As Jensen and Meckling explain, managers will seek to maximize their own utility (satisfaction, happiness, pleasure), which will not be the same as maximizing the firm's profits because managers obtain non-pecuniary benefits (such as on-the-job perquisites of expensive office furnishings), which are neither convertible to cash nor available to nonmanaging owners. In addition, managers may not expend maximum effort working on their job (that is, shirking is a nonpecuniary benefit).

Jensen and Meckling emphasize the organizational consequences of the agency problem: time and effort must be expended in fashioning institutions that align managers' incentives with owners' preferences. These expenditures are referred to as agency costs. Many features of corporate and securities laws fall into this cost category, such as audited financial statements. The separation of ownership and control does not imply that managers can operate firms in complete disregard of shareholder interests. Rather, institutions of corporate law, which are the focus of Chapter V, such as the board of directors, are devised to mitigate the agency problem. In addition, the many markets in which firms operate—product, labor, capital, and corporate control markets—constrain agency costs. For example, if managers shirk or consume excessive nonpecuniary benefits, the firm's production costs will rise and it will compete less effectively in its product market. This raises the cost of capital and places management at risk, for executives of poorly performing firms experience greater than normal turnover in their positions (see Chapter V, part A, note 5 and Chapter VI, part A, note 1). The role of the market for corporate control in reducing agency costs is explored in Chapter VI (see in particular the selections by Henry Manne and Michael Jensen).

Oliver Williamson focuses on complications for exchange that arise from uncertainty and two features of the human condition: opportunism and bounded rationality. Opportunism encompasses the possibility that individuals will act strategically to further their own interests at the expense of others. Bounded rationality refers both to limits on the information available to individuals when entering exchange relations and to limits on their ability to process information. Firms become necessary with the addition of another key variable, transaction-specific assets, assets necessary for an exchange transaction that are nonredeployable (that is, their value in the next-best use is significantly lower than their value in the specified exchange transaction). The owner of such an asset is vulnerable to exploitation: by realizing that the asset owner will suffer a loss if she does not transact with a particular party, that party can behave opportunistically and raise the transacting price. Williamson contends that in such a situation, simple contracts and markets will not do, and protective institutional arrangements, termed governance structures, are necessary to facilitate exchange. The corporation is one such device.

A real-world example should aid in understanding Williamson's analysis. Consider the owner of an aluminum refining plant located near a bauxite mine (the ore from which aluminum is produced). This plant is most valuable (production costs are lowest) if ore is purchased from the neighboring mine. Ore obtained from a more distant mine will be considerably more expensive because it must be transported: the refining process requires a large quantity of ore to extract a small quantity of aluminum and bauxite is a bulky product to ship. Consequently, once the refiner locates the plant near a specific mine, the mine owner can raise her price for ore (up to the refiner's costs of transporting ore from a more distant mine). The refining plant is a transaction-specific asset; it cannot be costlessly torn down and rebuilt elsewhere. Before locating the refining plant, the owner faced a competitive market in bauxite, but once built, the situation is fundamentally transformed, as it now matters with whom the refiner deals. It is much more profitable to transact with the adjoining mine owner than anyone else. The solution in the aluminum industry to this potential holdup problem is, indeed, a complex governance structure: vertical integration. Refining firms buy mines. See John Stuckey, *Vertical Integration and Joint Ventures in the Aluminum Industry* (Cambridge, Mass.: Harvard University Press, 1983).

Henry Hansmann extends the transaction cost analysis of firm organization by introducing as a key variable the cost of collective decision making by owners, in addition to the market and monitoring costs emphasized by Jensen and Meckling and Williamson. Homogeneity of interest across owners lessens decision costs. This insight is helpful in understanding why the vast majority of firms are owned by shareholders rather than workers.

Burton Malkiel provides an overview of modern portfolio theory, giving systematic content to the adage don't put all your eggs in one basket. Diversification reduces risk without a commensurate reduction in return, as the judiciously chosen portfolio eliminates risks that are idiosyncratic to specific firms. The investor who does not diversify pays a penalty; she bears unnecessary, and uncompensated, risk.

The selection from Stephen Ross, Randolph Westerfield, and Jeffrey Jaffe's textbook introduces the concept of an efficient market. A capital market is efficient if stock prices fully reflect all available information. As Ross, Westerfield, and Jaffe discuss, U.S. equity markets are extremely efficient. This insight is important for corporate law because an efficient capital market provides high-powered incentives for managers to act in the shareholders' interest. In an efficient market, decisions that fail to maximize stock value depress the firm's stock price, subjecting managers to an increased likelihood of replacement by either the board of directors or a change in control. Market efficiency is also important in assessing the regulation of securities markets, the topic of Chapter VII.

Theory of the Firm

Theory of the Firm: Managerial Behavior, Agency Costs, and Ownership Structure

MICHAEL C. JENSEN AND WILLIAM H. MECKLING

We define an agency relationship as a contract under which one or more persons (the principal[s]) engage another person (the agent) to perform some service on their behalf which involves delegating some decision-making authority to the agent. If both parties to the relationship are utility maximizers there is good reason to believe that the agent will not always act in the best interests of the principal. The *principal* can limit divergences from his interest by establishing appropriate incentives for the agent and by incurring monitoring costs designed to limit the aberrant activities of the agent. In addition in some situations it will pay the *agent* to expend resources (bonding costs) to guarantee that he will not take certain actions which would harm the principal or to ensure that the principal will be compensated if he does take such actions. However, it is generally impossible for the principal or the agent at zero cost to ensure that the agent will make optimal decisions from the principal's viewpoint. In most agency relationships the principal and the agent will incur positive monitoring and bonding costs (nonpecuniary as well as pecuniary), and in addition there will be some divergence between the agent's decisions and those decisions which would maximize the welfare of the principal. The dollar equivalent of the reduction in welfare experienced by the principal due to this divergence is also a cost of the agency relationship, and we refer to this latter cost as the "residual loss." We define *agency costs* as the sum of:

1. the monitoring expenditures by the principal,[1]
2. the bonding expenditures by the agent,
3. the residual loss.

. . .

Since the relationship between the stockholders and manager of a corporation fit the definition of a pure agency relationship it should be no surprise to

[1]As it is used in this article the term monitoring includes more than just measuring or observing the behavior of the agent. It includes efforts on the part of the principal to "control" the behavior of the agent through budget restrictions, compensation policies, operating rules, and so forth.

Reprinted by permission from 3 *Journal of Financial Economics* 305 (Amsterdam: Elsevier Science Pub., 1976).

discover that the issues associated with the "separation of ownership and control" in the modern diffuse ownership corporation are intimately associated with the general problem of agency. . . .

Overview

In this section we analyze the effect of outside equity on agency costs by comparing the behavior of a manager when he owns 100 percent of the residual claims on a firm to his behavior when he sells off a portion of those claims to outsiders. If a wholly owned firm is managed by the owner, he will make operating decisions which maximize his utility. These decisions will involve not only the benefits he derives from pecuniary returns but also the utility generated by various nonpecuniary aspects of his entrepreneurial activities such as the physical appointments of the office, . . . the level of employee discipline, the kind and amount of charitable contributions, personal relations ("love," "respect," etc.) with employees, a larger-than-optimal computer to play with, purchase of production inputs from friends, and so forth. The optimum mix (in the absence of taxes) of the various pecuniary and nonpecuniary benefits is achieved when the marginal utility derived from an additional dollar of expenditure (measured net of any productive effects) is equal for each nonpecuniary item and equal to the marginal utility derived from an additional dollar of after tax purchasing power (wealth).

If the owner-manager sells equity claims on the corporation which are identical to his (i.e., share proportionately in the profits of the firm and have limited liability) agency costs will be generated by the divergence between his interest and those of the outside shareholders, since he will then bear only a fraction of the costs of any nonpecuniary benefits he takes out in maximizing his own utility. If the manager owns only 95 percent of the stock, he will expend resources to the point where the marginal utility derived from a dollar's expenditure of the firm's resources on such items equals the marginal utility of an additional 95 cents in general purchasing power (i.e., *his* share of the wealth reduction) and not one dollar. Such activities, on his part, can be limited (but probably not eliminated) by the expenditure of resources on monitoring activities by the outside stockholders. . . . The owner will bear the entire wealth effects of these expected costs so long as the equity market anticipates these effects. Prospective minority shareholders will realize that the owner-manager's interests will diverge somewhat from theirs, hence the price which they will pay for shares will reflect the monitoring costs and the effect of the divergence between the manager's interest and theirs. Nevertheless, ignoring for the moment the possibility of borrowing against his wealth, the owner will find it desirable to bear these costs as long as the welfare increment he experiences from converting his claims on the firm into general purchasing power[2] is large enough to offset them.

[2]For use in consumption, for the diversification of his wealth, or more importantly, for the financing of "profitable" projects which he could not otherwise finance out of his personal wealth.

As the owner-manager's fraction of the equity falls, his fractional claim on the outcomes falls and this will tend to encourage him to appropriate larger amounts of the corporate resources in the form of perquisites. This also makes it desirable for the minority shareholders to expend more resources in monitoring his behavior. Thus, the wealth costs to the owner of obtaining additional cash in the equity markets rise as his fractional ownership falls.

We shall continue to characterize the agency conflict between the owner-manager and outside shareholders as deriving from the manager's tendency to appropriate perquisites out of the firm's resources for his own consumption. However, we do not mean to leave the impression that this is the only or even the most important source of conflict. Indeed, it is likely that the most important conflict arises from the fact that as the manager's ownership claim falls, his incentive to devote significant effort to creative activities such as searching out new profitable ventures falls. He may in fact avoid such ventures simply because it requires too much trouble or effort on his part to manage or to learn about new technologies. Avoidance of these personal costs and the anxieties that go with them also represent a source of on-the-job utility to him and it can result in the value of the firm being substantially lower than it otherwise could be. . . .

The Role of Monitoring and Bonding Activities in Reducing Agency Costs

In practice, it is usually possible by expending resources to alter the opportunity the owner-manager has for capturing nonpecuniary benefits. These methods include auditing, formal control systems, budget restrictions, and the establishment of incentive compensation systems which serve to more closely identify the manager's interests with those of the outside equity holders, and so forth. . . .

Since the current value of expected future monitoring expenditures by the outside equity holders reduce the value of any given claim on the firm to them dollar for dollar, the outside equity holders will take this into account in determining the maximum price they will pay for any given fraction of the firm's equity. . . . The entire increase in the value of the firm that accrues will be reflected in the owner's wealth, but his welfare will be increased by less than this because he forgoes some nonpecuniary benefits he previously enjoyed. . . .

It makes no difference who actually makes the monitoring expenditures— the owner bears the full amount of these costs as a wealth reduction in all cases. Suppose that the owner-manager could expend resources to guarantee to the outside equity holders that he would limit his activities which cost the firm [a specific amount]. We call these expenditures "bonding costs," and they would take such forms as contractual guarantees to have the financial accounts audited by a public account, explicit bonding against malfeasance on the part of the manager, and contractual limitations on the manager's decision-making power (which impose costs on the firm because they limit his ability to take full advantage of some profitable opportunities as well as limiting his ability to harm the stockholders while making himself better off).

If the incurrence of the bonding costs were entirely under the control of the manager, . . . he would incur them. . . . This would limit his consumption of perquisites . . . and the solution is exactly the same as if the outside equity holders had performed the monitoring. The manager finds it in his interest to incur these costs as long as the net increments in his wealth which they generate (by reducing the agency costs and therefore increasing the value of the firm) are more valuable than the perquisites given up. This optimum occurs [at the same point] in both cases under our assumption that the bonding expenditures yield the same opportunity set as the monitoring expenditures. In general, of course, it will pay the owner-manager to engage in bonding activities and to write contracts which allow monitoring as long as the marginal benefits of each are greater than their marginal cost. . . .

Pareto Optimality[3] and Agency Costs in Manager-Operated Firms

In general we expect to observe both bonding and external monitoring activities, and the incentives are such that the levels of these activities will satisfy the conditions of efficiency. They will not, however, result in the firm being run in a manner so as to maximize its value. The difference between . . . the efficient solution under zero monitoring and bonding costs (and therefore zero agency costs), and . . . the value of the firm given positive monitoring costs, are the total gross agency costs defined earlier in the introduction. These are the costs of the "separation of ownership and control." . . . The solutions outlined above to our highly simplified problem imply that agency costs will be positive as long as monitoring costs are positive—which they certainly are.

The reduced value of the firm caused by the manager's consumption of perquisites outlined above is "nonoptimal" or inefficient only in comparison to a world in which we could obtain compliance of the agent to the principal's wishes at zero cost or in comparison to a *hypothetical* world in which the agency costs were lower. But these costs (monitoring and bonding costs and "residual loss") are an unavoidable result of the agency relationship. Furthermore, since they are borne entirely by the decision maker (in this case the original owner) responsible for creating the relationship he has the incentives to see that they are minimized (because he captures the benefits from their reduction). Furthermore, these agency costs will be incurred only if the benefits to the owner-manager from their creation are great enough to outweigh them. In our current example these benefits arise from the availability of

---

[3]Pareto optimality or efficiency is a technical concept that refers to a situation in which no change can make one person better off without making some other person worse off [EDITOR'S NOTE].

profitable investments requiring capital investment in excess of the original owner's personal wealth. . . .

Factors Affecting the Size of the Divergence from Ideal Maximization

The magnitude of the agency costs discussed above will vary from firm to firm. It will depend on the tastes of managers, the ease with which they can exercise their own preferences as opposed to value maximization in decision making, and the costs of monitoring and bonding activities. The agency costs will also depend upon the cost of measuring the manager's (agent's) performance and evaluating it, the cost of devising and applying an index for compensating the manager which correlates with the owner's (principal's) welfare, and the cost of devising and enforcing specific behavioral rules or policies. Where the manager has less than a controlling interest in the firm, it will also depend upon the market for managers. Competition from other potential managers limits the costs of obtaining managerial services (including the extent to which a given manager can diverge from the idealized solution which would obtain if all monitoring and bonding costs were zero). The size of the divergence (the agency costs) will be directly related to the cost of replacing the manager. . . .

Risk and the Demand for Outside Financing

The model we have used to explain the existence of minority shareholders . . . in the capital structure of corporations implies that the owner-manager, if he resorts to any outside funding, will have his entire wealth invested in the firm. The reason is that he can thereby avoid the agency costs which additional outside funding impose. This suggests he would not resort to outside funding until he had invested 100 percent of his personal wealth in the firm—an implication which is not consistent with what we generally observe. Most owner-managers hold personal wealth in a variety of forms, and some have only a relatively small fraction of their wealth invested in the corporation they manage. Diversification on the part of owner-managers can be explained by risk aversion and optimal portfolio selection.

If the returns from assets are not perfectly correlated an individual can reduce the riskiness of the returns on his portfolio by dividing his wealth among many different assets, that is, by diversifying.[4] Thus a manager who invests all of his wealth in a single firm (his own) will generally bear a welfare loss (if he is risk averse) because he is bearing more risk than necessary. He will, of course, be willing to pay something to avoid this risk, and the costs he must bear to accomplish this diversification will be the agency costs outlined

[4]Diversification is explained in detail in the Malkiel selection in this chapter [EDITOR'S NOTE].

above. He will suffer a wealth loss as he reduces his fractional ownership because prospective shareholders will take into account the agency costs. Nevertheless, the manager's desire to avoid risk will contribute to his becoming a minority stockholder.

Transaction Cost Economics
OLIVER E. WILLIAMSON

Transaction cost economics adopts a contractual approach to the study of economic organization. Questions such as the following are germane: Why are there so many forms of organization? What main purpose is served by alternative modes of economic organization and best informs the study of these matters? Striking differences among labor markets, capital markets, intermediate product markets, corporate governance, regulation, and family organization notwithstanding, is it the case that a common theory of contract informs all? What core features—in human, technology, and process respects—does such a common theory of contract rely on? These queries go to the heart of the transaction cost economics research agenda. . . .

Background

Main Case

Economic organization services many purposes. Among those that have been ascribed by economists are monopoly and efficient risk bearing. Power and associational gains are sometimes held to be the main purposes of economic organization, especially by noneconomists. And some hold that "social institutions and arrangement . . . [are] the adventitious result of legal, historical, or political forces". . . .

The study of complex systems is facilitated by distinguishing core purposes from auxiliary purposes. Transaction cost economics subscribes to and develops the view that economizing is the core problem of economic organization. . . .

Behavioral Assumptions

Many economists treat behavioral assumptions as unimportant. This reflects a widely held opinion that the realism of the assumptions is unimportant and

Reprinted by permission from *Handbook of Industrial Organization*, R. Schmalensee and R. Willig, eds. (Amsterdam: Elsevier Science Pub., 1989), v. 1, p. 135.

that the fruitfulness of a theory turns on its implications. . . . But whereas transaction cost economics is prepared to be judged (comparatively) by the refutable implications which this approach uniquely affords, it also maintains that the behavioral assumptions are important—not least of all because they serve to delimit the study of contract to the feasible subset. . . .

Contracting man is distinguished from the orthodox conception of maximizing man in two respects. The first of these is the condition of bounded rationality. Second, contracting man is given to self-interest seeking of a deeper and more troublesome kind than his economic man predecessor. . . .

Transaction cost economics pairs the assumption of bounded rationality with a self-interest-seeking assumption that makes allowance for guile. Specifically, economic agents are permitted to disclose information in a selective and distorted manner. Calculated efforts to mislead, disguise, obfuscate, and confuse are thus admitted. This self-interest-seeking attribute is variously described as opportunism, moral hazard, and agency. . . .

Bounded rationality and opportunism serve both to refocus attention and help to distinguish between feasible and infeasible modes of contracting. Both impossibly complex and hopelessly naive modes of contracting are properly excluded from the feasible set. Thus:

1. Incomplete contracting. Although it is instructive and a great analytical convenience to assume that agents have the capacity to engage in comprehensive ex ante contracting (with or without private information), the condition of bounded rationality precludes this. All contracts within the feasible set are incomplete. Accordingly, the ex post side of a contract takes on special economic importance. The study of structures that facilitate gap filling, dispute settlement, adaptation, and the like thus become part of the problem of economic organization. Whereas such institutions play a central role in the transaction cost economics scheme of things, they are ignored (indeed, suppressed) by the fiction of comprehensive ex ante contracting.

2. Contract as promise. Another convenient concept of contract is to assume that economic agents will reliably fulfill their promises. Such stewardship behavior will not obtain, however, if economic agents are given to opportunism. Ex ante efforts to screen economic agents in terms of reliability and, even more, ex post safeguards to deter opportunism take on different economic significance as soon as the hazards of opportunism are granted. Institutional practices that were hitherto regarded as problematic are thus often seen to perform valued economizing purposes when their transaction cost features are assessed. . . .

Legal Centralism Versus Private Ordering

It is often assumed, sometimes tacitly, that property rights are well defined and that the courts dispense justice costlessly. . . .

If . . . the participants to a contract can often "devise more satisfactory solutions to their disputes than can professionals constrained to apply general rules on the basis of limited knowledge of the dispute," . . . then court ordering is better regarded as a background factor rather than the central forum for dispute resolution. Albeit useful for purposes of ultimate appeal, legal centralism (court ordering) gives way to private ordering. This is intimately connected to the incomplete contracting/ex post governance approach to which I refer above. . . .

Operationalizing Transaction Cost Economics

The Technology of Transacting

The economic counterpart of friction is transaction cost: for that subset of transactions where it is important to elicit cooperation,[1] do the parties to the exchange operate harmoniously, or are there frequent misunderstandings and conflicts that lead to delays, breakdowns, and other malfunctions? Transaction cost analysis entails an examination of the comparative costs of planning, adapting, and monitoring task completion under alternative governance structures.

Assessing the technology of transacting is facilitated by making the transaction the basic unit of analysis. The central question then becomes: What are the principal dimensions with respect to which transactions differ? Refutable implications are derived from the hypothesis that transactions, which differ in their attributes, are assigned to governance structures, which differ in their costs and competencies, in a discriminating—mainly transaction cost economizing—way.

The principal dimensions on which transaction cost economics presently relies for purposes of describing transactions are (1) the frequency with which they recur, (2) the degree and type of uncertainty to which they are subject, and (3) the condition of asset specificity. Although all are important, many of the refutable implications of transaction cost economics turn critically on this last.

Asset specificity. Asset specificity refers to the degree to which an asset can be redeployed to alternative uses and by alternative users without sacrifice of productive value. . . .

There are at least five kinds of asset specificity: (1) site specificity, as where successive stations are located in a cheek-by-jowl relation to each other so as to economize on inventory and transportation expenses; (2) physical asset specificity, such as specialized dies that are required to produce a component; (3) human asset specificity that arises in a learning-by-doing fashion; (4) dedicated

---

[1]The genius of neoclassical economics is that there are large numbers of transactions where conscious cooperation between traders is not necessary. The invisible hand works well if each party can go its own way—the buyer can secure product easily from alternative sources; the supplier can redeploy his assets without loss of productive value—with little cost to the other. Transaction cost economics is concerned with the frictions that obtain when bilateral dependency intrudes. This is not a trivial class of activity.

assets, which are discrete investments in general purpose plant that are made at the behest of a particular customer; and (5) brand name capital. . . . The organizational ramifications of each type of specificity differ. . . .

The fundamental transformation. Economists of all persuasions recognize that the terms upon which an initial bargain will be struck depend on whether noncollusive bids can be elicited from more than one qualified supplier. Monopolistic terms will obtain if there is only a single highly qualified supplier, while competitive terms will result if there are many. . . .

Transaction cost economics holds that a condition of large numbers bidding at the outset does not necessarily imply that a large numbers bidding condition will obtain thereafter. Whether ex post competition is fully efficacious or not depends on whether the good or service in question is supported by durable investments in transaction-specific human or physical assets. Where no such specialized investments are incurred, the initial winning bidder realizes no advantage over nonwinners. Although it may continue to supply for a long period of time, this is only because, in effect, it is continuously meeting competitive bids from qualified rivals. Rivals cannot be presumed to operate on a parity, however, once substantial investments in transaction-specific assets are put in place. Winners in these circumstances enjoy advantages over nonwinners, which is to say that parity at the renewal interval is upset. Accordingly, what was a large numbers bidding condition at the outset is effectively transformed into one of bilateral supply thereafter. The reason why significant reliance investments in durable, transaction-specific assets introduce contractual asymmetry between the winning bidder on the one hand and nonwinners on the other is because economic values would be sacrificed if the ongoing supply relation were to be terminated.

Faceless contracting is thereby supplanted by contracting in which the pairwise identity of the parties matters. Not only is the supplier unable to realize equivalent value were the specialized assets to be redeployed to other uses, but the buyer must induce potential suppliers to make similar specialized investments were he to seek least-cost supply from an outsider. The incentives of the parties to work things out rather than terminate are thus apparent. This has pervasive ramifications for the organization of economic activity.

A Simple Contractual Schema

The general approach. Assume that a good or service can be supplied by either of two alternative technologies. One is a general purpose technology, the other a special purpose technology. The special purpose technology requires greater investment in transaction-specific durable assets and is more efficient for servicing steady-state demands.

Using k as a measure of transaction-specific assets, transactions that use the general purpose technology are ones for which $k = 0$. When transactions use the special purpose technology, by contrast, a $k > 0$ condition exists. Assets here are specialized to the particular needs of the parties. Productive values would therefore be sacrificed if transactions of this kind were to be prematurely terminated. The bilateral monopoly condition described above and elaborated below applies to such transactions.

Whereas classical market contracting . . . suffices for transactions of the $k = 0$ kind, unassisted market governance poses hazards whenever nontrivial transaction-specific assets are placed at risk. Parties have an incentive to devise safeguards to protect investments in transactions of the latter kind. Let s denote the magnitude of any such safeguards. An $s = 0$ condition is one in which no safeguards are provided; a decision to provide safeguards is reflected by an $s > 0$ result.

Figure I.1 displays the three contracting outcomes corresponding to such a description. Associated with each node is a price. So as to facilitate comparisons between nodes, assume that suppliers (1) are risk neutral, (2) are prepared to supply under either technology, and (3) will accept any safeguard condition whatsoever so long as an expected break-even result can be projected. Thus, node A is the general purpose technology ($k = 0$) supply relation for which a break-even price of p_1 is projected. The node B contract is supported by transaction-specific assets ($k > 0$) for which no safeguard is offered ($s = 0$). The expected break-even price here is $\bar{p}$. The node C contract also employs the special purpose technology. But since the buyer at this node provides the supplier with a safeguard ($s > 0$), the break-even price, $\hat{p}$, at node C is less than $\bar{p}$.

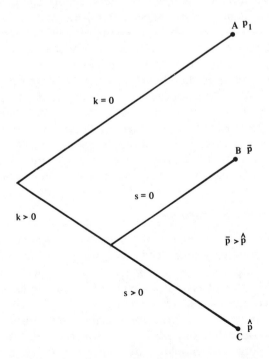

Figure I.1. A Simple Contracting Schema

The protective safeguards to which I refer normally take on one or more of three forms. The first is to realign incentives, which commonly involves some type of severance payment or penalty for premature termination. Albeit important and the central focus of much of the formal contracting literature, this is a very limited response. A second is to supplant court ordering by private ordering. Allowance is expressly made for contractual incompleteness; and a different forum for dispute resolution (of which arbitration is an example) is commonly provided. Third, the transactions may be embedded in a more complex trading network. The object here is to better assure continuity [of] purposes and facilitate adaptations. Expanding a trading relation from unilateral to bilateral exchange—through the concerted use, for example, of reciprocity—thereby to effect an equilibration of trading hazards, is one illustration. Recourse to collective decision making under some form of combined ownership is another. . . .

By way of summary, the nodes A, B, and C in the contractual schema set out in Figure I.1 have the following properties:

1. Transactions that are efficiently supported by general purpose assets $(k = 0)$ are located at node A and do not need protective governance structures. Discrete market contracting suffices. The world of competition obtains.
2. Transactions that involve significant investments of a transaction-specific kind $(k > 0)$ are ones for which the parties are effectively engaged in bilateral trade.
3. Transactions located at node B enjoy no safeguards $(s = 0)$, on which account the projected break-even supply price is great $(\bar{p} > \hat{p})$. Such transactions are apt to be unstable contractually. They may revert to node A (in which event the special purpose technology would be replaced by the general purpose $[k = 0]$ technology) or be relocated to node C (by introducing contractual safeguards that would encourage use of the $k > 0$ technology).
4. Transactions located at node C incorporate safeguards $(s > 0)$ and thus are protected against expropriation hazards.
5. Inasmuch as price and governance are linked, parties to a contract should not expect to have their cake (low price) and eat it too (no safeguard). More generally, it is important to study *contracting in its entirety*. Both the ex ante terms and the manner in which contracts are thereafter executed vary with the investment characteristics and the associated governance structures within which transactions are embedded.

Ownership of the Firm

HENRY HANSMANN

A Theoretical Framework

An Overview of the Theory

In principle, a firm could be owned by someone who is not a patron.[1] Such a firm's capital needs would be met entirely by borrowing; its other factors of production would likewise be purchased on the market, and its products would be sold on the market. The owner(s) would simply have the right to control the firm and to appropriate its (positive or negative) residual earnings. Such firms are rare, however. Ownership commonly is assigned to persons who have some other transactional relationship with the firm. The reason for this, evidently, is that the ownership relationship can be used to mitigate some of the costs that would otherwise attend these transactional relationships if they were managed through simple market contracting.

More particularly, market contracting can be especially costly in the presence of those conditions loosely called "market failure," such as market power or asymmetric information. In such circumstances, the total costs of transacting can sometimes be reduced by merging the purchasing and the selling party in an ownership relationship, hence eliminating the conflict of interest between buyer and seller that underlies or aggravates many of the avoidable costs of market contracting.

Ownership can itself involve substantial costs, however. Further, as we shall discuss below, these costs can be quite different for different classes of patrons. Efficiency will be best served if ownership is assigned so that total transaction costs for all patrons are minimized. This means minimizing the sum of both the costs of market contracting for those patrons who are not owners, and the costs of ownership for the class of patrons who are assigned ownership. . . .

The Costs of Market Contracting

Although a variety of factors can make market transactions costly, there are three characteristic types of problems that arise commonly and can often be mitigated by assigning ownership to the patrons involved.

---

[1]All persons who transact with a firm, as purchasers of its products or as suppliers of some factor of production, including capital, are referred to as patrons [EDITOR'S NOTE].

Market power. An obvious reason for assigning ownership to a given class of patrons is that the firm, owing to the relative absence of effective competition, has a degree of market power vis-à-vis those patrons. If, in such a situation, the patrons own the firm, they can avoid not only the efficiency losses that result from setting prices above marginal cost, but also the larger private costs that such prices would impose on the patrons.

Ex post market power ("Lock-in"). These problems appear where the patron must make substantial transaction-specific investments upon entering into the transactional relationship and where the situation is sufficiently complex that some elements of the transaction must initially be left unspecified and dealt with according to experience. Once such a transactional relationship has been entered into, the patron becomes locked in to a degree, losing the option of costless exit in case the firm seeks to renegotiate the terms of the transaction in its favor as events unfold. Ownership of the firm by the patron reduces the incentives for opportunistic behavior of this sort. This consideration is now widely recognized as an important incentive for vertical integration between individual firms. . . . It can also help explain why ownership is extended to whole classes of patrons.

Asymmetric information. Finally, contracting can also be costly when a firm has significantly better information than its patrons concerning the quality of performance that the firm offers or renders. Ownership by the patrons reduces the incentive for the firm to exploit such an information advantage. . . .

Costs of Ownership

The most significant of these costs can be grouped conveniently under three headings.

Monitoring. If a given class of patrons is to exercise effective control over the management of a firm, they must incur the costs of (1) becoming informed about the operations of the firm, (2) communicating among themselves for the purpose of exchanging information and making decisions, and (3) bringing their decisions to bear on the firm's management. These costs, which Jensen and Meckling . . . have labeled "monitoring" costs, can vary widely among different classes of patrons. They are most likely to be small, relative to the value of the patrons' transactions with the firm, where, for example, the patrons involved are relatively few in number, reside in geographic propinquity to each other and the firm, and transact regularly and repeatedly with the firm over a prolonged period of time for amounts that are a significant fraction of their budget. . . .

An equally important but less familiar point is that for a given class of patrons the costs of managerial opportunism may be worth bearing as an alternative to having no ownership at all. That is, just because a given class of patrons cannot monitor effectively, and thus cannot exercise much control beyond that which they would have simply by virtue of market transactions with the firm, it does not follow that there is no substantial gain to those patrons from having ownership of the firm. . . . By virtue of having ownership, the patrons in question are assured that there is no *other* group of owners to whom management is responsive. . . .

Collective decision making. When ownership of a firm is shared among a class of patrons, a method for collective decision making must be devised. Most commonly a voting mechanism of some sort is employed, with votes weighted by volume of patronage, although some cooperatives adhere to a one-member–one-vote scheme.

As methods for aggregating the preferences of a group of patrons, such collective choice mechanisms often involve substantial costs in comparison to market contracting. . . .

Although a variety of factors influence the magnitude of these costs, a fundamental consideration is the extent to which the patron-owners have divergent interests concerning the conduct of the firm's affairs. Where the patrons involved all have essentially identical interests—for example, where they all transact with the firm under similar circumstances for similar quantities of a single homogeneous commodity, . . . the costs associated with collective decision making are naturally small. Absent such circumstances, however, these costs may be large relative to those of market transactions. The costs can come in several different forms.

To begin with, even if no patron acts strategically such processes may yield decisions that are collectively inefficient in the sense that they do not maximize aggregate patron surplus. . . .

Further, the process of collective decision making itself can have high transaction costs in the face of heterogeneous interests. Because there is a strong incentive for individuals to form coalitions to shift benefits in their direction, efforts to form and break such coalitions may consume substantial effort. . . .

On the other hand, even where patrons diverge considerably in interest, the costs associated with collective decision making may be low if there is some simple and salient criterion for balancing their interests. For example, where it is easy to account separately for the net benefits bestowed on the firm by each individual patron, dividing up net returns according to such an accounting is likely to be both natural and uncontroversial even if the nature and the volume of the transactions with individual patrons differ substantially. . . .

Risk bearing. The preceding discussion has focused on the costs associated with the first element of ownership: the exercise of control. But costs are also associated with the second element of ownership: the receipt of compensation in the form of residual earnings. Most conspicuous among these is the cost of bearing the risk of the enterprise and is typically reflected in residual earnings. One class of a firm's patrons may be in a much better position than others to bear such risk, for example, through diversification. Assigning ownership to those patrons can then bring important economies. . . .

Another important consideration here, and one that has been little re-marked upon, is that market contracting with a given class of patrons itself sometimes *creates* a substantial degree of risk that can be avoided by assigning ownership to those patrons. This is particularly likely to be the case where the patrons must enter into a long-term relationship with the firm, so that the terms of the contract between them become a gamble on future contingencies. . . .

Investor-Owned Firms

Costs of Market Contracting

Because capital markets today are highly competitive, market power is rarely an incentive for lenders of capital to become owners of a firm to which they lend. Rather, problems of asymmetric information and lock-in provide the strongest incentive for assigning ownership to investors.

In theory a firm could borrow 100 percent of the capital it needs with the owners of the firm—whether they are another class of the firm's patrons or third parties who do not otherwise transact with the firm—investing no capital themselves. And if, in practice, the owners could be constrained by the terms of the loan contract to devote the borrowed funds only to the most efficient projects and to take for themselves only a specified rate of compensation until the loan had been repaid, this approach would be workable. But it is ex-tremely difficult to write and enforce such a contract. And without such contractual terms, the owners have an incentive to behave opportunistically, distributing to themselves dividends (or perquisites) that are unjustified by the firm's earnings or (what is harder to police) investing the proceeds of the loan in high-risk projects whose gains will go disproportionately to the owners and whose losses will fall disproportionately on the lenders.

The incentives of the owners to behave this way can be effectively curbed if they are made to post a bond for the full amount of the loan. If the proceeds of the loan are invested in assets that are not organization specific, this can be easily accomplished by giving the lenders either a lien on those assets or outright ownership of the assets (so that the loan becomes one in kind, that is, the assets are rented). Yet, as others have recognized, where the loan pro-ceeds are in some part invested in organization-specific assets—and this usu-ally will be the case—the problem is more difficult. A possible solution is to have the owners provide personal security for the loan by pledging personal assets or future income as collateral. This is, in fact, a common procedure, particularly in small firms. Where large-scale enterprise is involved, however, and the ownership class is numerous, this device is quite cumbersome. It is difficult for a lender to check the value of the numerous pledges of security, and it is expensive to foreclose on a large class of small guarantors in case of default. . . .

Costs of Ownership

Diversification of risk is a conspicuous advantage of investor ownership. Another great strength of investor-owned firms is the fact that the owners generally share a single, well-defined objective: to maximize the net present value of the firm's earnings per dollar invested. To be sure, differences in tax status or risk preference may lead investors to differ about the most appropriate financial policy for the firm. But even these differences can be eliminated to some extent if investors sort themselves among firms. . . .

The great liability of investor-owned firms, on the other hand, is that investors frequently are in a poor position to engage in meaningful supervision of the firm's management—particularly where, in order to obtain access to a large pool of capital and to diversify risk, the firm's capital is drawn from a numerous group of relatively small investors. . . .

We might, therefore, draw this conclusion from the success of such corporations: direct exercise of oversight and control by owners is not of decisive importance for the efficient conduct of enterprise; it is often worth trading off in favor of the other cost factors outlined above. Under this view, much of the protection that the investors in a widely held investor-owned firm have from opportunistic behavior on the part of the firm derives simply from the absence of a class of owners with interests contrary to theirs. But, as suggested earlier, this may be important protection and worth the costs of some managerial slack. . . .

Worker-Owned Firms

Worker-owned firms are the dominant form of organization in the service professions, such as law, accounting, investment banking, and management consulting. They are also relatively common in some other service industries, such as taxicabs and trash collection. . . . Outside the service sector, on the other hand, worker-owned firms are generally isolated and often short-lived entities, competing in industries in which investor-owned firms are clearly dominant. One of the few exceptions is plywood manufacturing; roughly two dozen plywood firms in the Pacific Northwest have long been operated, with considerable success, as labor cooperatives. . . .

Costs of Ownership

Workers in nearly all industries are in a very good position, in comparison with other classes of patrons, to monitor the management of the firm. The majority of their income typically comes from their work relationship with the firm; they are in daily contact with the firm's operations and are knowledgeable about some aspects of them; and they are easily organized for collective decision making. This is not to imply, of course, that the typical shop-floor worker necessarily knows much about the firm's marketing problems or capital investment program. Yet his or her opportunity and incentive to gain and use such information (or to locate, elect, and hold accountable representatives who will) is generally stronger than that of, say, the firm's customers or remote investors.

On the other hand, costs of risk bearing are often unfavorable to worker ownership. This is obviously true for firms with substantial amounts of firm-specific capital. . . . [W]orker-owners in such firms will face high costs of capital if they do not provide a significant fraction of this capital themselves. Yet investing heavily in the firm for which they work will cause their human capital and their savings, taken together, to be very poorly diversified—a problem of worker ownership that has frequently been noted. . . . It is not surprising, then, that those industries in which such firms are best established, such as law and accounting, are characterized by low amounts of organization-specific capital per worker. Yet there are many service industries, such as retailing and the construction trades, that are highly labor intensive but are nevertheless populated largely with investor-owned firms.

To be sure, risk bearing might also appear to be a comparative liability for worker-managed firms even in labor-intensive industry. Since workers generally cannot diversify their source of income by working for more than one firm at a time, it would seem advantageous to have the firm owned by investors, who would provide workers with job security and a contractually fixed wage. But, presumably because of the difficulty of writing workable long-term employment contracts, workers in fact generally bear substantial risk in the form of layoffs even in investor-owned firms. Consequently, the risk-bearing features of worker ownership seem unlikely to account for its rarity.

Rather, the truly striking feature that seems common to virtually all well-established worker-owned firms, and that seems most clearly to divide these firms from those that are investor owned, is the strong homogeneity of interest among the workers involved. In particular, what seems important is homogeneity of jobs and of skills: labor cooperatives appear to work best where all the workers who are also members of the cooperative perform essentially identical tasks within the firm. . . . [T]he plywood cooperatives typically follow a rigid principle of equal pay for all worker-owners. The manager of the firm often is not a member of the cooperative, but rather is hired by the worker-owners. Worker-owners are generally capable of undertaking any job in the plant other than that of manager, since only semiskilled labor is involved. Job assignments are made according to a bid system, with more senior workers generally given preference, and there is much rotation among jobs. . . . Such a system reinforces the equal pay rule and reduces conflicts of interest among workers: where all workers do, or will ultimately do, the same jobs, they will be affected similarly by any decision made by the firm.

To the extent that workers in worker-owned firms perform different jobs, it seems important to the viability of the firm that the returns to those jobs be separable. The reason, apparently, is that this permits a differential division of the firm's earnings with a minimum of friction. Thus some partners in law firms work longer hours, have greater skills, or bring in more new clients than others. Where such disparities are substantial, law firms sometimes use productivity-based formulas for dividing up earnings. Such formulas are feasible only where the returns to an individual worker's efforts are fairly easily observable, as they are in a law firm, in which such productivity measures as hours billed to individual clients are available. By contrast, it is hard to imagine how one would even design a productivity-based compensation formula for managers in most large business corporations, much less reach agreement on the terms of the formula among the different managers themselves.

Such considerations of homogeneity of interest are evidently an important reason why worker-owned firms appear, as remarked above, not where worker productivity is particularly difficult to monitor, but on the contrary in those industries in which worker output seems relatively easy to measure. Thus trash-collection crews, taxicab drivers, and service professionals are likely to form worker-owned firms, and not blue-collar or white-collar workers who work in large teams. It is, in fact, extremely difficult to find successful examples of worker-owned firms in which there is substantial hierarchy or division of labor among the worker-owners. . . .

Firms Without Owners: Nonprofit Enterprise

Sometimes the conflicts between the costs of market contracting and the costs of ownership are so strong that there is no class of individuals to whom ownership of a firm can be assigned without severe inefficiencies. In such situations, nonprofit firms, which effectively have no owners, often evolve as an expedient.

More specifically, two circumstances are commonly conjoined in those situations in which nonprofit firms emerge. First, there is an extreme problem of asymmetric information between the firm and some class of its patrons—usually a significant group of the firm's customers. As a consequence, assigning ownership to anyone other than that class of patrons would create both the incentive and the opportunity for the patrons to be severely exploited. But, second, those same patrons are so situated that the costs to them of exercising effective control over the firm are unacceptably large relative to the value of their transactions with the firm. The solution is to create a firm without owners—or, more accurately, to create a firm whose managers hold it in trust for its patrons. In essence, the nonprofit form abandons any benefits of full ownership in favor of stricter fiduciary constraints on management.

Most commonly, the problems of asymmetric information that render market contracting inadequate to protect the interests of patrons arise because the firm's patrons are either purchasing services to be delivered to third parties (such as food for starving children in Africa), so that the patrons cannot actually observe the firm's performance, or contributing toward the purchase of a public good (such as public broadcasting), so that the patrons cannot observe the marginal increment to the service purchased by their individual payment. (In both these cases, we usually refer to the patrons as donors.)

Notes and Questions

1. The agency problem discussed in Jensen and Meckling is also referred to as a moral hazard problem. This term comes from the insurance literature, and it indicates the problems for principals that arise when an agent's actions can affect event outcomes. For instance, a person who buys automobile insurance may take more risk driving or leaving her car unattended because she knows that if an accident or auto theft occurs she will be compensated for her loss by the insurance company. The problem for the insurance company is that, when a claim is filed, it cannot tell whether the insured took care. Similarly, in our context, the profitability of the firm will be affected by how hard the manager works, but if the firm performs poorly, it is possible that the manager worked hard but had bad luck. Insurance companies try to adjust for this perverse incentive problem by requiring insureds to bear some of the risk of loss through policy deductibles. The analogue in the corporation setting is incentive compensation, discussed in the selection by Clifford Smith and Ross Watts in Chapter V.

2. Jensen and Meckling develop their thesis by contrasting the choice of pecuniary and nonpecuniary benefit levels in the 100 percent owned and partially owned firm. Pecuniary benefits, a firm's cash flows, are received by all of the firm's owners and distributed proportionately across their shares, whereas nonpecuniary benefits, such as the perquisites of having nice office furnishings, by definition, cannot be shared by the nonmanager owners. The level of nonpecuniary benefits rises in the partially owned firm, compared to the sole-ownership firm, because the owner-manager's reduced stock ownership causes a change in the relative prices of pecuniary and nonpecuniary benefits. Under 100 percent ownership there is a dollar for dollar trade-off between pecuniary and nonpecuniary benefits, so that the owner-manager internalizes fully the financial loss from taking a nonpecuniary benefit, but if she sells part of her holdings, say 50 percent, for every dollar shift from pecuniary to nonpecuniary benefit, she loses only fifty cents on the dollar (the financial loss is shared equally with the other equity investors), while she retains the full dollar of nonpecuniary benefit for herself. A positive level of nonpecuniary benefits in the 100 percent owned firm is not inefficient: the owner is obtaining utility from the benefits. For example, she is willing to take less cash as a return on her investment in exchange for a more pleasant workplace environment.

If the pecuniary value of the sole-owner firm is $1 million and the owner decides to sell 50 percent of her shares, will outside investors pay $500,000 (half of the firm's value) for their interest? Jensen and Meckling's answer is no. The investors pay an amount less than $500,000 because they correctly anticipate that the sole owner's trade-off of pecuniary and nonpecuniary benefits is changing from $1 to $.50 and that she will now take an inefficiently large nonpecuniary benefit from the firm's perspective (i.e., she takes more than the amount she chose under sole ownership), which by the nature of the trade-off reduces the firm's pecuniary value below $1 million.

3. The owner-manager bears all of the agency costs of selling shares to the public because Jensen and Meckling are operating in a world of rational expectations: investors' expectations about the future are correct. To put it another way, their expectations are confirmed by events. Investors expect the sole owner-manager to increase her consumption of nonpecuniary benefits when she sells her stock, and correspondingly, they pay less for the shares than they are worth at the current (sole owner) level of nonpecuniary benefits (less than $500,000 in the example in note 2). After the shares are sold, the owner-manager does, in fact, increase her consumption of nonpecuniary benefits to the level the outside investors predicted.

Adding Williamson's assumption of bounded rationality need not alter the conclusion that the selling owner bears the agency costs. If the buyers cannot predict what trade-off the manager will make with respect to nonpecuniary benefits upon the stock sale, then they may very well discount the shares far more heavily than the seller's optimal consumption of nonpecuniary benefits would warrant. The seller then has an incentive to bond her choice of nonpecuniary benefits to increase the cash she will obtain for her shares; otherwise, she will be forced to take more nonpecuniary benefits than she desires upon selling the shares because of the lower price she receives. The seller will not bear all of the agency costs if the buyers underestimate the level of nonpecuniary benefits that she will take once they purchase the shares.

4. The selling owner will desire to minimize the decline in firm value caused by the shift to partial ownership. She could, for instance, expend resources to precommit to taking a given level of nonpecuniary benefits. In addition, the new investors may expend resources to monitor the owner-manager's nonpecuniary benefit consumption. Audited financial statements are an example of an expenditure that parties to firms undertake to reduce agency costs. The difference between what Jensen and Meckling term bonding and monitoring costs is simply who bears the *initial* expense, selling owner or new investors, respectively; the selling owner ultimately bears the cost in either case, under the same reasoning as outlined in note 3 above.

5. Jensen and Meckling's set-up of the agency problem raises an important question, why deviate from the sole-owner situation? Both of Jensen and Meckling's answers to this question, capital needs arising from personal wealth constraints or a desire for personal wealth diversification, are facilitated by an investment with the corporate characteristics of limited liability and free transferability of shares. The residual agency loss—the decrease in total value of the firm upon partial ownership—must be at least equal to the utility gained by the original owner from the cash received for her shares. Otherwise, she would not be better off selling the shares and the firm would remain with 100 percent ownership.

A further important question is if we wish to deviate from sole ownership, what contracts will arise? While Jensen and Meckling are not concerned with specifying optimal compensation contracts for agents, there is now an extensive formal economic literature addressing this issue. For an excellent introduction to these models, see Paul Milgrom and John Roberts, *Economics, Organization and Management* (Englewood Cliffs, N.J.: Prentice Hall, 1992). Jensen and Meckling are instead interested in considering what other types of expenditures can be undertaken to reduce the agency problem. Can provisions of corporation codes be understood as devices to reduce agency costs? See Chapters III and V.

6. How useful do you think the principal-agent formulation is in predicting organizational arrangements? See Mark A. Wolfson, "Empirical Evidence of Incentive Problems and Their Mitigation in Oil and Gas Tax Shelter Programs," in J. W. Pratt and R. J. Zeckhauser, eds., *Principals and Agents: The Structure of Business* 101 (Boston: Harvard Business School, 1985); and the selection by William Sahlman in Chapter IV.

7. Berle and Means had an optimistic view of the significance of the separation of ownership and control. They maintained (or hoped) that corporate managers would be public-spirited and would manage the corporation as statesmen, orchestrating the firm to meet public needs. See Roberta Romano, "Metapolitics and Corporate Law Reform," 36 *Stanford Law Review* 923 (1984). Is this a plausible conception of the corporation, or of human behavior as we know it? If you were investing your money in a firm, would you want management to have the option of ignoring your wishes? We will have the opportunity to appraise the realism of Berle and Means's aspiration when examining certain state takeover statutes, which instruct directors to consider the interests of constituencies other than shareholders, in Chapters V and VI.

8. The classical question for transaction cost economics, first posed by Ronald Coase, is why are there firms, that is, when does a business enterprise choose to purchase an input in the market and when does it choose to produce it internally? Coase, "The Nature of the Firm," 4 *Economica* 386 (1937). The trade-off in the decision to bring transactions into the firm (such as the backward vertical integration, discussed in this chapter's introduction, of the aluminum producer who purchases the bauxite mine) is between avoiding a supplier's holdup power to extract the value of transaction-specific assets and losing the high-powered incentives of markets, that is, the incentives for cost-effective production that are created by the disciplining effect of market competition. For a more detailed presentation of transaction cost economics, see Oliver E. Williamson, *The Economic Institutions of Capitalism* (New York: Free Press, 1985).

9. Williamson offers a hostage model as an example of nonstandard commercial contracting that provides safeguards for investments in transaction-specific assets (node C in Figure I.1). In this situation, while only one party has to invest in transaction-specific assets to complete the transaction, the other

party also makes a transaction-specific investment (the hostage) that is forfeited if he breaches the agreement. A key feature of a Williamsonian hostage is that its value is worth more to the hostage giver than to the hostage taker. Otherwise, a hostage taker will have an incentive to renege on contractual performance in order to keep the hostage.

The hostage concept was inspired by the exchange of hostages by kings in medieval times to guarantee the peace. A whimsical example of a Williamsonian hostage appears in Gilbert and Sullivan's *Princess Ida*. Princess Ida refuses to marry Prince Hilarion, to whom she was betrothed as a child to ensure comity between their families' kingdoms, and has instead established a women's university. When King Gama fails to produce the princess at the appointed time, Hilarion goes to the university in order to appeal to Ida in person (in part, at Gama's suggestion). His father, King Hildebrand, takes Gama and his three sons as hostages for Hilarion's safe return.

10. Williamson and Jensen and Meckling employ different vocabulary to describe related phenomena. For example, Williamson's opportunism includes the moral hazard problem of concern to Jensen and Meckling. In addition, Jensen and Meckling's bonding costs share features with Williamsonian hostages: bonding expenditures are the means by which an owner-agent precommits not to exploit outside investors, although such investments need not be worth more to the insider than the outsiders. However, their core explanations of corporate organization differ. Williamson emphasizes that the corporate form is a mechanism to reduce transaction costs, whereas for Jensen and Meckling, risk shifting is the critical element. Although Hansmann emphasizes the costs of collective decision making as the principal reason for the absence of worker-owned manufacturing firms, he does mention a natural extension of Jensen and Meckling's risk-based explanation for outside owners in closely held firms to public corporations: a division of labor derived from differences in attitudes toward risk between risk-loving capital providers and workers who dislike risk. Burton Malkiel's discussion of risk and portfolio theory that immediately follows in part B of this chapter presents the intuition motivating such an explanation of the firm.

B

Theory of Capital Markets

From *A Random Walk Down Wall Street*

BURTON G. MALKIEL

Defining Risk: The Dispersion of Returns

Risk is the chance that expected security returns will not materialize and, in particular, that the securities you hold will fall in price.

Once academics accepted the idea that risk for investors is related to the chance of disappointment in achieving expected security returns, a natural measure suggested itself—the probable variability or dispersion of future returns. Thus, financial risk has generally been defined as the variance or standard deviation of returns. Being long-winded, we use the accompanying exhibit to illustrate what we mean. A security whose returns are not likely to depart much, if at all, from its average (or expected) return is said to carry little or no risk. A security whose returns from year to year are likely to be quite volatile (and for which sharp losses are typical in some years) is said to be risky.

Exhibit

Expected Return and Variance: Measures of Reward and Risk

This simple example will illustrate the concept of expected return and variance and how they are measured. Suppose you buy a stock from which you expect overall returns (including both dividends and price changes) under different economic conditions, as shown in Table I.1. If, on average, a third of past years have been "normal," another third characterized by rapid growth, and the remaining third characterized by "stagflation," it might be reasonable to take these relative frequencies of past events and treat them as our best guesses (probabilities) of the likelihood of future business conditions. We could then say that an investor's *expected return* is 10 percent. A third of the time the investor gets 30 percent, another third 10 percent, and the rest of the

Reprinted from *A Random Walk Down Wall Street, Including a Life-Cycle Guide to Personal Investing,* Fifth Edition, by Burton G. Malkiel, by permission of the author and W. W. Norton & Company, Inc. Copyright © 1990, 1985, 1981, 1975, 1973 by W. W. Norton & Company, Inc.

Table I.1.

Business Conditions	Probability of Occurrence	Expected Return
Normal economic conditions	1 chance in 3	10 percent
Rapid real growth	1 chance in 3	30 percent
Recession with inflation (stagflation)	1 chance in 3	−10 percent

time he suffers a 10 percent loss. This means that, *on average,* his yearly return will turn out to be 10 percent.

$$\text{Expected Return} = \tfrac{1}{3}(0.30) + \tfrac{1}{3}(0.10) + \tfrac{1}{3}(-0.10) = 0.10.$$

The yearly returns will be quite variable, however, ranging from a 30 percent gain to a 10 percent loss. The "variance" is a measure of the dispersion of returns. It is defined as the average squared deviation of each possible return from its average (or expected) value, which we just saw was 10 percent.

$$\begin{aligned} \text{Variance} &= \tfrac{1}{3}(0.30 - 0.10)^2 + \tfrac{1}{3}(0.10 - 0.10)^2 + \tfrac{1}{3}(-0.10 - 0.10)^2 \\ &= \tfrac{1}{3}(0.20)^2 + \tfrac{1}{3}(0.00)^2 + \tfrac{1}{3}(-0.20)^2 = 0.0267. \end{aligned}$$

The square root of the variance is called the *standard deviation.* In this example, the standard deviation equals 0.1634. . . .

Documenting Risk: A Long-Run Study

One of the best-documented propositions in the field of finance is that, on average, investors have received higher rates of return for bearing greater risk. The most thorough study has been done by Roger Ibbotson and Rex Sinquefield. Their data cover the period 1926 through 1988. The results are shown in Table I.2. . . .

A quick glance shows that over long periods of time, common stocks have, on average, provided relatively generous total rates of return. These returns, including dividends and capital gains, have exceeded by a substantial margin the returns from long-term corporate bonds. The stock returns have also tended to be well in excess of the inflation rate as measured by the annual rate of increase in consumer prices. Thus, stocks have also tended to provide positive "real" rates of return, that is, returns after washing out the effects of inflation. The data show, however, that common-stock returns are highly variable, as indicated by the standard deviation and the range of annual returns, shown in adjacent columns of the table. Returns from equities have ranged from a gain of over 50 percent (in 1933) to a loss of almost the same magnitude (in 1931). Clearly, the extra returns that have been available to investors from stocks have come at the expense of assuming considerably higher risk. Note that small company stocks have provided an even higher rate of return since 1926 but the dispersion (standard deviation) of those returns

Table I.2. Selected Performance Statistics, 1926–1988

Series	Annual (Geometric Mean Rate of Return)	Number of Years Returns Are Positive	Number of Years Returns Are Negative	Highest Annual Return (and Year)	Lowest Annual Return (and Year)	Standard Deviation of Annual Returns	Distribution
Common stocks	10.0%	44	19	54.0% (1933)	−43.3% (1931)	20.9	
Small company stocks	12.3	43	20	142.9 (1933)	−49.8 (1931)	35.6	
Long-term corporate bonds	5.0	48	15	43.8 (1982)	−8.1 (1969)	8.4	
U.S. Treasury bills	3.5	62	1	14.7 (1981)	−0.0 (1940)	3.3	
Consumer price index	3.1	53	10	18.2 (1946)	−10.3 (1932)	4.8	

−50%   0%   50%

Source: © Stocks, Bonds, Bills and Inflation 1992 Yearbook,™ Ibbotson Associates, Inc., Chicago (annually updates work by Roger G. Ibbotson and Rex A. Sinquefield). All rights reserved. Reproduced with permission.

has been even larger than for equities in general. Again, we see that higher returns have been associated with higher risks. . . .

The patterns evident in Ibbotson and Sinquefield's table also appear when the returns and risks of individual stock portfolios are compared. Indeed . . . the differences that exist in the returns from different funds can be explained almost entirely by differences in the risk they have taken. However, given the rate of return they seek, there are ways in which investors can reduce the risks they take. This brings us to the subject of modern portfolio theory, which has revolutionized the investment thinking of professionals.

Reducing Risk: Modern Portfolio Theory (MPT)

Portfolio theory begins with the premise that all investors . . . are risk averse. They want high returns and guaranteed outcomes. The theory tells investors how to combine stocks in their portfolios to give them the least risk possible, consistent with the return they seek. It also gives a rigorous mathematical justification for the time-honored investment maxim that diversification is a sensible strategy for individuals who like to reduce their risks.

The theory was invented in the 1950s by Harry Markowitz. . . . What Markowitz discovered was that portfolios of risky (volatile) stocks might be put together in such a way that the portfolio as a whole would actually be less risky than any one of the individual stocks in it. . . .

Let's suppose we have an island economy with only two businesses. The first is a large resort with beaches, tennis courts, a golf course, and the like. The second is a manufacturer of umbrellas. Weather affects the fortunes of both. During sunny seasons the resort does a booming business and umbrella sales plummet. During rainy seasons the resort owner does very poorly, while the umbrella manufacturer enjoys high sales and large profits. Table I.3 shows some hypothetical returns for the two businesses during the different seasons.

Suppose that, on average, one-half the seasons are sunny and one-half are rainy (i.e., the probability of a sunny or rainy season is ½). An investor who bought stock in the umbrella manufacturer would find that half the time he earned a 50 percent return and half the time he lost 25 percent of his investment. On average, he would earn a return of 12½ percent. This is what we have called the investor's *expected return*. Similarly, investment in the resort would produce the same results. Investing in either one of these businesses would be fairly risky, however, because the results are quite variable and there could be several sunny or rainy seasons in a row.

Table I.3.

	Umbrella Manufacturer	Resort Owner
Rainy season	50%	−25%
Sunny season	−25%	50%

Suppose, however, that instead of buying only one security an investor with two dollars diversified and put half his money in the umbrella manufacturer's and half in the resort owner's business. In sunny seasons, a one-dollar investment in the resort would produce a fifty-cent return, while a one-dollar investment in the umbrella manufacturer would lose twenty-five cents. The investor's total return would be twenty-five cents (fifty cents minus twenty-five cents), which is 12½ percent of his total investment of two dollars.

Note that during rainy seasons exactly the same thing happens—only the names are changed. Investment in the umbrella manufacturer produces a good 50 percent return while the investment in the resort loses 25 percent. Again, however, the diversified investor makes a 12½ percent return on his total investment.

This simple illustration points out the basic advantage of diversification. Whatever happens to the weather, and thus to the island economy, by diversifying investments over both of the firms an investor is sure of making a 12½ percent return each year. The trick that made the game work was that while both companies were risky (returns were variable from year to year), the companies were affected differently by weather conditions. (In statistical terms, the two companies had a negative covariance.[1]) As long as there is some lack of parallelism in the fortunes of the individual companies in the economy, diversification will always reduce risk. In the present case, where there is a perfect negative relationship between the companies' fortunes (one always does well when the other does poorly), diversification can totally eliminate risk.

Of course, there is always a rub, and the rub in this case is that the fortunes of most companies move pretty much in tandem. When there is a recession and people are unemployed, they may buy neither summer vacations nor umbrellas. Therefore, one should not expect in practice to get the neat kind of total risk elimination just shown. Nevertheless, since company fortunes don't always move completely in parallel, investment in a diversified portfolio of stocks is likely to be less risky than investment in one or two single securities. . . . The point to realize in setting up a portfolio is that while the variability (variance) of the returns from individual stocks is important, even more

---

[1]Statisticians use the term *covariance* to measure what I have called the degree of parallelism between the returns of the two securities. If we let R stand for the actual return from the resort and $\bar{R}$ be the expected or average return, while U stands for the actual return from the umbrella manufacturer and $\bar{U}$ is the average return, we define the covariance between U and R (or COV_{UR}) as follows:

$$COV_{UR} = \text{Prob. rain } (U, \text{ if rain} - \bar{U}) \ (R, \text{ if rain} - \bar{R}) + \text{Prob. sun}$$
$$(U, \text{ if sun} - \bar{U}) \ (R, \text{ if sun} - \bar{R}).$$

From the preceding table of returns and assumed probabilities we can fill in the relevant numbers:

$$COV_{UR} = \tfrac{1}{2}(0.50 - 0.125)(-0.25 - 0.125) + \tfrac{1}{2}(-0.25 - 0.125)(0.50 - 0.125)$$
$$= -0.141.$$

Whenever the returns from two securities move in tandem (when one goes up the other always goes up), the covariance number will be a large positive number. If the returns are completely out of phase, as in the present example, the two securities are said to have negative covariance.

important in judging the risk of a portfolio is covariance, the extent to which the securities move in parallel. It is this covariance that plays the critical role in Markowitz's portfolio theory.

True diversification depends on having stocks in your portfolio that are not all dependent on the same economic variables (consumer spending, business investment, housing construction, etc.). Wise investors will diversify their portfolios not by names or industries but by the determinants that influence the fluctuations of various securities. . . .

Modeling Risk: The Capital-Asset Pricing Model (CAPM)

Portfolio theory has important implications for how stocks are actually valued. If investors seek to reduce risk in anything like the manner Harry Markowitz described, the stock market will tend to reflect these risk-reducing activities. This brings us to what is called the "capital-asset pricing model," a creation devised by Stanford professor William Sharpe, the late Harvard professor John Lintner, and others.

I've mentioned that the reason diversification cannot usually produce the miracle of risk elimination, as it did in my mythical island economy, is that usually stocks tend to move up and down together. Still, diversification is worthwhile—it can eliminate some risks. What Sharpe and Lintner did was to focus directly on what part of a security's risk can be eliminated by diversification and what part can't. . . .

Systematic risk, also called market risk, captures the reaction of individual stocks (or portfolios) to general market swings. Some stocks and portfolios tend to be very sensitive to market movements. Others are more stable. This relative volatility or sensitivity to market moves can be estimated on the basis of the past record, and is popularly known by the Greek letter beta. . . . Basically, beta is the numerical description of systematic risk. Despite the mathematical manipulations involved, the basic idea behind the beta measurement is one of putting some precise numbers on the subjective feelings money managers have had for years. The beta calculation is essentially a comparison between the movements of an individual stock (or portfolio) and the movements of the market as a whole.

The calculation begins by assigning a beta of 1 to a broad market index, such as the NYSE [New York Stock Exchange] index or the S&P [Standard & Poors] 500. If a stock has a beta of 2, then on average it swings twice as far as the market. If the market goes up 10 percent, the stock rises 20 percent. If a stock has a beta of 0.5, it tends to be more stable than the market (it will go up or down 5 percent when the market rises or declines 10 percent). . . .

Now the important thing to realize is that *systematic risk cannot be eliminated by diversification.* It is precisely because all stocks move more or less in tandem (a large share of their variability is systematic) that even diversified stock portfolios are risky. Indeed, if you diversified perfectly by buying a share in the S&P index (which by definition has a beta of 1) you would still have quite variable (risky) returns because the market as a whole fluctuates widely.

Unsystematic risk is the variability in stock prices (and therefore, in returns from stocks) that results from factors peculiar to an individual company. Receipt of a large new contract, the finding of mineral resources on the company's property, labor difficulties, the discovery that the corporation's treasurer has had his hand in the company till—all can make a stock's price move independently of the market. The risk associated with such variability is precisely the kind that diversification can reduce. The whole point of portfolio theory is that, to the extent that stocks don't move in tandem all the time, variations in the returns from any one security will tend to be washed away or smoothed out by complementary variation in the returns from other securities. . . . [To illustrate] the important relationship between diversification and total risk, [s]uppose we randomly select securities for our portfolio that tend on average to be just as volatile as the market (the average betas for the securities in our portfolio will always be equal to 1). . . . [A]s we add more and more securities the total risk of our portfolio declines, especially at the start.

When ten securities are selected for our portfolio, a good deal of the unsystematic risk is eliminated, and additional diversification yields little further risk reduction. By the time twenty well-diversified securities are in the portfolio, the unsystematic risk is substantially eliminated and our portfolio (with a beta of 1) will tend to move up and down essentially in tandem with the market. . . .

Now comes the key step in the argument. Both financial theorists and practitioners agree that investors should be compensated for taking on more risk by a higher expected return. Stock prices must therefore adjust to offer higher returns where more risk is perceived, to ensure that all securities are held by someone. Obviously, risk-averse investors wouldn't buy securities with extra risk without the expectation of extra reward. But not all of the risk of individual securities is relevant in determining the premium for bearing risk. The unsystematic part of the total risk is easily eliminated by adequate diversification. So there is no reason to think that investors will be compensated with a risk premium for bearing unsystematic risk. The only part of total risk that investors will get paid for bearing is systematic risk, the risk that diversification cannot help. Thus, the capital-asset pricing model says that returns (and, therefore, risk premiums) for any stock (or portfolio) will be related to beta, the systematic risk that cannot be diversified away.

The proposition that risk and reward are related is not new. Finance specialists have agreed for years that investors do need to be compensated for taking on more risk. What is different about the new investment technology is the definition and measurement of risk. Before the advent of the capital-asset pricing model, it was believed that the return on each security was related to the total risk inherent in that security. It was believed that the return from a security varied with the instability of that security's particular performance, that is, with the variability or standard deviation of the returns it produced.

The new theory says that the *total* risk of each individual security is irrelevant. It is only the systematic component of that total instability that is relevant for valuation. . . . Because stocks can be combined in portfolios to eliminate specific risk, only the undiversifiable or systematic risk will command a risk premium. Investors will not get paid for bearing risks that can be diversified away. This is the basic logic behind the capital-asset pricing model.

In a big fat nutshell, the proof of the capital-asset pricing model (henceforth to be known as CAPM because we economists love to use letter abbreviations) can be stated as follows:

> If investors did get an extra return (a risk premium) for bearing unsystematic risk, it would turn out that diversified portfolios made up of stocks with large amounts of unsystematic risk would give larger returns than equally risky portfolios of stocks with less unsystematic risk. Investors would snap at the chance to have these higher returns, bidding up the prices of stocks with large unsystematic risk and selling stocks with equivalent betas but lower unsystematic risk. This process would continue until the prospective returns of stocks with the same betas were equalized and no risk premium could be obtained for bearing unsystematic risk. Any other result would be inconsistent with the existence of an efficient market.

The key relationship of the theory is shown in Figure I.2. As the systematic risk (beta) of an individual stock (or portfolio) increases, so does the return an investor can expect. If an investor's portfolio has a beta of zero, as might be the case if all his funds were invested in a bank savings certificate (beta would be zero since the returns from the certificate would not vary at all with swings in the stock market), the investor would receive some modest rate of return, which is generally called the risk-free rate of interest. As the individual takes on more risk, however, the return should increase. If the investor holds a portfolio with a beta of 1 (as, for example, holding a share in one of the broad stock market averages) his return will equal the general return from common stocks. This return has over long periods of time exceeded the risk-free rate of interest, but the investment is a risky one. In certain periods the return is much less than the risk-free rate and involves taking substantial losses. This, as we have said, is precisely what is meant by risk.

The diagram shows that a number of different expected returns are possible simply by adjusting the beta of the portfolio. For example, suppose the investor put half of his money in a savings certificate and half in a share of the market averages. In this case he would receive a return midway between the risk-free return and the return from the market and his portfolio would have an average beta of 0.5.[2] The CAPM then asserts very simply that to get a higher average long-run rate of return you should just increase the beta of your portfolio. An investor can get a portfolio with a beta larger than 1 either by buying high-beta stocks or by purchasing a portfolio with average volatility on margin. . . .

[2]In general, the beta of a portfolio is simply the weighted average of the betas of its component parts.

Rate of Return

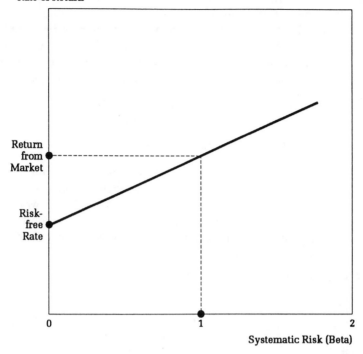

Figure I.2. Risk and Return According to the Capital-Asset Pricing Model

Note: Those who remember their high school algebra will recall that any straight line can be written as an equation. The equation for the straight line in the diagram is

Rate of Return = Risk-free Rate + Beta (Return from Market − Risk-free Rate).

Alternately, the equation can be written as an expression for the risk premium, that is, the rate of return on the portfolio or stock over and above the risk-free rate of interest:

Rate of Return − Risk-free Rate = Beta (Return from Market − Risk-free Rate).

The equation says that the risk premium you get on any stock or portfolio increases directly with the beta value you assume. Some readers may wonder what relationship beta has to the covariance concept that was so critical in our discussion of portfolio theory. The beta for any security is essentially the same thing as the covariance between that security and the market index as measured on the basis of past experience.

Anyone can theorize about how security markets work, and the capital-asset pricing model is just another theory. The really important question is: Does it work? . . .

By the early 1980s, according to a *Wall Street Journal* article, beta had become so popular that it underlay the investment rationale for $65 billion in U.S. pension funds. Beta also appeared to provide a method of evaluating a portfolio manager's performance. If the realized return is larger than that predicted by the overall portfolio beta, the manager is said to have produced a positive alpha. Lots of money in the market sought out the manager who could deliver the largest alpha.

But is beta a useful measure of risk? Do high-beta portfolios always fall farther in bear markets than low-beta ones? Is it true that high-beta portfolios will provide larger long-term returns than lower-beta ones, as the capital-asset pricing model suggests? Do present methods of calculating beta on the basis of past history give any useful information about future betas? Does beta alone summarize a security's total systematic risk, or do we need to consider other factors as well? . . .

Batting for Beta: The Supporting Evidence

Tests of the capital-asset pricing model have tried to ascertain if security returns are in fact directly related to beta, as the theory asserts. . . .

The enthusiasm for beta and for the CAPM in which it is wrapped has been fueled by charts, such as Figure I.3, that show the relationship over a twenty-year period between the performance of a large number of profession-

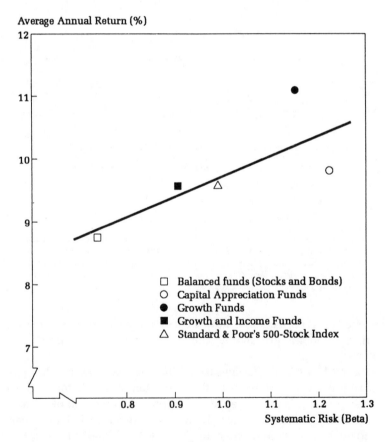

Figure I.3. Average Annual Return Versus Risk: Selected Mutual Funds (20 Years, 1969–1988)

ally managed funds and the beta measure of relative volatility. It is because the numbers are averages of many funds that the relationship between risk and reward is so tight. Still, the results appear to be quite consistent with the theory. The portfolio returns have varied positively with beta in (almost) a straight-line manner, so that over the long pull, high-beta portfolios have provided larger total returns than low-risk ones.

Figures I.4 and I.5 break down the twenty-year performance into two subperiods: (1) the fifteen years when the market went up and (2) the five years when it went down. Again, the relationship is exactly as predicted by the theory. In "up" years, the high-beta portfolios well outdistanced the low-beta ones. (Since the market was up on average from 1969 through 1988, this same relationship held over the whole period.) In "down" years, however, the high-beta portfolios did considerably worse than the low-volatility ones. It was the high-beta portfolios that took the real drubbings in the bear market periods of the 1970s. Of course, this is precisely what we mean by the concept of risk, and this is why betas for diversified portfolios appear to be useful risk measures.

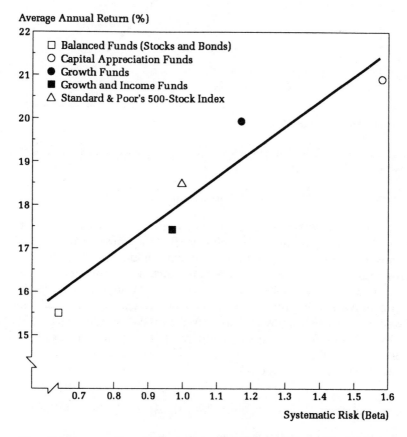

Figure I.4. Average Annual Return Versus Risk: Selected Mutual Funds (15 "Up" Years, 1969–1988)

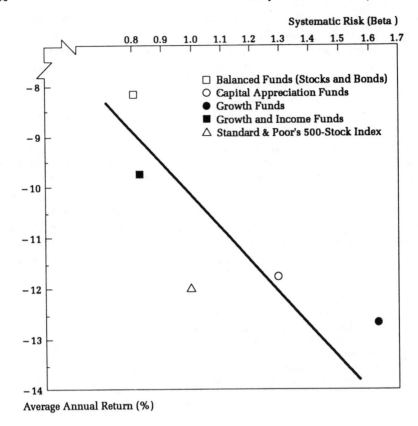

Figure I.5. Average Annual Return Versus Risk: Selected Mutual Funds (5 "Down" Years, 1969–1988)

Being Bearish on Beta: Some Disquieting Results

Like just about everything in life, beta may work well some of the time, but it certainly doesn't live up to its press billings all of the time. Burrowing away at the statistical base of the capital-asset pricing model, the beta bears have uncovered major flaws. The evidence contradicting this fundamental part of the new investment technology has sent some practitioners and academics off in search of ways to improve the CAPM. And even some institutional investors who in the past swore by the model began disavowing it altogether by the late 1980s. In order to understand this reaction, we need to examine the academic studies that led to beta's fall from grace, at least in the minds of some academics and professionals.

Academic Attack 1: Theory Does Not Measure Up to Practice

Recall that the CAPM could be reduced to a very simple formula:

Rate of Return = Risk-free Rate + Beta (Return from Market
− Risk-free Rate).

Thus, a security with a zero beta should give a return exactly equal to the risk-free rate. Unfortunately, the actual results don't come out that way.

This damning accusation is the finding from an exhaustive study of all the stocks on the New York Stock Exchange over a thirty-five-year period. The securities were grouped into ten portfolios of equal size, according to their beta measures for the year. Thus, Portfolio I consisted of the 10 percent of the NYSE securities with the highest betas. Portfolio II contained the 10 percent with the second-highest betas, and so forth. Figure I.6 shows the relation between the average monthly return and the beta for each of the ten different portfolios (shown by the black dots on the chart) over the entire period. The market portfolio is denoted by an open circle, and the solid line is a line of best fit (a regression line) drawn through the dots. The dashed line connects the average risk-free rate of return with the rate of return on the market portfolio. This is the theoretical relationship of the CAPM that was described above.

If the CAPM were absolutely correct, the theoretical and the actual relationship would be one and the same. But practice, as can quickly be seen, is not

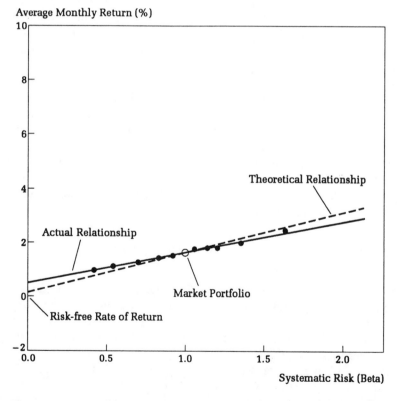

Figure I.6. Systematic Risk (Beta) Versus Average Monthly Return for Ten Different-Risk Portfolios, and the Market Portfolio, for 1931–1965

Source: Black, Jensen, and Scholes, "The Capital Asset Pricing Model: Some Empirical Tests," in *Studies in the Theory of Capital Markets,* ed. Jensen (New York: Praeger 1972). Reprinted with permission.

represented by the same line as theory on the chart. Note particularly the difference between the rate of return on an actual zero-beta common stock or portfolio of stocks and the risk-free rate. From the chart, it is clear that the measured zero-beta rate of return exceeds the risk-free rate. Since the zero-beta portfolio and a portfolio of riskless assets such as Treasury bills have the same systematic risk (beta), this result implies that something besides a beta measure of risk is being valued in the market. It appears that some unsystematic (or at least some nonbeta) risk makes the return higher for the zero-beta portfolio.

Furthermore, the actual risk–return relationship (examined by Black, Jensen, and Scholes) appears to be flatter than that predicted by the CAPM; low-risk stocks earn higher returns, and high-risk stocks earn lower returns, than the theory predicts. . . .

Academic Attack 2: Beta Is a Fickle Short-Term Performer (and Sometimes It Fails to Work for a Full Decade)

The divergence of theory from evidence is even more striking in the short run: For some short periods, it may happen that risk and return are *negatively* related. In 1972, for example, which was an "up" market year, it turned out that safer (lower-beta) stocks went up *more* than the more volatile securities. . . .

Black, Jensen, and Scholes found a similar type of anomaly for the entire period from April 1957 through December 1965. . . .

Not only does the zero-beta return exceed the riskless rate here, but during this period of nearly nine years, securities with higher risk produced *lower* returns than less-risky (lower-beta) securities. Substantial deviations from the relationship predicted by the CAPM were also found for many subperiods.

The experience of the 1980s provided even more dramatic evidence of the folly of relying on beta measures to predict realized rates of return. It turned out that for the entire decade of the 1980s realized mutual fund returns bore no relationship to their beta measures of risk. . . .

Academic Attack 3: Estimated Betas Are Unstable

Another problem the theory encounters is the instability of measured betas. . . . Any changes in the economy, in the characteristics of an individual company, or in the competitive situation facing the company can be expected to change the sensitivity of the company's stock to market fluctuations. It would be surprising to discover that betas of individual stocks did not vary widely over time. In fact, they do vary. . . .

Marshall Blume, a professor at the Wharton School of Finance, conducted several tests of the stability of historical beta estimates. He found that the smaller the number of securities in the portfolio, the weaker the relationship between portfolio betas for consecutive periods. . . . Better predictive power is obtained from betas calculated for portfolios containing larger numbers of stocks. Thanks to the law of large numbers, a number of inaccurate beta estimates on individual stocks can be combined to form a much more accurate estimate of the risk of the portfolio as a whole. While the beta estimates for some securities will be much too high, the estimates for many others will be too low. . . .

Academic Attack 4: Beta Is Easily Rolled Over

Perhaps the most devastating criticism of beta has been delivered by Richard Roll. . . .

Roll says that it is impossible to observe the market's return. Because, in principle, the market includes *all* stocks, a variety of other financial instruments, and even nonmarketable assets such as an individual's investment in education, the S&P index (or any other index used to represent the market) is a very imperfect market proxy at best. And when we measure market risk using an imperfect proxy, we may obtain a quite imperfect estimate of market sensitivity. Roll showed that by changing the market index against which betas are measured, one can obtain quite different measures of the risk levels of individual stocks or portfolios. As a consequence, one would make very different predictions about the expected returns from the stocks or portfolios. He further demonstrated that by changing market indexes (from, say, the S&P 500 to the much broader Wilshire 5,000) one could actually reverse the risk-adjusted performance rankings (alphas) of fund managers. But if betas differ according to the market proxy that they are measured against, and if you never can get a measure of the "true" market portfolio, then, in effect, the CAPM has not been (and cannot be) adequately tested. . . .

Searching for the Investment Grail

The evidence that supports the efficiency of capital markets and the existence of what is usually a positive relationship between measured risk and return is far too abundant for anyone to reject the new investment technology out of hand. And since academics and practitioners have already made substantial progress in building better theories of the risk–return relationship, the practical consequence of the failures of the CAPM is likely to be *more* discriminating risk measurement, with the use of even more quantitative tools in risk analysis—not less. In this section, I will give the flavor of some of the new approaches to security pricing that have been developed as alternatives to the CAPM, and will present their practical meaning to investment analysts.

The Quant[3] Quest for Better Measures of Risk: Arbitrage Pricing Theory

One of the pioneers in the field of risk measurement is the Yale School of Management's finance wunderkind, Stephen Ross. Ross has developed a new

---

[3]"Quant" is the Wall Street nickname for the quantitatively inclined financial analyst who devotes attention largely to the new investment technology [of mathematical models such as the CAPM or APT].

theory of pricing in the capital markets called "APT," or arbitrage pricing theory. APT has had wide influence both in the academic community and in the practical world of portfolio management. To understand the logic of the newest APT work on risk measurement, one must remember the correct insight underlying the CAPM: The only risk that investors should be compensated for bearing is the risk that cannot be diversified away. Only systematic risk will command a risk premium in the market. But the systematic elements of risk in particular stocks and portfolios may be too complicated to be capturable by a measure of beta—the tendency of the stocks to move more or less than the market. This is especially so since any particular stock index is a very imperfect representative of the general market. Hence, many quants now feel that beta may fail to capture a number of important systematic elements of risk.

Let's take a look at several of these other systematic risk elements. Changes in national income, for one, may affect returns from individual stocks in a systematic way. This was shown in our illustration of a simple island economy, in the preceding section. Also, changes in national income mirror changes in the personal income of individuals, and the systematic relationship between security returns and salary income can be expected to have a significant effect on individual behavior. For example, the laborer in a Ford plant will find a holding of Ford common stock particularly risky, since job layoffs and poor returns from Ford stock are likely to occur at the same time. Changes in national income may also reflect changes in other forms of property income and may therefore be relevant for institutional portfolio managers as well.

Changes in interest rates also systematically affect the returns from individual stocks and are important nondiversifiable risk elements. To the extent that stocks tend to suffer as interest rates go up, equities are a risky investment, and those stocks that are particularly vulnerable to increases in the general level of interest rates are especially risky. Thus, many stocks and fixed-income investments will tend to move in parallel, and these stocks will not be helpful in reducing the risk of a bond portfolio. Since fixed-income securities are a major part of the portfolios of many institutional investors, this systematic risk factor is particularly important for some of the largest investors in the market. Clearly, then, investors who think of risk in its broadest and most meaningful sense will be sensitive to the tendency of certain stocks to be particularly affected by changes in interest rates.

Changes in the rate of inflation will similarly tend to have a systematic influence on the returns from common stocks. This is so for at least two reasons. First, an increase in the rate of inflation tends to increase interest rates and thus tends to lower the prices of equities, as just discussed. Second, the increase in inflation may squeeze profit margins for certain groups of companies—public utilities, for example, which often find that rate increases lag behind increases in costs. On the other hand, inflation may benefit the prices of common stocks in the natural resource industries. Thus, again there are important systematic relationships between stock returns and economic variables that may not be captured adequately by a simple beta measure of risk.

Statistical tests of the influence on security returns of several systematic risk variables have shown promising results. Better explanations than those given by the CAPM can be obtained for the variation in returns among different securities by using, in addition to the traditional beta measure of risk, a number of systematic risk variables, such as sensitivity to changes in national income, in interest rates, and in the rate of inflation. Of course, the evidence supporting many risk-factor models of security pricing has only begun to accumulate. It is not yet certain how these new theories will stand up to more extensive examination. Still, the preliminary results are definitely encouraging.

From *Corporate Finance*

STEPHEN A. ROSS, RANDOLPH W. WESTERFIELD, AND JEFFREY F. JAFFE

A Description of Efficient Capital Markets

An efficient capital market is one in which stock prices fully reflect available information. To illustrate how an efficient market works, suppose the F-stop Camera Corporation (FCC) is attempting to develop a camera that will double the speed of the auto-focusing system now available. FCC believes this research has positive NPV.[1] The value of the new auto-focusing system will depend on demand for cameras at the time of the discovery, as well as on many other factors.

Now consider a share of stock in FCC. What determines the willingness of investors to hold shares of FCC at a particular price? One important factor is the probability that FCC will be the company to develop the new auto-focusing system first. In an efficient market we would expect the price of the shares of FCC to increase if this probability increases.

Suppose a well-known engineer is hired by FCC to help develop the new auto-focusing system. In an efficient market, what will happen to FCC's share price when this is announced? If the well-known scientist is paid a salary that fully reflects his or her contribution to the firm, the price of the stock will not necessarily change. Suppose, instead, that hiring the scientist is a positive

[1]NPV, the net present value of an investment, is the present value of the project's future cash flows (which is determined by answering the question how much money must be put in a bank today to receive the future cash flow, and then adjusting that answer for risk) minus the present value of its costs [EDITOR'S NOTE].

NPV transaction. In this case, the price of shares in FCC will increase because the market for scientists is imperfect and FCC can pay the scientist a salary below his or her true value to the company.

When will the increase in the price of FCC's shares take place? Assume that the hiring announcement is made in a press release on Wednesday morning. In an efficient market, the price of shares in FCC will immediately adjust to this new information. Investors should not be able to buy the stock on Wednesday afternoon and make a profit on Thursday. This would imply that it took the stock market a day to realize the implication of the FCC press release. The *efficient-market hypothesis* predicts that the price of shares of FCC stock on Wednesday afternoon will already reflect the information contained in the Wednesday morning press release.

The efficient-market hypothesis (EMH) has implications for investors and for firms.

> Because information is reflected in prices immediately, investors should only expect to obtain a normal rate of return. Awareness of information when it is released does an investor no good. The price adjusts before the investor has time to trade on it.
>
> Firms should expect to receive the fair value for securities that they sell. *Fair* means that the price they receive for the securities they issue is the present value. Thus, valuable financing opportunities that arise from fooling investors are unavailable in efficient capital markets.

Some people spend their entire careers trying to pick stocks that will outperform the average. For any given stock, they can learn not only what has happened in the past to the stock price and dividends, but also what the company earnings have been, how much debt it owes, what taxes it pays, what businesses it is in, what market share it has for its products and how well it is doing in each of its businesses, what new investments it has planned, how sensitive it is to the economy, and so on.

If you want to learn about a given company and its stock, an enormous amount of information is available to you. The preceding list only scratches the surface; we haven't even included the information that only insiders know. *Inside information* is information possessed by people in special positions inside the company, such as the major officers of the company or people farther down in the company who might be aware of some special discovery or new development.

Not only is there a lot to know about any given company, there is also a powerful motive for doing so, the profit motive. If you know more about a company than other investors in the marketplace, you can profit from that knowledge by investing in the company's stock if you have good news or selling it if you have bad news.

There are other ways to use your information. If you could convince investors that you have reliable information about the fortunes of companies, you might start a newsletter and sell investors that information. . . .

The logical consequence of all of this information being available, studied, sold, and used in an effort to make profits from stock market trading is that the market becomes *efficient.* A market is efficient with respect to information if there is no way to make unusual or excess profits by using that information. When a market is efficient with respect to information, we say that prices *incorporate* the information. Without knowing anything special about a stock, an investor in an efficient market expects to earn an equilibrium required return from an investment, and a company expects to pay an equilibrium cost of capital.

Example. Suppose IBM announces it has invented a microprocessor that will make its computer thirty times faster than existing computers. The price of a share of IBM should increase immediately to a new equilibrium level.

Figure I.7 presents three possible adjustments in stock prices. The solid line represents the path taken by the stock in an efficient market. In this case the price adjusts immediately to the new information so that no further changes take place in the price of the stock. The broken line depicts a delayed reaction. Here it takes the market eight to ten days to fully absorb the informa-

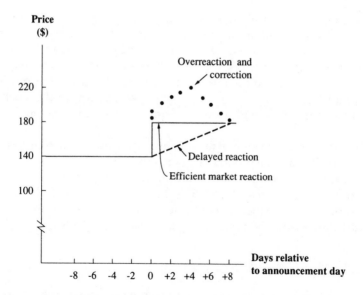

Figure I.7. Reaction of Stock Price to New Information in Efficient and Inefficient Markets

Efficient market reaction: The price instantaneously adjusts to and fully reflects new information; there is no tendency for subsequent increases and decreases.

Delayed reaction: The price partially adjusts to the new information; eight days elapse before the price completely reflects the new information.

Overreaction: The price overadjusts to the new information; there is a bubble in the price sequence.

tion. Finally, the dotted line illustrates an overreaction and subsequent correction back to the true price. The broken line and the dotted line show the paths that the stock price might take in an inefficient market. If the price of the stock takes several days to adjust, trading profits would be available to investors who bought at the date of the announcement and sold once the price settled back to the equilibrium. . . .

The Different Types of Efficiency

In our previous discussion, we assumed that the market responds immediately to all available information. In actuality, certain information may affect stock prices more quickly than other information. To handle differential response rates, researchers separate information into different types. The most common classification system speaks of three types: information on past prices, publicly available information, and all information. The effect of these three information sets on prices is examined below.

The Weak Form

A capital market is said to be *weakly efficient* or to satisfy weak-form efficiency if it fully incorporates the information in past stock prices. . . .

Weak-form efficiency is about the weakest type of efficiency that we would expect a financial market to display because historical price information is the easiest kind of information about a stock to acquire. If it were possible to make extraordinary profits simply by finding the patterns in the stock price movements, everyone would do it, and any profits would disappear in the scramble. . . .

By denying that future market movements can be predicted from past movements, we are denying the profitability of a host of techniques falling under the heading of technical analysis. Furthermore, we are denigrating the work of all of their followers, who are called technical analysts. The term technical analysis, when applied to the stock market, refers among other things to attempts to predict the future from the patterns of past price movements. . . .

The Semistrong and Strong Forms

If weak-form efficiency is controversial, even more contentious are the two stronger types of efficiency, semistrong-form efficiency and strong-form efficiency. A market is semistrong-form efficient if prices reflect (incorporate) all publicly available information, including information such as published accounting statements for the firm as well as historical price information. A market is strong-form efficient if prices reflect all information, public or private.

The information set of past prices is a subset of the information set of publicly available information, which in turn is a subset of all information. . . . Thus, strong-form efficiency implies semistrong-form efficiency, and semistrong-form efficiency implies weak-form efficiency. The distinction between semistrong-form efficiency and weak-form efficiency is that semistrong-form efficiency requires not only that the market be efficient with respect to historical price information, but that all of the information available to the public be reflected in price. . . .

At the furthest end of the spectrum is strong-form efficiency, which incorporates the other two types of efficiency. This form says that anything that is pertinent to the value of the stock and that is known to at least one investor is, in fact, fully incorporated into the stock value. A strict believer in strong-form efficiency would deny that an insider who knew whether a company mining operation had struck gold could profit from that information. Such a devotee of the strong-form efficient-market hypothesis might argue that as soon as the insider tried to trade on his or her information, the market would recognize what was happening, and the price would shoot up before he or she could buy any of the stock. Alternatively, sometimes believers in strong-form efficiency take the view that there are no such things as secrets and that as soon as the gold is discovered, the secret gets out.

Are the hypotheses of semistrong-form efficiency and strong-form efficiency good descriptions of how markets work? Expert opinion is divided here. The evidence in support of semistrong-form efficiency is, of course, more compelling than that in support of strong-form efficiency, and for many purposes it seems reasonable to assume that the market is semistrong-form efficient. The extreme of strong-form efficiency seems more difficult to accept. Before we look at the evidence on market efficiency, we will summarize our thinking on the versions of the efficient-market hypothesis in terms of basic economic arguments.

One reason to expect that markets are weak-form efficient is that it is so cheap and easy to find patterns in stock prices. Anyone who can program a computer and knows a little bit of statistics can search for such patterns. It stands to reason that if there were such patterns, people would find and exploit them, in the process causing them to disappear.

Semistrong-form efficiency, though, uses much more sophisticated information and reasoning than weak-form efficiency. An investor must be skilled at economics and statistics, and steeped in the idiosyncrasies of individual industries and companies and their products. Furthermore, to acquire and use such skills requires talent, ability, and time. In the jargon of the economist, such an effort is costly, and the ability to be successful at it is probably in scarce supply.

As for strong-form efficiency, this is just farther down the road than semistrong-form efficiency. It is difficult to believe that the market is so efficient that someone with true and valuable inside information cannot prosper by using it. It is also difficult to find direct evidence concerning strong-form efficiency. What we have tends to be unfavorable to this hypothesis of market efficiency.

Some Common Misconceptions about the Efficient-Market Hypothesis

No idea in finance has attracted as much attention as that of efficient markets, and not all of the attention has been flattering. To a certain extent this is because much of the criticism has been based on a misunderstanding of what the hypothesis does and does not say. We illustrate three misconceptions below.

The efficacy of dart throwing. When the notion of market efficiency was first publicized and debated in the popular financial press, it was often characterized by the following quote: ". . . throwing darts at the financial page will produce a portfolio that can be expected to do as well as any managed by professional security analysts." This is almost, but not quite, true.

All the efficient-market hypothesis really says is that, on average, the manager will not be able to achieve an abnormal or excess return. The excess return is defined with respect to some benchmark expected return that comes from the SML.[2] The investor must still decide how risky a portfolio he or she wants and what expected return it will normally have. A random dart thrower might wind up with all of the darts sticking into one or two high-risk stocks that deal in genetic engineering. Would you really want all of your stock investments in two such stocks? (Beware, though—a professional portfolio manager could do the same.)

The failure to understand this has often led to a confusion about market efficiency. For example, sometimes it is wrongly argued that market efficiency means that it does not matter what you do because the efficiency of the market will protect the unwary. However, someone once remarked. "The efficient market protects the sheep from the wolves, but nothing can protect the sheep from themselves."

What efficiency does say is that the price that a firm will obtain when it sells a share of its stock is a fair price in the sense that it reflects the value of that stock given the information that is available about it. Shareholders need not worry that they are paying too much for a stock with a low dividend or some other characteristic, because the market has already incorporated it into the price. We sometimes say that the information has been *priced out.*

—

Price fluctuations. Much of the public is skeptical of efficiency because stock prices fluctuate from day to day. However, this price movement is in no way inconsistent with efficiency, because a stock in an efficient market adjusts to

[2]SML, the Security Market Line, is the upward sloping line described by the CAPM, which relates the expected return on a security to its beta (see Figure 1.2 in the Malkiel selection). All securities must lie on the line in equilibrium. For example, if a security with a beta of 1 was below the line, then investors could buy the market portfolio (which is on the line) and obtain a higher return for the same risk. No one would want to hold the other security and its price would fall as demand declined, thus raising the security's expected return, until it lay on the SML [EDITOR'S NOTE].

new information by changing price. In fact, the absence of price movements in a changing world might suggest an inefficiency.

Stockholder disinterest. Many laypersons are skeptical that the market price can be efficient if only a fraction of the outstanding shares changes hands on any given day. However, the number of traders in a stock on a given day is generally far less than the number of people following the stock. This is true because an individual will trade only when his appraisal of the value of the stock differs enough from the market price to justify incurring brokerage commissions and other transactions costs. Furthermore, even if the number of traders following a stock is small relative to the number of outstanding share-holders, the stock can be expected to be efficiently priced as long as a number of interested traders use the publicly available information. That is, the stock price can reflect the available information even if many stockholders never follow the stock and are not considering trading in the near future, and even if some stockholders trade with little or no information. Thus, the empirical findings suggesting that the stock market is predominantly efficient need not be surprising. . . .

The Evidence

The record on the efficient-market hypothesis is extensive, and in large measure it is reassuring to advocates of the efficiency of markets. The studies done by academicians fall into broad categories. First, there is evidence as to whether changes of stock prices are random. Second are event studies. Third is the record of professionally managed investment firms.

The Weak Form

The random-walk hypothesis . . . implies that a stock's price movement in the past is unrelated to its price movement in the future. . . .

Financial economists frequently speak of serial correlation, which involves only one security. This is the correlation between the current return on a security and the return on the same security over a later period. A positive coefficient of serial correlation for a particular stock indicates a tendency toward *continuation*. That is, a higher-than-average return today is likely to be followed by higher-than-average returns in the future. Similarly, a lower-than-average return today is likely to be followed by lower-than-average returns in the future.

A negative coefficient of serial correlation for a particular stock indicates a tendency toward *reversal*. A higher-than-average return today is likely to be followed by lower-than-average returns in the future. Similarly, a lower-than-average return today is likely to be followed by higher-than-average returns in the future. Both significantly positive and significantly negative serial-correlation coefficients are indications of market inefficiencies; in either case, returns today can be used to predict future returns.

Serial correlation coefficients for stock returns near zero would be consistent with the random-walk hypothesis. Thus, a current stock return that is higher than average is as likely to be followed by lower-than-average returns as by higher-than-average returns. Similarly, a current stock return that is lower than average is as likely to be followed by higher-than-average returns as by lower-than-average returns.

Figure I.8 shows the serial correlation for daily stock price changes for nine stock markets. These coefficients indicate whether or not there are relationships between yesterday's return and today's return. For example, Germany's coefficient of 0.08 is slightly positive, implying that a higher-than-average return today makes a higher-than-average return tomorrow slightly more likely. Conversely, Belgium's coefficient is slightly negative, implying that a lower-than-average return today makes a higher-than-average return tomorrow slightly more likely.

However, because correlation coefficients can, in principle, vary between −1 and 1, the reported coefficients are quite small. In fact, the coefficients are so small relative to both estimation errors and to transactions costs that the results are generally considered to be consistent with weak-form efficiency.

The weak form of the efficient-market hypothesis has been tested in many

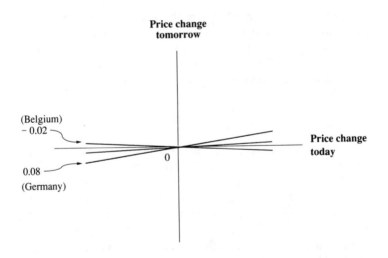

Figure I.8. Testing the Random-walk Hypothesis

Germany's coefficient of 0.08 is slightly positive, implying that a positive return today makes a positive return tomorrow slightly more likely. Belgium's coefficient is negative, implying that a negative return today makes a positive return tomorrow slightly more likely. However, the coefficients are so small relative to estimation error and to transaction costs that the results are generally considered to be consistent with efficient capital markets.

Source: Data from B. Solnik, "A Note on the Validity of the Random Walk for European Stock Prices," *Journal of Finance* (December 1973). Used with permission.

Simulated market levels for 52 weeks

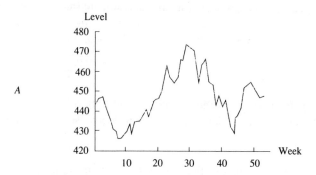

A

Friday closing levels: December 30, 1955, to December 28, 1956
Dow Jones Industrial Index

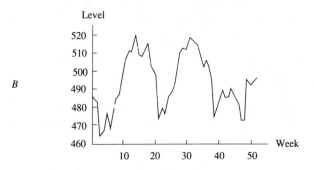

B

Figure I.9. Simulated and Actual Stock Price Returns.
A. Simulated market levels for 52 weeks.
B. Friday closing levels from December 30, 1955, to December 28, 1956—Dow Jones Industrial Index.

The two graphs display similar patterns. Because the upper graph is randomly generated, it is hard to argue that the bottom graph is nonrandom.

Source: From H. Roberts, "Stock Market Patterns and Financial Analysis: Methodological Suggestions," *Journal of Finance* (March 1959). Used with permission.

other ways as well. Our view of the literature is that the evidence, taken as a whole, is strongly consistent with weak-form efficiency.

This finding raises an interesting thought: If price changes are truly random, why do so many believe that prices follow patterns? The work of Harry Roberts suggests that most people simply do not know what randomness looks like. For example, consider Figure I.9. The top graph was generated by computer using random numbers and [the] equation [for a random walk]. Because of this, it must follow a random walk. Yet, we have found that people examining the chart continue to see patterns. Different people will see different patterns and will forecast different future price movements. However, in our experience, viewers are all quite confident of the patterns they see.

Next, consider the bottom graph, which tracks actual movements in the Dow Jones Index. This graph may look quite nonrandom to some, suggesting weak-form inefficiency. However, it also bears a close visual resemblance to the simulated series above, and statistical tests indicate that it indeed behaves like a purely random series. Thus, in our opinion, people claiming to see patterns in stock price data are probably seeing optical illusions.

The Semistrong Form

The semistrong form of the efficient-market hypothesis implies that prices should reflect all publicly available information. We present two types of tests of this form.

Event studies. A way to think of the tests of the semistrong form is to examine the following system of relationships:

$$\text{Information released at time } t - 1 \rightarrow AR_{t-1}$$
$$\text{Information released at time } t \rightarrow AR_t$$
$$\text{Information released at time } t + 1 \rightarrow AR_{t+1}$$

where AR stands for a stock's abnormal return and where the arrows indicate that the return in any time period is related only to the information released during that period. The abnormal return (AR) of a given stock on a particular day can be measured by subtracting the market's return on the same day (R_m)—as measured by the S&P (Standard & Poor's) composite index—from the actual return (R) of the stock on that day:[3]

$$AR = R - R_m$$

According to the efficient-market hypothesis, a stock's abnormal return at time t, AR_t, should reflect the release of information at the same time, t. Any information released before then, though, should have no effect on abnormal returns in this period, because all of its influence should have been felt before. In other words, an efficient market would already have incorporated previous information into prices. Because a stock's return today cannot depend on what the market does not yet know, the information that will be known only in the future cannot influence the stock's return either. Hence the arrows point in the direction that is shown, with information in any one time period affecting only that period's abnormal return. *Event studies* are statistical studies that examine whether the arrows are as shown or whether the release of information influences returns on other days.

One of the first event studies was conducted by Fama, Fisher, Jensen, and Roll, who studied 940 stock splits. Figure I.10 shows [a redrawn] plot of the *cumulative abnormal return* (CAR) for the stock split sample. Compare the CAR with the plots in Figure I.7. Positive abnormal returns were observed before the stock split, probably because firms tend to split in good times. In

---

[3]The abnormal return can also be measured by using the market model. In this case the abnormal return is

$$AR = R - (\alpha + \beta R_m)$$

Cumulative Abnormal Returns

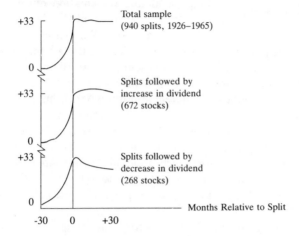

Figure I.10. Abnormal Returns for Companies Announcing Stock Splits.

Cumulative abnormal returns rise prior to month of split. Very likely this occurs because splits take place in good times, that is, they take place *following* a rise in stock price. Abnormal returns are flat after month of split, a finding consistent with efficient capital markets.

Source: Redrawn from E. F. Fama, L. Fisher, M. C. Jensen, and R. Roll, "The Adjustment of Stock Prices to New Information," *International Economic Review* 10 (February 1969), pp. 1–31. Reproduced with permission.

addition, positive abnormal returns were observed around the time the split was announced. Fama et al. suggested that stock splits released information to the market, perhaps as signals of future dividend increases. After the split they observed no further tendency for the CAR to increase. This is consistent with efficient financial markets. To see this, note that investors could profit by buying stock on the split date, if the CAR continued to rise after that date.

Over the years this type of methodology has been applied to a large number of events. Announcements of dividends, earnings, mergers, capital expenditures, and new issues of stock are a few examples of the vast literature in the area. Although there are exceptions, the event-study tests generally support the view that the market is semistrong-form (and therefore also weak-form) efficient. In fact, the tests even tend to support the view that the market is gifted with a certain amount of foresight. By this we mean that news tends to leak out and be reflected in stock prices even before the official release of the information.

Tests of market efficiency can be found in the oddest places. The price of frozen orange juice depends to a large extent on the weather in Orlando, Florida, where many of the oranges that are frozen for juice are grown. One researcher found that he could actually use frozen-orange-juice prices to improve the U.S. Weather Bureau's forecast of the temperature for the following night! Clearly the market knows something that the weather forecasters do not. . . .

The record of mutual funds.   If the market is efficient in the semistrong form, then no matter what publicly available information mutual-fund managers rely on to pick stocks, their average returns should be the same as those of the average investor in the market as a whole. We can test efficiency, then, by comparing the performance of these professionals with that of a market index.

As you might imagine, the studies differ in their particular samples of mutual-fund performance, and they differ in their use of statistics. Because the object of the studies is to detect abnormal performance, it is not surprising that they employ different theories for determining what normal performance is. But the general conclusion of all of the theories is the same, that there is no evidence that the funds outperform suitably selected indices. The most common index of market performance is the S&P composite index. Ignoring some small costs of trading, a big investor could achieve the S&P 500 return by buying the same stocks in the same proportion as the index, that is, by *buying the index.*

The general picture that emerges from these studies is illustrated in Figure I.11. This figure plots the average return and beta of different mutual fund managers. The funds were divided into groups based on their investment objectives. In this study the average mutual fund manager underperformed

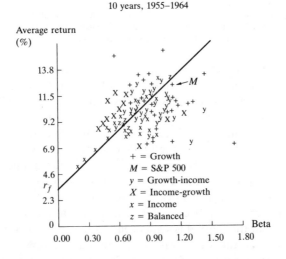

Figure I.11. Jensen Index for Mutual Funds

The graph depicts average return and beta for many different mutual funds over period from 1955 through 1964. The solid line depicts the return that a naive investor can earn from randomly selecting securities with a given beta. The evidence suggests that about half of the mutual funds outperform the line and about half underperform the line. The evidence is consistent with market efficiency.

Source: From M. Jensen, "Risks, the Pricing of Capital Assets, and the Evaluation of Investment Performance," *Journal of Business* 42 (April 1969), p. 2. Copyright © 1969 The University of Chicago. Reproduced by permission.

the S&P composite index (denoted M in Figure I.11). (One mutual fund manager whose fund was recently beaten by the S&P composite index was heard to remark, "If the S&P were an athlete, it would be accused of taking steroids."[4])

Perhaps nothing rankles successful stock market investors more than to have some professor tell them that they are not necessarily smart, just lucky. However, the view that mutual fund managers, on average, have no special ability to beat the stock market indices is largely supported by the evidence. This does not mean that no individual investor can beat the market average or that he or she lacks a special insight, only that proof seems difficult to find.

By and large, mutual fund managers rely on publicly available information. Thus the finding that they do not outperform the market indices is consistent with semistrong-form and weak-form efficiency. This does not imply that mutual funds are bad investments for individuals. Though these funds fail to achieve better returns than some indices of the market, they do permit the investor to buy a portfolio that has a large number of stocks in it (the phrase "a well-diversified portfolio" is often used). They might also be very good at providing a variety of services such as keeping custody and records of all of the stocks.

Some contrary views. Although the bulk of the evidence supports the view that markets are efficient, we would not be fair if we did not note the existence of contrary results. A number of researchers have argued that the particular statistical tests that have been used are so weak that even an inefficient market would pass them. For example, some have pointed out that looking to see if prices are serially correlated is not enough; instead we should be looking at much more complicated kinds of dependence.

These arguments tend to bog down in a morass of statistics, but one particularly interesting direction suggested by these critics is the variability of

---

[4]Sometimes when confronted with this sort of evidence, practitioners take the view that although in any given year the S&P composite and other market indices may do very well, nevertheless some managers are able to outperform the index year after year. That may well be true, but it is not necessarily evidence against the efficient-market view. Rather it may be an example of what statisticians refer to as *survivorship bias.*

Suppose we look at a sample of the twenty largest mutual funds as of December 1988 and we ask how these funds performed over the past ten years. Would you be surprised to discover that these funds had outperformed the stock market indices? Would this constitute evidence against market efficiency?

The answer to the first question is that you should not be surprised to find that these funds had been good performers. After all, you are looking at twenty funds that survived and, since you are looking at the twenty biggest, you are looking at the funds that have thrived. It is a bit like looking at the five tallest people in a high school class and going back to when they were in the seventh grade and asking how tall they were then. You wouldn't be surprised to find that, on average, they were tall then, too.

Even if mutual fund managers had no ability to pick stocks at all, if you go backward and look at the records of the twenty biggest funds today, you would still find that they were strong performers in the past. The proper way to study performance is to go back into the past, pick any sample of managers, and follow them forward in time.

stock returns. If the market is efficient, a stock's price should change with the arrival of information. The variance of stock returns, then, should be related to the amount of information. Examining this proposition, Shiller and others have unearthed some new evidence on efficiency, and much of it remains difficult to fit into our story. They conclude that the variance of stock prices is too large for efficient markets.

Some of financial economics' most enigmatic empirical findings concern seasonalities. Keim has found that firms with small market capitalizations have abnormally high returns on the first five trading days of January.[5] French found that the returns on stocks on Friday are abnormally high and on Monday they are negative. It is difficult to imagine any reasonable model of equilibrium consistent with the efficient-market hypothesis that could also be consistent with the Keim and French results.

In addition, the stock market crash of October 19, 1987, is extremely puzzling. The market dropped between 20 percent and 25 percent on a Monday following a weekend during which little surprising news was released. A drop of this magnitude for no apparent reason is not consistent with market efficiency. Because the crash of 1929 is still an enigma, it is doubtful that the more recent debacle will be explained anytime soon. The comments of an eminent historian in the 1970s are apt here: When asked what, in his opinion, the effect of the French Revolution of 1789 was, he replied that it was too early to tell. . . .

The Strong Form

Even the strongest adherents to the efficient-market hypothesis would not be surprised to find that markets are inefficient in the strong form. After all, if an individual has information that no one else has, it is likely that he can profit from it.

One group of studies of strong-form efficiency investigates insider trading. Insiders in firms have access to information that is not generally available. But if the strong form of the efficient-market hypothesis holds, they should not be able to profit by trading on their information. A government agency, the Securities and Exchange Commission, requires insiders in companies to reveal any trading they might do in their own company's stock. By examining the record of such trades, we can see whether they made abnormal returns. A number of studies support the view that these trades were abnormally profitable. Thus, strong-form efficiency does not seem to be substantiated by the evidence.

---

[5]Market capitalization is the price per share of stock multiplied by the number of shares outstanding.

Notes and Questions

1. As Malkiel indicates, portfolio theory is founded on the behavioral assumption that investors dislike risk. A risk-averse investor will not accept a fair bet, which is defined as a gamble whose cost to play equals its expected value. An example will clarify the concept. Consider the following gamble: a fair coin is flipped and if it comes up heads you receive $500 and if tails you pay $100. The expected value of this gamble is .5(500) + .5(−100) = $200. A risk-averse investor will accept this bet only if she can pay less than $200 to play. In fact, she would be willing to pay a premium to avoid risk. A risk-loving investor enjoys taking risk and is willing to pay more than $200 to play the game. This investor enjoys visiting Las Vegas as well. A risk-neutral investor is defined as an individual who accepts fair bets (she is willing to take our gamble at the $200 price.) Economists equate risk aversion with a declining marginal utility for wealth, that is, a person's second dollar provides less utility than her first, and a dollar loss provides more disutility or pain than a dollar gain provides in utility or pleasure. Risk-averse investors are concerned, then, not only with an investment's return but also with its risk. There is considerable evidence in support of the assumption of risk aversion: individuals buy insurance and hold diversified portfolios.

2. The insight from portfolio theory that only market or systematic risk is priced, has several implications for corporate law. It implies that diversified shareholders will not benefit from a merger whose objective is to reduce firm-specific risk. The selection by Yakov Amihud and Baruch Lev in Chapter VI is an elaboration of this idea. It further suggests that the oft-noted absence of cases holding directors liable for breach of the duty of care (the fiduciary duty of directors to act with reasonable skill, diligence, and care), see Joseph Bishop, "Sitting Ducks and Decoy Ducks: New Trends in the Indemnification of Corporate Directors and Officers," 77 *Yale Law Journal* 1078 (1968), may not be of great concern because diversification protects shareholders from the vagaries of a particular firm's performance due to management's negligence.

3. Malkiel's proof of the CAPM involves what is referred to as a no-arbitrage condition for market equilibrium. If portfolio prices do not adjust immediately, then a profitable opportunity exists for arbitrage, which is defined as buying something at one price and simultaneously selling it, or something equivalent, at a higher price. A more colloquial term for an arbitrage opportunity is a money machine (for every dollar you put in, such a machine gives you more than a dollar back). In competitive financial markets, investors will flock to exploit an arbitrage opportunity, and their demand for the cheaply priced product and supply of the dear product eliminates the price differential. As a consequence, investors' ability to make arbitrage profits in such markets is small and fleeting. As a well-known finance text puts it, in

well-functioning capital markets, "there is no such thing as a money machine." Richard A. Brealey and Stewart C. Myers, *Principles of Corporate Finance,* 4th ed., 31 (New York: McGraw-Hill 1992).

4. Ross, Westerfield, and Jaffe refer to event studies, statistical studies used to test the semistrong form of the efficient-market hypothesis. They are also an important technique for policy analysis in corporate law. As Ross, Westerfield, and Jaffe indicate, event-study topics are quite varied, from actions taken by managers, such as stock splits, to exogenous events, such as deaths of CEOs, or events that are a mix of both (caused by managers and natural events or actions by third parties), such as product recalls. The technique has also been used to evaluate the impact of new statutes and regulation on firms. For a useful discussion of the methodology's adaptation to such events, see G. William Schwert, "Using Financial Data to Measure Effects of Financial Regulation," 24 *Journal of Law and Economics* 121 (1981). Finally the event-study methodology has proved quite useful in securities litigation (see Chapter VII). An event study is included in Chapter VI (the selection by Jonathan Karpoff and Paul Malatesta), along with selections that collect and review the findings of numerous event studies (the selections by Michael Jensen and Richard Ruback and by Gregg Jarrell, James Brickley, and Jeffry Netter).

The event study's importance is in measuring; by providing an anchor for determining value, it eliminates reliance on ad hoc judgments about the impact of specific information ("events") on stock prices. We begin with a model of how stock prices are normally generated, such as the CAPM. This enables us to determine whether a piece of information (an "event"), such as the announcement of a stock split, affects the stock price, for we can ask whether stock returns in the event period are different from what our pricing model predicted. Because, as Malkiel discusses, the CAPM is a linear model, the statistical technique employed is linear regression analysis, which fits the best line to the data. For an excellent introduction to the linear regression model, see Robert S. Pindyck and Daniel L. Rubinfeld, *Econometric Models & Economic Forecasts,* 3d ed. (New York: McGraw-Hill, 1991).

Event studies often use a pricing model somewhat simpler than the CAPM, known as the market model, which simply adjusts the stock's return, R_i for the return on the market, R_m:

$$R_{it} = a_i + b_i R_{mt} + \epsilon_{it}$$

Actual returns for the stock and the market portfolio are used to estimate the intercept and slope parameters over an extended period of time prior to the event date (the subscript t for the variables indexes time). These estimates are then used to predict the stock price on the event date. Stephen Brown and Jerold Warner, in a comprehensive review of event-study methodology, indicate that the market model performs as well as other models, including the CAPM. Brown and Warner, "Using Daily Stock Returns: The Case of Event Studies," 14 *Journal of Financial Economics* 3 (1985); Brown and Warner,

"Measuring Security Price Performance," 8 *Journal of Financial Economics* 205 (1980).

The expected value of the error term, ϵ_i (the regression's residual), is zero. This reflects the assumption that the pricing equation is expected to hold exactly. The residual is the difference between actual returns and predicted returns, and if our model is correct, there should be no difference between the two. This property is key for testing whether an event has an impact on stock price (and thus whether the efficient-market hypothesis holds for this information set). If the event is viewed positively by investors (cash flows are expected to increase), then the residual will be positive rather than zero (the predicted price will understate the actual price because the information set has changed). Similarly, if the event is expected to have a negative impact on cash flows, then the residual will be negative (the predicted price will overstate the actual price). This is a straightforward application of the efficient-market hypothesis: the event releases new information that immediately affects the stock price and is picked up in the error term because it measures the difference between the actual and predicted return on the event date. Another name for the residual is, therefore, the abnormal return; it is the change in the stock price beyond what is normally expected.

In most event studies, a researcher creates a portfolio of firms experiencing the event and investigates the return to the portfolio, which is an average of the returns of the individual stocks. The error term in this context is therefore called an average residual. A portfolio of firms is more desirable than one firm to isolate the stock price effect of an event because although there may be additional firm-specific information, unrelated to the event, that affects one firm's return on the event date, the effects of extraneous information on individual firms will cancel each other out when there are many firms experiencing the same event on different calendar days. If the event date is uncertain, the researcher uses an interval as the event date and sums the average residuals up over that period; the sum is called a cumulative average residual.

Because the world is uncertain, expected returns will differ from actual returns. We therefore cannot expect the residual to equal zero as the pricing equation predicts, even in the absence of an event. Thus, when we obtain a positive or negative value for the residual on the event date, we must test whether it is statistically significantly different from the expected value of zero. Few residuals plot as dramatically as those in the event study of stock splits discussed in the Ross, Westerfield, and Jaffe excerpt, where visual inspection makes significance self-evident. Instead, a test statistic is created using the average residual, called a *t*-statistic, in which the data value is tested for the likelihood that its true value is zero. Statisticians, by convention, accept a 5 percent significance level as an acceptable level of risk of error in a test. This means that the researcher is willing to tolerate a 5 percent risk that she will reject the hypothesis that the abnormal return is zero when it truly is zero. As a rule of thumb, a *t*-statistic greater than 2 is statistically significant (i.e., there is only a 5 percent chance when the *t*-statistic is greater than 2 that the abnormal return is zero).

5. Eugene Fama, who coined the three types of market efficiency detailed in Ross, Westerfield, and Jaffe, recently suggested an alternative classification: tests for return predictability (which include weak-form efficiency tests and new work on forecasting returns with variables such as dividend yields and interest rates); event studies (semistrong-form market-efficiency tests); and tests for private information (strong-form efficiency tests). Fama, "Efficient Capital Markets: II," 46 *Journal of Finance* 1575 (1991). He notes that the controversy over market efficiency involves the first category of tests, which are tests of asset-pricing models such as the CAPM. Several studies have uncovered anomalies, many of which are noted in Ross, Westerfield, and Jaffe, which cast doubt on the pricing models' predictability of returns; others have found that returns can be predicted from variables such as dividend yields. While Fama cautions that it is possible that the apparent predictability of returns may be spurious, the findings do indicate that this class of tests are subject to a joint-hypothesis test problem, that a test of a pricing model cannot be distinguished from a test of market efficiency.

Fama characterizes as uncontroversial the implication of the second category of research, event studies: the market is efficient (stock prices adjust within a day of an information event). The criticism of the first class of studies is inapposite to event studies because use of daily return data permits precise measurement of stock price responses and mitigates the joint-hypothesis problem. When an information event can be dated precisely and the stock price effect is large, statistical inferences are robust across asset-pricing models. Id. at 1601, 1607. Fama concludes that "event studies are the cleanest evidence we have on efficiency. . . . With few exceptions, the evidence is supportive." Id. at 1602.

6. Helpful discussions of the efficient-market hypothesis can be found in Ronald J. Gilson and Reinier H. Kraakman, "The Mechanisms of Market Efficiency," 70 *Virginia Law Review* 549 (1984), which relates different forms of the hypothesis to the means by which information is transmitted into the market; and Jeffrey N. Gordon and Lewis A. Kornhauser, "Efficient Markets, Costly Information, and Securities Research," 60 *New York University Law Review* 761 (1985), which critiques the hypothesis and its use in policy analysis.

II

Legal Characteristics of the Corporation: Limited Liability

Four characteristics distinguish corporations from the other principal forms of business organizations (proprietorships and partnerships): limited liability for investors, free transferability of shares, perpetual life, and centralized management. The formal distinctions often blur for small businesses. Lenders, for example, typically require personal guarantees from controlling shareholders of close corporations. In addition, limited partners' liability is, like shareholders, limited to their investment. But for large public corporations, the formal distinctions are important organizational features, as they facilitate the development of thick capital markets for shares. Of the four characteristics, commentators' attention has focused most on limited liability and whether it is crucial for the corporate form. It is, correspondingly, the focus of this chapter.

Jensen and Meckling's explanation, in Chapter I, for why a sole owner would choose to sell part of her equity interest depends upon limited liability: restricting personal financial liability to an individual's capital investment in the firm facilitates portfolio diversification, which reduces the risk of corporate investment. This lowers the cost of capital. The reason why limited liability is a subject for debate despite this clear-cut benefit is because it permits investors to externalize risk. Since shareholders do not have to pay creditors in full if the business does poorly and cannot repay its debts, they do not bear the full costs of doing business. Corporations are, accordingly, likely to take too much risk, as the equity owners capture all the benefits from the upside potential of a risky project, while the downside risk is shared with creditors.

This incentive problem is analogous to that posed by the changing trade-off between pecuniary and nonpecuniary benefits for an owner-manager in Jensen and Meckling's move from 100 percent to partial ownership of the firm.

The externality problem of limited liability is not severe for voluntary creditors, such as lenders, because they will include the risk of nonpayment in the contract price (by, for example, charging a higher rate of interest). Equity thus bears the cost of limited liability, just as the selling shareholder bears the agency costs of outside ownership in Jensen and Meckling's analysis. The individuals who bear uncompensated risk under limited liability are instead a firm's involuntary creditors, such as tort victims, for they cannot contract with the firm in advance of their injury. For the most part, all sides in the debate over limited liability recognize the competing concerns; the disagreement is over whether the benefits of limited liability in reducing the cost of capital outweigh the costs of externalizing risk.

Frank Easterbrook and Daniel Fischel review a number of arguments in support of limited liability and add their own extension, which is related to the agency problem, that limited liability reduces monitoring costs. Susan Woodward develops the analysis in support of limited liability from a transaction cost perspective, with a twist. She maintains that limited liability is simply a necessary concomitant to free transferability of shares, and it is the latter that is essential for raising corporate capital, not the former.

Henry Hansmann and Reinier Kraakman contend that the consequences from eliminating corporate limited liability for the capital market, and hence for the corporate form, would not be severe. They advocate a pro rata rule of unlimited liability, under which each investor's liability is proportional to her stock ownership. They would, however, permit participants in the firm to contract for limited liability as between themselves. This is because such contracts do not affect the concern underlying the proposal, preventing uncompensated transfers of business risk to involuntary creditors.

Limited Liability and the Corporation

FRANK H. EASTERBROOK AND DANIEL R. FISCHEL

Limited liability is a fundamental principle of corporate law. Yet liability has never been absolutely limited. Courts occasionally allow creditors to "pierce the corporate veil," which means that shareholders must satisfy creditors' claims. . . . The rule of limited liability means that the investors in the corporation are not liable for more than the amount they invest. A person who pays $100 for stock risks that $100, but no more. . . . Henry Manne, in an important contribution, argues that the modern publicly held corporation with many small shareholders could not exist without limited liability. If investors could be required to supply unlimited amounts of additional capital, wealthy people would be reluctant to make small investments. Every share of stock would place all of their personal assets at risk. To guard against this risk, the investor would reduce the number of different firms he holds and monitor each more closely.

Manne's insight is powerful but incomplete. Limited liability does not eliminate the risk of business failure but rather shifts some of the risk to creditors. The creditors can invest in T[reasury]-bills and other riskless securities, and they will not make risky investments in firms unless offered more interest, which comes out of the shareholders' returns. Why are the increased returns demanded by creditors not exactly offset by the lower returns paid to shareholders? Manne's analysis does not explain why creditors bear as much risk as they do.

There are several reasons why the value of the firm might be maximized if creditors bear a substantial portion of the risk of business failure. Richard Posner maintains that creditors might be appropriate risk bearers because they are less risk averse than stockholders or have superior information. We find this implausible. Creditors are generally more risk averse than stockholders; why else do creditors arrange for the equity claimants to bear the most risk? . . .

The possibility that creditors might be the superior risk bearers because of superior information has considerably more appeal, but it cannot completely explain limited liability. . . . [T]hough creditors may sometimes possess superior information, this will not always be true. To the contrary, we expect creditors to know less. The equity investors have the residual claim. They stand to gain or lose almost the whole value of modest fluctuations in the fortunes of the firm. The residual claimants therefore have incentives to invest in the amount of monitoring likely to produce these gains (or avoid the losses), net of the costs of monitoring. Debt claimants, protected by the

Reprinted by permission from 52 *University of Chicago Law Review* 89 (1985).

"equity cushion," are more likely to be ignorant. They might do more monitoring if debt claims were more concentrated then equity claims, so that there would be less free riding on information, but no data show dramatic differences in the concentration of holdings. . . . [W]e know from the survival of large corporations that the costs generated by agency relations are outweighed by the gains from separation and specialization of function. Limited liability reduces the costs of this separation and specialization.

First, limited liability decreases the need to monitor. All investors risk losing wealth because of the actions of agents. They could monitor these agents more closely. The more risk they bear, the more they will monitor. But beyond a point more monitoring is not worth the cost. Moreover, specialized risk bearing implies that many investors will have diversified holdings. Only a small portion of their wealth will be invested in any one firm. These diversified investors have neither the expertise nor the incentive to monitor the actions of specialized agents. Limited liability makes diversification and passivity a more rational strategy and so potentially reduces the cost of operating the corporation. . . .

Second, limited liability reduces the costs of monitoring other shareholders. Under a rule exposing equity investors to additional liability, the greater the wealth of other shareholders, the lower the probability that any one shareholder's assets will be needed to pay a judgment. Thus existing shareholders would have incentives to engage in costly monitoring of other shareholders to ensure that they do not transfer assets to others or sell to others with less wealth. Limited liability makes the identity of other shareholders irrelevant and thus avoids these costs.

Third, by promoting free transfer of shares, limited liability gives managers incentives to act efficiently. . . . The ability of individual investors to sell creates new opportunities for investors as a group and thus constrains agents' actions. So long as shares are tied to votes, poorly run firms will attract new investors who can assemble large blocs at a discount and install new managerial teams. This potential for displacement gives existing managers incentives to operate efficiently in order to keep share prices high. . . .

Fourth, limited liability makes it possible for market prices to impound additional information about the value of firms. With unlimited liability, shares would not be homogeneous commodities, so they would no longer have one market price. Investors would therefore be required to expend greater resources analyzing the prospects of the firm in order to know whether "the price is right." . . .

Fifth . . . limited liability allows more efficient diversification. Investors can minimize risk by owning a diversified portfolio of assets. Firms can raise capital at lower costs because investors need to bear the special risk associated with nondiversified holdings. This is true, though, only under a rule of limited liability or some good substitute. Diversification would increase rather than reduce risk under a rule of unlimited liability. If any one firm went bankrupt, an investor could lose his entire wealth. The rational strategy under unlimited liability, therefore, would be to minimize the number of securities held. As a result, investors would be forced to bear risk that could have been avoided by diversification, and the cost to firms of raising capital would rise. . . .

Because limited liability increases the probability that there will be insufficient assets to pay creditors' claims, shareholders of a firm reap all of the benefits of risky activities but do not bear all of the costs. These are borne in part by creditors. Critics of limited liability have focused on this moral hazard—the incentive created by limited liability to transfer the cost of risky activities to creditors—as a justification for substantial modification of the doctrine.

Externalization of risk imposes social costs and thus is undesirable. The implications of this point, however, are unclear, both because modifying limited liability has its costs and because moral hazard would exist without limited liability. The social loss from reducing investment in certain types of projects—a consequence of seriously modifying limited liability—might far exceed the gains from reducing moral hazard. Too, even the abolition of limited liability would not eliminate the moral hazard problem. The incentive to engage in overly risky activities is a general phenomenon that exists whenever a person or firm has insufficient assets to cover its expected liabilities. Although the problem of moral hazard may be more severe under limited liability, it will exist under any rule. . . .

Employees, consumers, trade creditors, and lenders are voluntary creditors. The compensation they demand will be a function of the risk they face. One risk is the possibility of nonpayment because of limited liability. Another is the prospect, common to all debtor-creditor relations, that after the terms of the transaction are set the debtor will take increased risk, to the detriment of the lender.

So long as these risks are known, the firm pays for the freedom to engage in risky activities. . . . There is no "externality." Voluntary creditors receive compensation in advance for the risk that the firm will be unable to meet its obligations. . . .

When corporations must pay for the right to engage in risky activities, they will tend to undertake projects only where social benefits equal social costs at the margin. Where high transactions costs prohibit those affected by risky activities from charging an appropriate risk premium, however, the probability that firms with limited liability will undertake projects with an inefficiently high level of risk increases. Firms capture the benefits from such activities while bearing only some of the costs; other costs are shifted to involuntary creditors. This is a real cost of limited liability, but its magnitude is reduced by corporations' incentives to insure. . . . The common explanation for insurance is risk aversion. . . .

Corporate purchase of insurance seems inconsistent with this explanation. Investors can diversify, and this is a cheap way to reduce risk. Limited liability facilitates this diversification. Thus investors should not be willing to pay insurers to reduce risk. Why buy something you already have for free?

But not all who enter into contracts with a firm have the same ability to minimize risk by diversification. Human capital, for example, is notoriously difficult to diversify. Managers who have firm-specific investments of human capital cannot diversify the risk of business failure. To the contrary, investors want managers' fortunes tied to the fate of the firms under their control, and so they induce managers to bear extra costs if these firms fail, and they offer disproportionate rewards for success. The possibility of bankruptcy also represents a real cost to those with firm-specific investments of human capital, and firms must compensate those who bear this risk. The purchase of insurance in amounts greater than the amount of the firm's capital is one method of reducing the amount that the firm must pay. A firm with insurance against tort claims is less likely to become bankrupt, and thus less likely to impose costs on managers and other employees. Insurance thus induces people to make firm-specific investments of human capital.

Whether the purchase of insurance in amounts greater than the size of the firm's capital will reduce the incentive to engage in overly risky activities is a complex question. Before purchasing insurance, the firm's investors have the full amount of their investment at risk. After the purchase, the investors have much less at risk—in the limit, nothing at risk. The insurance company now bears the risk of business failure caused by tortious conduct. Thus the purchase of insurance might be thought to reduce the managers' incentives to take care, incentives already too low because of the existence of limited liability. The purchase of insurance has the effect, however, of creating a contract creditor where none may have existed before. The insured corporation must pay (through higher premiums) for engaging in risky activities. Because the firm will now bear the costs of engaging in risky projects, it will tend to equate social benefits and costs when making investment decisions.

Our argument is not that firms' incentive to purchase insurance eliminates the possibility that firms will engage in excessively risky activities. There may well be situations where firms will decide not to insure even if insurance is available. If potential losses are extremely large, the premium paid to risk-averse managers with firm-specific investments of human capital to compensate them for bearing the risk of bankruptcy might be less than the premium that would have to be paid to the insurance company. In other words, the limited liability of the manager (particularly in a world where discharge in bankruptcy is available) coupled with the limited liability of the firm may cause firms faced with large expected liabilities to involuntary creditors to pay managers a premium and simultaneously decrease their capitalization. Moreover, there is no guarantee that insurance will be offered if risks are highly correlated. Even when offered, the insurance will exclude very large losses. On the other hand, our discussion of firms' incentives to insure suggests that firms will insure in some situations where people and partnerships would not. A corporation with a fleet of trucks might insure for an amount in excess of its capital, for example, while an individual owner with few assets might insure for a small amount or not insure at all.

Piercing the Corporate Veil

Courts have allowed creditors in some situations to reach the assets of share-holders. . . . The cases may be understood, at least roughly, as attempts to balance the benefits of limited liability against its costs. Courts are more likely to allow creditors to reach the assets of shareholders where limited liability provides minimal gains from improved liquidity and diversification, while creating a high probability that a firm will engage in a socially excessive level of risk taking. . . .

Close Versus Public Corporations

Almost every case in which a court has allowed creditors to reach the assets of shareholders has involved a close corporation. The distinction between close and public corporations is supported by economic logic. In close corporations, there is much less separation between management and risk bearing. This has profound implications for the role of limited liability. Because those who supply capital in close corporations typically are also involved in decision making, limited liability does not reduce monitoring costs. Other benefits of limited liability in public corporations—facilitating efficient risk bearing and monitoring by the capital market—also are absent for close corporations. . . .

Coporate Versus Personal Shareholders

The other major category of piercing cases involves parent-subsidiary combinations, where creditors of the subsidiary attempt to reach assets of the parent. Courts' greater willingness to allow creditors to reach the assets of corporate as opposed to personal shareholders is again consistent with economic principles.

Allowing creditors to reach the assets of parent corporations does not create unlimited liability for any people. Thus the benefits of diversification, liquidity, and monitoring by the capital market are unaffected. Moreover, the moral hazard problem is probably greater in parent-subsidiary situations because subsidiaries have less incentive to insure. . . .

It does not follow that parent and affiliate corporations always should be liable for the debts of those in which they hold stock. Far from it. Such general liability would give unaffiliated firms a competitive advantage. Think of the taxicab business. Taxi firms may incorporate each cab or put just a few cabs in a firm. If courts routinely pierced this arrangement and put the assets of the full venture at risk for the accidents of each cab, then "true" single-cab firms would have lower costs of operation because they alone could cut off liability. That would create a perverse incentive because, as we have emphasized, larger firms are apt to carry more insurance. Potential victims of torts would not gain from a legal rule that promoted corporate disintegration. As a result, courts properly disregard the corporate form only when the corporate arrangement has increased risks over what they would be if firms generally were organized as separate ventures.

Contracts Versus Torts, and the Fraud or Misrepresentation Exception

Courts are more willing to disregard the corporate veil in tort than in contract cases. The rationale for this distinction follows directly from the economics of moral hazard—where corporations must pay for the risk faced by creditors as a result of limited liability, they are less likely to engage in activities with social costs that exceed their social benefits. Contract creditors, in other words, are compensated ex ante for the increased risk of default ex post. Tort creditors, by contrast, are not compensated.

This distinction between contract and tort creditors breaks down when the debtor engages in fraud or misrepresentation. For the costs of excessive risk taking to be fully internalized, creditors must be able to assess the risk of default accurately. If the creditor is misled into believing that the risk of default is lower than it actually is, the creditor will not demand adequate compensation. This will lead to an excessive amount of risk taking by firms, because some of the costs are now shifted to creditors.

Courts have responded to this problem by allowing creditors to go beyond the assets of the corporation in cases of fraud or misrepresentation. . . . In all these situations, creditors are unable to assess the risk of default accurately and thus the probability that the firm will engage in excessively risky activities is increased.

Alternative Methods for Reducing the Problem of Moral Hazard

Piercing the corporate veil is one of several methods for decreasing the incentive created by limited liability to engage in overly risky activities. We briefly analyze in this section the costs and benefits of . . . other methods of decreasing this risk: minimum-capital requirements, mandatory insurance, [and] managerial liability. . . .

The lower a firm's capitalization, the higher the probability that it will engage in excessively risky activities. Legislatively imposed minimum-capitalization requirements are one method of internalizing the costs of risk taking. But such regulations have problems of their own. One is the obvious administrative cost associated with determining what amount of capital firms should raise. Another is the cost of error. If capital requirements are set too high, this will impede new entry and permit the existing firms to charge monopoly prices. Still another is the question of how firms must satisfy their capitalization requirements. For such requirements to be effective, the corporation must post a bond equal to its highest expected liability or hold sufficient funds in the corporate treasury and invest them in risk-free assets. The total held in this way will far exceed the expected risk created by firms as a group (because not all firms go bankrupt or incur the maximum possible loss). Under either alternative, the rate of return on equity investments will decrease. Thus at the margin people will shift capital away from equity investment in risky industries. This too represents a social cost.

Mandatory-insurance requirements are similar in some respects to minimum-capitalization requirements. Both involve administrative costs and may act as barriers to entry. Whether mandatory insurance poses a greater barrier to entry is difficult to determine. New firms, with less experience than existing ones, have higher risks, as insurers see things, and must therefore pay higher premiums. If these premiums are less than the cost of self-insurance, then mandatory insurance facilitates new entry (compared with mimimum-capitalization rules). On the other hand, some firms, particularly new firms, might find it difficult to obtain insurance at all. The effect of mandatory insurance on new firms, therefore, might be greater than minimum-capitalization requirements.

One important difference between the two regimes is the effect each has on firms' incentives to engage in excessively risky activities. Minimum-capitalization requirements decrease this incentive. Mandatory-insurance requirements, by contrast, may increase or decrease the level of risky activities, depending on insurers' ability to monitor. Where the insurer is unable to monitor, the level of risky activities will increase because of the existence of the insurance. . . .

One method of minimizing the incentive to engage in overly risky activities while avoiding the administrative costs of minimum-capitalization requirements or mandatory insurance is to impose liability on managers as well as enterprises. Managerial liability is an additional risk for which firms must compensate the managers. From the firm's perspective, however, there are problems with paying managers for bearing risk. Because of their inability to diversify their human capital, managers are inefficient risk bearers. Thus firms have incentives to undo managerial liability by providing managers with indemnification or insurance.

This risk shifting does not, however, defeat the purpose of managerial liability. If only the firm may be held liable, the value-maximizing strategy for a firm with few firm-specific investments of human capital may well be to maintain assets less than expected liabilities. If managers may also be held liable, they have incentives to monitor the firm's capitalization and insurance, because they bear the costs if risk shifting is not complete. Thus the value-maximizing strategy under managerial liability is for firms to self-insure by increasing their capitalization or to purchase insurance, whichever is cheaper. In either case, the incentive to engage in overly risky activities is reduced.

The problem with managerial liability is that risk shifting may not work perfectly. It is unlikely, for example, that managers who are liable for mass torts, with huge but uncertain expected liabilities, could shift all of this risk. Because of the huge amounts involved and the difficulty of monitoring, insurers are unwilling to assume the highest possible expected liability. To the extent that risk is not completely shifted, a legal rule of managerial liability creates risk for a group with a comparative disadvantage in bearing that risk. This inefficiency leads to both an increase in the competitive wage for managers and a shift away from risky activities. And there is no guarantee that the social costs of this shift away from risky activities will not exceed the social costs of the excessively risky activities in the absence of managerial liability.

Limited Liability in the Theory of the Firm

SUSAN E. WOODWARD

Most of the explanations for the prevalence of limited liability appeal to risk aversion. While I will agree that risk aversion may motivate limited liability in small, closely held firms, I will argue that for the large, publicly traded firm, for which limited liability is ubiquitous, risk aversion is neither necessary nor sufficient to explain the presence of limited liability. The explanation lies rather in the lower information and transaction costs associated with limited liability.

In particular, the most important feature of limited liability is that it accommodates transferable shares. Any extension of liability beyond the assets of the firm to the personal (extra-firm) assets of the shareholders must, in order to be enforceable, impair transferability of shares.

Little effort is necessary to motivate the desirability of transferable shares. When numerous investors cooperate in large, long-lived investment projects, transferability of the shares in these projects makes them much more desirable than they would otherwise be. Investors who wish to sell (or buy) shares to accommodate their intertemporal consumption plans, to revise their portfolios to achieve desired risk profiles, or to indulge changes in their beliefs about the enterprise, can do so without interfering with the management of the firm's physical assets so long as the shares are transferable. It is the transferability of the shares that allows for the separation of consumption/production decisions and consumption/investment decisions, the importance of which we laboriously inculcate in courses in elementary price theory.

The connection between transferability and liability assignment is easily seen by imagining a firm with unconditionally saleable shares which tries to extend liability to the shareholders (where the liable party is the holder of the share at the time request for resources is made). Were bankruptcy to threaten such a firm, any shareholders with assets worth the creditor's pursuit could simply sell their shares rather than pay up. The only willing buyers of the

Reprinted by permission from *Journal of Institutional and Theoretical Economics* v. 141, pp. 601ff, J.C.B. Mohr (Paul Siebeck) publisher, 1985.

shares would be those whose wealth is too small for the creditors to bother pursuing. So whenever the creditors try to reach beyond the assets of the firm to the other assets of the shareholders, the shareholders will be found a group with no assets worth the cost of pursuit. If the shares are unconditionally saleable and liability extends only to the current holder of the share, extended liability simply cannot be enforced.

Thus, an easy answer to why firms limit liability is that freely transferable shares result in de facto limited liability. Making limited liability explicit simply saves transaction costs associated with the scramble of transfers when bankruptcy is imminent.

But between the extremes of perfectly transferable shares and perfectly inalienable shares (wherein shareholders could not escape specific liability) lies an array of restrictions on transfer which could make extended liability feasible. Thus, task remains to explain why firms do not choose some interior combination of liability to shareholders and transferability of shares.

The explanation lies in understanding the costs imposed by extended liability. Any effective extension of liability makes the cash flows both to creditors and to shareholders depend on the personal wealth of each shareholder. This dependence creates incentives on the part of both creditors and shareholders to (1) invest in information about the shareholders, and also to (2) make side contracts in an attempt to control each others' behavior. Limited liability, by eliminating the dependence of firm credit on shareholder wealth, can lower the transaction and information costs for all parties connected with mutual investment projects, especially those with numerous shareholders.

Against the benefits of transferability conferred by limited liability we must consider the costs of limiting liability in order to understand the stylized facts. These costs are agency costs which arise when equity holders do not bear the full consequences of their decisions, as the equity holders of an indebted, limited liability firm do not. . . .

The magnitude of the costs of extending liability to shareholders depends on the specific nature of the shareholder's responsibilities. The most common form of liability holds the shareholders jointly and severally liable for the debts of the firm. If the firm's assets are insufficient to meet its liabilities, each shareholder is initially responsible for a fraction of the debts proportionate to his shareholdings. If the other stockholders are unable to meet their obligations, each remaining shareholder can be held responsible for the obligations of the others. A milder form of extended liability holds shareholders responsible for only a fraction of firm debts proportionate to their shareholdings, leaving those debts unmet by individual shareholders as losses to the creditors, not the other shareholders. An even milder form (referred to in legal discussions as assessability), which prevailed on some U.S. bank shares even into the twentieth century, holds shareholders liable only for a specified amount per share.

All of these extended liability rules give rise to costly activity associated with enforcement and with anticipating enforcement. . . . The resources creditors will expend pursuing each shareholder will of course depend on what they anticipate they can recover. Under strong rules, such as joint and several liability, richer shareholders will be pursued more vigorously than they will be with weaker rules such as assessability. Shareholders themselves will have no motivation to pursue each other with a proportionate liability or assessability rule. With joint and several liability, however, the richer shareholders (who are most likely to be pursued by the creditors) will pursue poorer ones for their share.

But likely the greater costs of extending liability are not those associated with collections ex post of a bankruptcy, but with the dealings ex ante attempting to assure these collections take place.

Prior to and in anticipation of a bankruptcy, creditors and shareholders alike are motivated to examine and assure the solvency of each shareholder, regardless of the form of the extended liability. The issue is not simply whether the shareholders will be able to pay their share in a bankruptcy, but also that even if no bankruptcy actually occurs, the mere likelihood of it implies that every stockholder's wealth influences the firm's credit, and consequently makes each shareholder care about who the other shareholders are and how they manage their other assets. Even if the liability rule is just assessability (liability only for a specified amount per share), the prospects of each shareholder to meet the assessment affect the firm's credit and, as a result, the wealth of all other shareholders. Thus, the creditors (in order to assess the firm's credit) and the shareholders (in order to determine the firm's value) have incentives to invest real resources to secure information about the amount and composition of shareholder wealth—information unrelated to the firm's productive activities. . . .

The interest in shareholder wealth which accompanies extended liability also motivates shareholders and creditors to invest in making side contracts with one another to resolve the conflicts of interest that extended liability creates. One way to prevent a wealthy shareholder from selling to an impecunious one is to restrict the set of potential owners of the stock to those with at least some minimum wealth. This may improve the firm's credit, but it could also lower the demand price for the firm's shares and make the cost of capital higher. Another solution is to require any shareholder who could not bond wealth of a particular amount and composition to carry insurance. The investigation to determine the insurance premium and the enforcement of the purchase of insurance are themselves costly. Yet another suggestion is that the firm carry insurance on behalf of the shareholders.

Even this solution (insurance purchased by the firm) would not, however, resolve all conflicts of interest, as poor shareholders and rich shareholders would not agree on the level of insurance that ought to be carried. If the firm buys the insurance the rich shareholders and poor shareholders bear equally (prorated by shareholdings) the cost of the insurance. But the rich shareholders have more assets to protect from extended liability than do the poor ones, and will likely desire a higher level of coverage. Even if the shareholders are unanimous on the probabilities of various states of the world, and consequently, how much should be invested in, say, finding oil, they will still not, because of their different personal situations, agree on the level of liability insurance to be carried. . . .

Another device to prevent shareholders from escaping liability on transferable shares is to hold liable those who owned the shares when the liability was created rather than those who own it when the bankruptcy occurs. Essentially this liability assignment makes inalienable the potential debts created during one's tenure of ownership. By assigning liability in this way, new shareholders, who control the firm, could appropriate wealth from former shareholders by managing old assets in a more risky way. . . . The former shareholders would have even less opportunity than the creditors, who at least have continuing transactions with the firm, to influence how assets are managed. . . .

The lesson of this section is that the costs of extended liability rise with the number of shareholders. . . . Thus, we have at least a partial explanation for the first of our stylized facts: publicly traded firms are more likely to limit liability because it is for these firms that extended liability is more costly.

Toward Unlimited Shareholder Liability
for Corporate Torts

HENRY HANSMANN AND REINIER KRAAKMAN

Designing an Unlimited Liability Rule

The threshold issue is whether unlimited liability is feasible for public corporations. Several commentators have questioned whether any unlimited liability regime can obtain meaningful recovery from widely dispersed shareholders without sacrificing the liquidity of the securities market. We believe that such a regime is clearly feasible: a well-crafted rule of unlimited liability would neither impair the marketability of securities nor impose excessive collection costs.

It might be argued that, even if feasible, unlimited liability in tort for publicly traded corporations is unnecessary, since relatively few publicly traded firms have been bankrupted by tort liability. There are, however, important reasons for extending an unlimited liability regime to publicly traded firms. First, the paucity of tort judgments bankrupting such firms may underrepresent the frequency with which publicly traded firms cause tort damages exceeding their net worth, since tort victims suing under a regime of limited liability have a strong incentive to accept a settlement for less than the full value of the firm. Further, . . . the threat of tort liability exceeding the net

Reprinted by permission of The Yale Law Journal Company and Fred. B. Rothman & Company from *The Yale Law Journal,* Vol. 100, pp. 1879–1934 (1991).

assets of such firms is likely to increase in the future. Finally, and most importantly, any effort to extend unlimited liability to closely held firms without including publicly traded firms as well would invite corporations that are currently closely held to avoid unlimited liability simply by selling some portion of their stock to the public.

When Should Liability Attach to Shareholders?

An administrable rule of unlimited liability for the public firm must specify both a measure of shareholder liability and the point at which freely trading stock imposes personal liability on its holders. . . . The most plausible measure of shareholder liability is a rule of pro rata liability for any excess tort damages that the firm's estate fails to satisfy. The more difficult problem is selecting the timing rule that determines which shareholders become liable after a tort occurs.

The choice of a timing rule in determining when excess liability attaches to shareholders inevitably involves a conflict between administrative complexity and opportunities for evasion. An "occurrence" rule, under which liability attaches to persons who are shareholders at the time a tort occurs, is the most difficult rule for shareholders to evade but also the most difficult to administer. A "judgment" rule that attaches residual liability only to those persons who are shareholders at the time of judgment is the simplest to administer but creates widespread opportunities for evasion. A variety of considerations suggest that a reasonable compromise is a modified version of a "claims-made" rule, which attaches liability to persons who are shareholders when—or somewhat before—the tort claim is filed. We style this modified claims-made rule an "information-based" rule. . . .

At first glance, a simple claims-made rule might seem to answer the drawbacks of an occurrence rule. A claims-made rule would permit shareholders to escape liability for undiscovered torts merely by selling their shares. Under a claims-made rule, moreover, shareholders who did incur a risk of personal liability by virtue of a tort filing would usually receive clear notice of that contingent liability even though any actual judgment against these shareholders would follow only much later, after an unfavorable resolution of the tort action and a subsequent bankruptcy proceeding. For the same reason, the administrative task of apportioning liability among shareholders would also be greatly simplified under a claims-made rule.

Nevertheless, a simple claims-made rule would not be satisfactory in every case. Where claims accumulate gradually, as they did in the asbestos cases, a simple claims-made rule might invite massive selling at the time the first claims are filed.[1] Indeed, even when a catastrophic tort occurs in a single and

---

[1]This problem of wholesale evasion of shareholder liability through the market has been analyzed by Woodward . . . . If personal liability for excess tort damages could be transferred by sale of the shares until a relatively late date (e.g., until the time of judgment or even bankruptcy), then there would always be a strong incentive, once it became apparent that the expected value of tort

well-publicized moment, as with the Bhopal disaster, shareholders might succeed in exploiting the interval between this moment and the time when the first claims are filed to dispose of their shares opportunistically. In addition, a simple claims-made rule might also leave residual uncertainty as to the liability of particular shareholders when, for example, claims filed at different times were later consolidated for trial. Applying a claims-made rule mechanically in these cases would recreate the administrative complexity of an occurrence rule in determining which claimants could ultimately collect against which shareholders.

For these reasons, we propose a modified form of the claims-made rule—an "information-based rule"—under which liability would attach to shareholders at the earliest of the following moments: (1) when the tort claims in question were filed; (2) when the corporation's management first became aware that, with high probability, such claims would be filed; or (3) when the corporation dissolved without leaving a contractual successor. This information-based rule would fix liability before shareholders could evade responsibility for tort damages, without creating the uncertainties and complexities that would attend an occurrence rule. . . .

The information-based rule we propose would permit shares of an affected company to resume trading unencumbered by outstanding personal liability shortly after the first filings in a major tort action, or even earlier if management announced a liability date. Thus, new investors and old shareholders could trade in a liquid market without reallocating liability for actual or anticipated tort claims, which would remain firmly attached to the old shareholders who had held stock as of the liability date. This is not to say, however, that old shareholders would be unable to shift their contingent personal liability. We would expect a specialized market in retroactive insurance to arise shortly after the adoption of an unlimited liability regime. Such a market would permit old shareholders to insure against the contingency that a tort judgment might ultimately exceed the net value of the firm. . . .

The Costs of Collection

A second inquiry bearing on the feasibility of unlimited liability is whether recovering from numerous public shareholders would be prohibitively costly even in the absence of opportunistic efforts to disperse share ownership. Very large collection costs would make unlimited liability less attractive not only because they would be wasteful, but also because they would lower settlement values and hence reduce the deterrent effect of tort rules. We believe that collection costs are unlikely to be prohibitive in this sense. . . .

There is no doubt that unlimited liability, as we have described it, would increase the cost of equity. Indeed, the *purpose* of unlimited liability is to

claims exceeded the firm's net worth, for shareholders to seek to create "homemade limited liability" by selling, or simply giving, shares to insolvent or dispersed persons from whom collection would be difficult.

make share prices reflect tort costs. Yet the literature suggests that, beyond internalizing tort losses, unlimited liability might generate additional costs by (1) impairing the market's capacity to diversify risk and to value shares, (2) altering the identities and investment strategies of shareholders, and (3) inducing market participants to monitor excessively. The magnitude of these additional costs, however, turns chiefly on the choice between a joint and several or a pro rata liability rule. If shareholders faced joint and several personal liability for all corporate debts, whether arising in contract or tort, these costs might be very large. But . . . they should be much smaller if shareholder liability is restricted to tort judgments and governed by a pro rata rule. . . .

Unlimited liability would bar today's low-cost strategies of evasion by self-incorporation or the partitioning of assets into corporate subsidiaries. It would leave open, however, the three liability evasion strategies that are presently available for unincorporated individuals or partnerships: (1) hiding personal assets or shifting shares; (2) shifting from equity to debt financing; and (3) disaggregating industry. All three of these evasion strategies operate by placing equity interests in the hands of judgment-proof individuals. . . . Each of these strategies would not only blunt safety incentives under an unlimited liability regime but would also generate other efficiency costs, ranging from the reallocation of individual assets to the loss of integration gains. Yet the magnitude of these effects and the ease with which they might be countered are empirical questions. . . . [T]here is good reason to believe that they would not overshadow the efficiency gains that unlimited shareholder liability would introduce, given the seeming implausibility of a wholesale shift of assets into the hands of high rollers who happen to have just the right assets, preferences, and access to credit necessary to become repositories of risky equity. . . .

With unlimited liability, it will remain to the courts to determine which costs are efficiently and equitably borne by a corporation and its shareholders and which are not. Some costs associated with corporate activity should be left on victims or their insurers rather than borne by corporations or, particularly, their shareholders. But there is no reason to believe that this will always be the case, as the prevailing limited liability regime necessarily presumes. . . . [C]ourts should appropriately consider the structure of particular corporate defendants in determining the extent of their tort liability under an unlimited liability regime. For example, when the defendant corporation is the wholly owned subsidiary of a large parent corporation, the prospect that a judgment might exceed the corporation's net assets and thus spill over onto its parent shareholder should generally not, in itself, affect the size of the judgment. When the firm's shareholders are individuals, however, the prospect of shareholder liability might sometimes be a reason to temper the amount of the damages assessed. In some cases, the transaction costs occasioned by individual shareholder liability may not justify whatever added deterrence or insurance benefits come from assessing damages that exceed the firm's assets. Moreover, among firms with individual shareholders, it may often be worthwhile to distinguish between small closely held firms and publicly held firms. For example, corporate liability that is justified on the grounds of risk bearing (insurance) is more sensibly imposed on individual shareholders in publicly held firms than on shareholders in privately held firms, since public shareholders presumably have better-diversified investments. Similarly, smaller judgments against closely held firms will often be justified for purposes of deterrence, since liability is likely to deter risk-averse shareholders with concentrated stockholdings more readily than diversified shareholders in public corporations. . . .

We do not want to exaggerate our faith in tort law as a means of controlling behavior. It is a very rough and costly mechanism. But it usefully discourages the most severe forms of opportunistic cost externalization. Moreover, if any class of actors is likely to respond rationally to the deterrence incentives created by tort law, it is corporations and their shareholders. Similarly, if tort law is to have any role in shifting risks to low-cost insurers, then using it to shift risks to the equity market makes sense. Consequently, allowing corporations to avoid tort liability through the simple device of limited liability seems, at the very least, highly suspect.

Limited Liability for Contract Creditors

The case against limited liability in tort does not extend to contract. . . . Limited liability for contractual debts simply permits the owners and creditors of a firm to allocate the risks of the enterprise between themselves in whatever fashion is the most efficient. . . .

Which Creditors Are Involuntary?

Because limited liability should be retained as the background rule for contract creditors, it would be necessary, if unlimited liability were adopted for tort claims, to distinguish clearly between claims against a corporation that arise in tort and claims that arise in contract. The obvious difficulties lie in areas, such as products liability and workplace injuries, where, although the victim had a contractual relationship with the firm prior to the injury, the courts have been inclined to classify the injury as a tort.

These difficulties do not, however, seem serious. The critical question is whether the victim was able, prior to the injury, to assess the risks she took in dealing with the firm and to decline to deal if those risks seemed excessive in comparison with the net advantages she otherwise derived from the transaction. In other words, the question is whether the victim can reasonably be understood to have contracted with the firm in substantial awareness of the risks of injury involved.[2] If so, then the liability should be considered contractual, and limited liability should be considered a background term of the

---

[2]Of course, this question should not generally be asked of each individual victim, but rather for categories of victims.

contract, to be respected unless specifically waived. If not, the victim should be considered an involuntary creditor and the corporation's shareholders should not be permitted to invoke limited liability.

Notes and Questions

1. How convincing are Easterbrook and Fischel's criticisms of the alternatives, besides judicial veil-piercing, for reducing the externalization of risk of limited liability of minimum capital requirements, mandatory insurance, and personal liability for managers? What are the comparative costs and benefits of these proposals versus Hansmann and Kraakman's approach?

2. From the shareholders' perspective, a limited liability regime will be equivalent to a regime of unlimited liability with insurance. This replaces one contractual setting with another, as the insurance company, rather than the creditors, bears the residual (bankruptcy) risk. Paul Halpern, Michael Trebilcock, and Stuart Turnbull suggest, however, that insurance markets will not exist in many contexts. In particular, insurance is unlikely to be available for small high-risk firms, where the moral hazard problem is most severe. See Halpern, Trebilcock, and Turnbull, "An Economic Analysis of Limited Liability in Corporation Law," 30 *University of Toronto Law Journal* 117 (1980). Note that such firms' owners typically must provide personal guarantees to borrow in a limited liability regime.

Consider the difference between these two regimes for involuntary creditors. First, under unlimited liability with insurance, the corporation's insurance will be available to compensate these creditors, whereas under limited liability, compensation is less likely as the debtholders' unreturned capital may not cover their claims, especially because some debt is superior to tort claims. For a review of priorities and related issues in bankruptcy law, see Thomas H. Jackson, *The Logic and Limits of Bankruptcy Law* (Cambridge, Mass.: Harvard University Press, 1986). Note, however, that Easterbrook and Fischel maintain that incentives to acquire adequate insurance still hold under limited liability because of the separation of ownership and control.

Second, when shareholders obtain insurance against third-party claims, premiums are priced in the market. The "insurance" they obtain under limited liability will not, however, be fully priced because the providers of debt capital are not, as earlier noted, insurers of involuntary creditors. Thus, shareholders will be induced to take more care under a regime of unlimited liability with insurance.

In an unlimited liability regime, who insures the insurance companies? The diversification principles discussed in Chapter I apply most generally to insurance: when the risks of loss are independent across firms, the insurer, by pooling risk, can diversify it away. Most U.S. property and casualty insurers are incorporated and thus operate under the limited liability regime. Lloyds of London, the oldest and most celebrated insurance company, is organized as a partnership, whose members have unlimited liability. In this regard, an interesting empirical finding is that Lloyds purchases significantly more reinsurance (insurance for liabilities under the primary policies that it has underwritten) than incorporated insurance companies. See David Mayers and Clifford W. Smith, Jr., "On the Corporate Demand for Insurance: Evidence from the Reinsurance Market," 63 *Journal of Business* 19 (1990). With some syndicate members facing catastrophic losses and its capital base shrinking in the early 1990s, Lloyd's announced plans to reorganize so as to limit members' liability.

3. Liability insurance contracts have policy limits (ceilings on the dollar amount of loss that is covered). Would you expect a business's insurance policy limit to vary depending upon whether or not it is incorporated? How might Easterbrook and Fischel explain this contractual feature? Gur Huberman, David Mayers, and Clifford Smith show that optimal insurance policies include upper limits on coverage when liability is limited in bankruptcy. Huberman, Mayers, and Smith, "Optimal Insurance Policy Indemnity Schedules," 14 *Bell Journal of Economics* 415 (1983). As they note, even individual liability is not truly unlimited: bankruptcy laws permit debtors to retain some of their wealth and the bankruptcy discharge prevents attachment of future earnings.

4. Does Hansmann and Kraakman's modified regime of unlimited liability answer Woodward's concerns? Are they unduly optimistic about the costs of a pro rata unlimited liability system? What effect would such a regime have on the incentives for concentrating ownership in large blocks? Would this mitigate or exacerbate the agency problem? How small do you think collection costs would be in practice? For a procedural perspective on Hansmann and Kraakman, see Janet Cooper Alexander, "Unlimited Shareholder Liability Through a Procedural Lens," 106 *Harvard Law Review* 387 (1992), which points out barriers to their proposal, of constitutional dimension, concerning a tort forum state's ability to obtain personal jurisdiction over out-of-state shareholders and concerning altering the choice of law rule, known as the internal affairs doctrine, that the law of the state of incorporation governs shareholder rights and obligations. Cooper maintains that these barriers raise considerably the costs of administering an unlimited liability system and concludes that the procedural difficulties can only be overcome through federal legislation. The materials in Chapter III, part A, address the general question of federalism in corporate law: Are the states, as opposed to the federal government, the appropriate sovereign?

5. Hansmann and Kraakman contend that opposition to unlimited liability because it will increase the cost of equity capital is misguided since increased costs are appropriate to ensure that corporations fully internalize

liability risk. How useful is their proposal if stock prices do not significantly decrease, as they anticipate, with a liability regime change? Would unchanged prices reflect an inefficient capital market, or a capital market extraordinarily adaptive to regulatory change that will arbitrage (defined at Chapter I, part B, note 3) away the purported benefits of unlimited liability? Consider Joseph Grundfest's scenario: Foreign investors, who are U.S. judgment-proof, hold stock in firms likely to run into liability problems while U.S. citizens hold less risky firms' shares; then investors reallocate the risk and return of their equity investments by trading innovative financial instruments known as derivative products, that is, foreigners sell (buy) and U.S. citizens buy (sell) options on a risky-firm (safe-firm) index. Grundfest, "The Limited Future of Unlimited Liability: A Capital Markets Perspective," 102 *Yale Law Journal* 387 (1992). Grundfest concludes that the alternative approaches to internalizing business risk listed in note 1, which directly regulate firms, are more viable solutions than unlimited liability because demand and supply in markets for goods and services are far more inelastic than demand and supply in capital markets.

6. Hansmann and Kraakman's concern, risk spreading for tort losses, and Easterbrook and Fischel's and Woodward's concern, stock market liquidity, provide contradictory guidance to courts distinguishing between close and public corporation defendants in piercing cases. Can these competing concerns be reconciled in a useable criterion for distinguishing between corporate defendants?

7. Although most commentators agree that limited liability is problematic for involuntary creditors of the firm, such as tort victims, courts do not appear to share this concern. In a study of 1,600 piercing-the-veil cases, Robert Thompson finds that courts pierce the veil more frequently in contract rather than tort cases (42 percent compared to 31 percent). This difference is statistically significant at a level of less than 1 percent. The number of tort claims is also far less than contract claims (226 compared to 900 cases). Thompson, "Piercing the Corporate Veil: An Empirical Study," 76 *Cornell Law Review* 1036 (1991). The successful contract cases appear to be those in which there is a strong scent of fraud.

None of the successful piercing cases involved a public corporation, except when a corporate group was involved and the courts pierced the veil to hold a parent company liable for its subsidiary's debts. There were 777 cases involving close corporations, 637 corporate group cases, and only 9 public corporation cases, and in none of these 9 cases did the court pierce the veil. Thus, the only individuals who were deprived of the protection of limited liability were the owners of close corporations. The likelihood of piercing, however, decreased as the number of shareholders of the close corporation increased: for corporations with only one shareholder, the court pierced the veil in approximately 50 percent of the cases, whereas for firms with more than three shareholders, the veil was pierced in only 35 percent of the cases.

How consistent are these data with Easterbook and Fischel's and Woodward's explanations of limited liability? Would consistency between theory and evidence demonstrate that the common law is efficient? The efficiency of the common law has been the source of considerable debate since the idea was propounded by Richard Posner. For his assessment of the debate, see Posner, "A Reply to Some Recent Criticisms of the Efficiency Theory of the Common Law," 9 *Hofstra Law Review* 775 (1981).

Does the apparent scarcity of tort plaintiffs and public corporation defendants in piercing cases suggest that Hansmann and Kraakman are concerned with a nonissue? Or is there a problem of selection bias in Thompson's data set? His data consist of reported cases, which are only a fraction of all claims because many claims are settled or dropped prior to trial. If claims against public corporations are more likely to be settled or dropped than litigated, compared to claims against close corporations, then they would not be picked up in his study. In addition, to the extent that public corporations are more likely to carry insurance than close corporations, there will be fewer piercing claims filed against them. Finally, if inefficient rules are litigated more frequently because they are more costly to society (see Robert Cooter and Thomas Ulen, *Law and Economics* 492–96 [Glenview, Ill.: Scott, Foresman, 1988]) and if limited liability is inefficient for close corporations but not public corporations, then we would also expect to obtain the skewed distribution of cases that Thompson finds. The latter two explanations provide, however, only pyrrhic support for Hansmann and Kraakman because they are inconsistent with an externality problem. For an analysis of selection bias problems in empirical research based on litigated cases, see George Priest and Benjamin Klein, "The Selection of Disputes for Litigation," 13 *Journal of Legal Studies* 1 (1984).

III

The Production of Corporation Laws

Corporation codes can be viewed as products, whose producers are states and consumers, private corporations. The readings in this chapter focus on the structure of corporation laws—enabling versus mandatory—and the dynamics by which they are produced.

A central question for corporate law concerns the competitive process by which state corporation codes are produced: Is there any reason to suppose that the provisions in state codes benefit shareholders? Roberta Romano reviews the positions in what is a long-standing debate in corporate law, whether competition among the states for the corporate charter business is a "race for the bottom," and the empirical evidence, which tends to support the federal system rather than its critics. She also offers an explanation of the dominant position of Delaware in the chartering market, which draws upon Williamson's hostage analysis, in Chapter I, of transaction-specific assets. Jonathan Macey and Geoffrey Miller emphasize a different concern in the calculus, the actors in the political process, whose interests may affect the contours of corporation codes as much as code consumers' interests.

Frank Easterbrook and Daniel Fischel address another foundational question that concerns the states' choice of product attributes: What is the role of corporate law? As they explain, state corporation codes reflect the contractual context of the corporation by taking an enabling approach. Corporation codes provide a standard-form contract, which firms can tailor quite extensively to meet their particular needs. The policy issue posed, given this context, concerns the scope of an enabling regime: Should all provisions in corporation codes be optional, or is there a role for mandatory corporate laws?

Easterbrook and Fischel emphasize the important role of efficient capital markets in maintaining an enabling structure. In an efficient market, corporate charter terms are priced and, consequently, investors will not bear the cost of harmful provisions (they get what they pay for). This weakens the case for mandatory rules. Jeffrey Gordon, however, offers several justifications for mandatory rules that do not depend on market failure in the pricing of charter terms. His most compelling argument involves a problem that is also of concern to Easterbrook and Fischel, changes in corporate organization undertaken midstream, for here capitalization of corporate changes in stock prices is of little solace to the shareholders, as they already own the stock. Shareholders, not managers, bear the cost of unexpected value-decreasing decisions that occur midstream (postinvestment). Because shareholders must approve changes that amend the corporate charter, the midstream opportunism problem would seem to be a nonissue, why would shareholders approve an amendment that is adverse to their interest? The response is that shareholders are rationally ignorant. They face a collective action problem in which the costs of gathering information on an amendment's price effects are greater than their pro rata share of the benefit obtained from an informed vote.

John Coffee reminds us that there is another actor besides the legislature that affects the configuration of corporate laws: courts. He suggests that, from the perspective of shareholder protection, active judicial review is a substitute for mandatory rules. Concerns over institutional competence, however, limit the potential for benefits from judicial activism. These concerns are most forcefully conveyed in the application of the business judgment rule, through which courts hesitate to second-guess managers' decisions, even when there is evidence that the decision will not maximize shareholder value. See Fred S. McChesney and William J. Carney, "The Theft of Time, Inc?" *Regulation* 78 (Spring 1991).

State Competition for Corporate Charters

The State Competition Debate in Corporate Law

ROBERTA ROMANO

A perennial issue in corporate law reform is the desirability of a federal system. For notwithstanding the invasive growth of regulation by the national government, principally through the federal securities laws, corporate law is still the domain of the states. While no two corporation codes are identical, there is substantial uniformity across the states. Provisions typically spread in a discernible *S*-shaped pattern, as one state amends its code in response to another state's innovation. The revision process is often analogized in the academic literature to market competition, in which states compete to provide firms with a product, corporate charters, in order to obtain franchise tax revenues. This characterization is the centerpiece of the federalism debate in corporate law—whether competition, and hence a federal system, benefits shareholders. The hero—or culprit—in the debate is Delaware, the most successful state in the market for corporate charters. . . .

The State Competition Literature

The Classic Positions Revisited

The foundation of the federalism debate in corporate law is that revenues derived from franchise taxes provide a powerful incentive for state legislatures to implement corporation codes that will maintain the number of domiciled corporations, if not lure new firms to incorporate in their state. All participants in the debate believe that the income produced by the chartering business spurs states to enact laws that firms desire. This behavioral assumption is plausible: There is a positive linear relation between the percentage of total revenues that states obtain from franchise taxes and states' responsiveness to firms in their corporate codes. The more dependent a state is on income from

Published originally in 8 *Cardozo Law Review* 709 (1987). Reprinted by permission.

franchise tax revenues, the more responsive is its corporation code. The potential revenue from this tax source can be substantial for a small state. Delaware's franchise tax revenue averaged 15.8 percent of its total revenue from 1960 to 1980, and while it is impossible to generate a precise figure, this income considerably outdistances the cost of operating its chartering business.

Given the shared assertion that revenues compel states to be responsive to firms' demands for legislation, the crux of the dispute is, therefore, whether this responsiveness is for the better. Because of the separation of ownership and control in the management of many large public corporations, when a firm's managers propose a reincorporation or urge the enactment of a statute, no less the adoption of a charter provision, we are concerned about whether they are maximizing the value of the firm. This is the classic agency problem, which goes to the heart of corporation law . . . : How do principals—the shareholders—ensure that their agents—the managers—behave faithfully?

Advocates of a national corporation law have termed state competition a race for the bottom because they believe that managers' discretion is unfettered, enabling them to promote laws that are detrimental to shareholders' welfare. They base this conclusion on a characterization of the statutes and case law of Delaware—which is the most frequent location for a reincorporation—as excessively permissive, by which they mean tilted toward management. Proponents of the current federal system, however, question this phrasing of the issue, typically viewing the agency problem as trivial. They maintain that the many markets in which firms operate—the product, capital, and labor markets—constrain managers to further the shareholders' interests. Accordingly, in their view, conflict between investors and managers over the content of state laws is largely illusory, and the laws that are promulgated can best be explained as mechanisms for maximizing equity share prices.

The initial articulation of the market argument in the state competition debate was by Ralph Winter. Responding to William Cary, who launched the first salvo in the modern debate, Winter contended that if management chose a state whose laws were adverse to the shareholders' interests, the value of the firm's stock would decline relative to stock in a comparable firm incorporated in a state with value-maximizing laws, as investors would require a higher return on capital to finance the business operating under the inferior legal regime. This impact in the capital market would affect managers by threatening their jobs. Either the lower stock price would attract a takeover artist who could turn a profit by acquiring the firm and relocating it in a state with superior laws, or the firm would go bankrupt by being undercut in its product market by rivals whose cost of capital would be lower because they were incorporated in value-maximizing states. In either scenario, in order to maintain their positions, managers are compelled, by natural selection, to seek the state whose laws are most favorable to shareholders.

Winter's critique is devastating to Cary's analysis because Cary completely overlooked the interaction of markets on managers' incentives. Yet Cary's position cannot be entirely dismissed: More sophisticated proponents of national chartering can move to another line of attack by maintaining that there is a true difference in opinion that turns upon Cary's and Winter's assessments of the disciplining effect of markets on managers. Winter can assume away the agency problem because of his view that the capital market is efficient such that information concerning the impact of different legal regimes is publicly available and fully assimilated into stock prices. In contrast, support for national chartering presupposes a market that is, at best, only weakly efficient, such that it does not digest information concerning legal rules. In addition, even if stock prices accurately reflect the value of different legal regimes, if product markets are not competitive or the costs of takeovers are substantial, then a manager's livelihood may not be jeopardized by the choice of a non-value-maximizing incorporation state. When the debate is phrased in this way, the disagreement is over an empirical question concerning market efficiency, for which, in principle, there is a clean answer.

To be sure, Cary sees a failure not only in financial and product markets, but also in local politics. His recommendation of national standards for corporations implies that the political process at the national level differs fundamentally from that of the states. Cary considers the flaw in Delaware's code to be a function of that state's desire for revenue and the close personal connections between Delaware legislators, judges, and corporate law firms. The national government certainly would not be as sensitive to franchise tax dollars as would a small state, and practically speaking, there would be no competing sovereigns to attract dissatisfied corporations.

But even if we grant Cary's premise that the states' responsiveness is the source of the problem, the elimination of intergovernmental competition is not necessarily the cure. The hitch in Cary's position is that he leaves unexplained why national legislators in pursuit of reelection would be less susceptible to the political influence of managers for "pet" statutes than state legislators. For why should diffuse and unorganized shareholders be appreciably better able to communicate their views to Congress when they cannot do so to state legislatures? There are countless pieces of legislation produced by pork barrel politics in Congress—the tax code is perhaps the most notable example—and there is no convincing reason to believe that firms' managers would be any less skilled at protecting their interests when it comes to a federal corporation code. . . .

A Transaction Cost Explanation of the Market for Corporate Charters

Why is Delaware the destination state of choice? The transaction cost explanation of the corporate charter market provides a different perspective on state competition. Delaware's persistent large market share is maintained by a first-mover advantage created by the reciprocal relation that develops between the chartering state and firms due to their substantial investment in assets that are specific to the chartering transaction. . . . How does [Williamson's] analysis apply to the corporate charter market? Because the transactions between a firm and its incorporation state extend over a long period of time and reincorporation is not costless, relocation makes a firm vulnerable to exploitation by the state. In particular, the state may charge a premium for incorporation and then alter its code or simply not implement the latest innovations, to the firm's detriment, knowing that the firm cannot quickly migrate again without incurring additional expenses. Hence, due to this nonsimultaneity in performance, a state with a favorable corporation code must guarantee its code's continued responsiveness to be successful in the corporate charter market.

Of all the states, Delaware is best positioned to credibly commit itself to responsiveness. First, its very success in the incorporation business serves, ironically, to constrain its behavior: The high proportion of total revenue it derives from franchise taxes guarantees continued responsiveness because it has so much to lose. For unlike states less dependent on franchise revenues, Delaware has no readily available alternative source to which it can turn in order to maintain expenditures. It cannot afford to lose firms to other states by failing to keep its code up-to-date. In this way, Delaware offers itself as a hostage by its reliance on franchise taxes to finance its expenditures.

Second, an additional institutional mechanism warranting responsiveness is Delaware's constitutional provision mandating that all changes in the corporation code be adopted by a two-thirds vote of both houses of the state legislature. This makes it difficult to renege on provisions already in the code and, correspondingly, on the overall policy of being responsive to firms. While the provision would appear to make future changes equally difficult, if firms are risk averse when it comes to corporation codes, they might favor a maximin strategy in which the constitutional provision would be desirable, since it helps to ensure that the legal regime will never be worse than it is at the time of incorporation. This provision thus complements Delaware's high proportionate franchise tax, for while the constitution is backward-looking, limiting radical revamping of the code, the incentives provided by the franchise tax revenue are forward-looking, as the state reacts to the high proportion of franchise tax revenues in the past by maintaining its responsiveness to incremental change in the future.

Third, Delaware has invested in assets that have no use outside of the chartering business. These assets, which can best be characterized as legal capital, consist of a store of legal precedents forming a comprehensive body of case law, judicial expertise in corporate law, and administrative expertise in the rapid processing of corporate filings. These features are not as easily duplicated by other states as the provisions of a corporation code because of the start-up costs in developing expertise and the dynamic precedent-based nature of adjudication by courts.

The combination of these factors—the high proportion of franchise tax revenue, the constitutional supermajority requirement, and the investments in legal capital—create an intangible asset with hostage-like qualities, a reputation for responsiveness, that firms weigh in their incorporation decision. The large number of firms already incorporated in Delaware further solidifies its commanding position in the market by giving it a first-mover advantage.

There is safety in numbers—the more firms there are the higher the level of franchise tax paid and the more the state relies on its incorporation business for revenue, which provides the incentive to behave responsively. In addition, the large number of firms makes it more likely that any particular issue will be litigated and decided in Delaware, providing a sound basis for corporate planning. This attracts even more firms for the more responsive a state and the more settled its law, the cheaper it is for a firm to operate under that legal regime. The first-mover advantage is self-sustaining because the more firms there are paying a franchise tax, the greater the return Delaware earns on its reputation for responsiveness, and the stronger its incentive to not engage in an end-game strategy of exploiting firms that would damage, if not destroy, its investment in a reputation.

This brings us to the demand side of the market, which also aids Delaware in maintaining an edge. There is a third party affected by the incorporation system, legal counsel, and the features of Delaware's legal regime that are attractive to firms—a well-developed case law with a pool of handy precedents and a means for rapidly obtaining a legal opinion on any issue—are also advantageous to corporate lawyers. For these features of Delaware law lower the cost of furnishing advice to clients. This is especially important for outside counsel, who service firms that are headquartered in different states, and who are instrumental in choosing the incorporation state. They realize cost savings by having clients operate under one legal regime. In addition to encouraging the choice of Delaware as the incorporation site for clients, specialization also provides an incentive for advising firms to remain in Delaware, because moving will diminish the attorney's human capital. Counsel's desire to recoup the investment in mastery of the institutional detail of Delaware law ties firms reciprocally to Delaware, just as Delaware is tied to firms.

Human capital is important in another way. Delaware's stake in the chartering business exceeds the revenues it receives from the franchise tax. A number of its citizens specialize in providing services to nonresident Delaware corporations. Accordingly, it is in the interest of those individuals that Delaware be responsive to corporations so that the demand for their services does not decline. Delaware's supermajority constitutional provision therefore serves an important function aside from credibly precommiting it to be responsive to firms: It protects the value of these individuals' personal investments by making it more difficult for a political realignment in the state to alter the long-standing course of corporate responsiveness.

This transaction-specific human capital, which creates a "mutual reliance relation" between firms and Delaware, joins the parties in long-term cooperation because of their reciprocal vulnerability and cements Delaware's market position, as it makes it difficult for a rival state to compete successfully. Another state cannot simply offer corporations the same code at a lower price and attract the marginal firm because a switch would increase operating and legal costs, and more importantly, the state cannot provide a credible commitment of superior service. In particular, a rival state cannot place itself in the same vulnerable position as Delaware because it starts from a low franchise tax ratio and has not yet invested in legal capital. In order for a state to begin to compete, a significant number of firms would have to agree to move to it in concert. But there is no incentive for corporations to move to another state so long as Delaware continues to cooperate, and there are powerful incentives for Delaware to continue to do so. . . .

Event Studies as Arbiters of the Debate

The debate over the efficacy of state corporation codes essentially boils down to an empirically testable hypothesis: whether managers or shareholders benefit from the market for corporate charters. If we could identify the beneficiaries, then fashioning a political consensus regarding the optimal level of government regulation would be straightforward. The best available means of generating information bearing on this issue is to examine the impact of reincorporations on stock prices, for a change in equity value conveys investors' assessment of the event's expected effect on shareholder wealth. A stock price increase upon a firm's reincorporation would mean that investors expect the change in incorporation state to increase the firm's future cash flows, and from this it could be concluded that shareholders benefit from a move. Similarly, a decline in stock price would indicate the anticipation that shareholder welfare will be diminished by the move and confirm the managerialist position.

Several event studies have been performed that bear on the state competition debate. Researchers have addressed the issue directly by investigating the impact of reincorporating, and indirectly by looking at the effect of state court decisions. . . . None of the studies support the managerialist position, for none found a negative effect on stock price. Rather, to the extent that they can be used to buttress any position, it is the value-maximizing view associated with Ralph Winter.

Event Studies on State Competition

Peter Dodd and Richard Leftwich, in the first empirical study concerning state competition, found statistically significant positive abnormal returns to the stock of reincorporating firms over the two-year period preceding the reincorporation. The returns around the event date were not, however, significant. While this finding undermines Cary's position, it is difficult to assert that it bolsters Winter's view because the period of abnormal returns is so far before the announcement of the move that it is possible the abnormal returns are due to some other factor affecting the firms.

I sought to refine the Dodd and Leftwich study by partitioning the portfolio of reincorporating firms according to the reasons for which the reincorporation was undertaken, and by using daily rather than monthly stock price data. I found that firms which reincorporated in order to embark on merger-and-acquisition programs, as well as the aggregate portfolio of reincorporating firms, experienced statistically significant positive abnormal returns on and around the event date. The signs of the cumulative average residuals for the other groups were also positive, although they were not significant. This finding creates further difficulty for the Cary thesis and provides more clear-cut support for Winter's value-maximizing interpretation of state competition. . . .

Weiss and White examined another theme in the literature to get at the crux of the state competition debate: Who is helped out by Delaware court decisions? They investigated the effect of seven Delaware opinions that they characterized as reversals or departures from existing corporate law rules. They hypothesized that if the decisions benefited shareholders, firms would experience abnormal positive returns, and if not, there would be negative returns. They found no statistically significant abnormal returns earned by Delaware firms, and the signs of the residuals were not consistent with any particular thesis.

Toward an Interest Group Theory of Delaware Corporate Law

JONATHAN R. MACEY AND GEOFFREY P. MILLER

[We posit] the existence of two groups within the state whose interests are differentially affected depending on the nature of the legal rules supplied to firms incorporated there: (1) the taxpayers and groups allied with them; and (2) the Delaware bar. We envisage the supply side as a political equilibrium in which each group obtains desired legal rules depending on its political influence. A principal hypothesis emerging from this model is that the bar is the most important interest group within this equilibrium. Thus, the rules that Delaware supplies often can be viewed as attempts to maximize revenues to the bar, and more particularly to an elite cadre of Wilmington lawyers who practice corporate law in the state. . . .

From the standpoint of shareholders and corporate managers, both franchise fees and the panoply of indirect costs are largely interchangeable. These decision makers are ordinarily indifferent to whether the firm must pay an extra dollar to the state fisc in franchise fees, to a lawyer for defense of a lawsuit arising under Delaware law, or to someone else. As far as the firm is concerned, the dollar is a cost of Delaware incorporation regardless of the identity of its recipient.

To the recipients, however, the identity of the payee is crucial. It makes a great deal of difference to Delaware taxpayers, lawyers, and corporation

Published originally in 65 *Texas Law Review* 469–523 (1987). Copyright 1987 by the Texas Law Review Association. Reprinted by permission.

service firms whether the dollar is paid in franchise fees, legal fees, or for some other purpose. This observation is crucial . . . for it suggests that the shape of Delaware corporate law may be affected significantly by interest group rivalries, as different factions within the state vie to capture the lion's share of the revenues for themselves.

The trade-off between direct and indirect costs, however, is not simply a matter of allocating gains among different groups within Delaware. Although the direct costs of Delaware incorporation are all captured by the state's taxpayers in the form of revenues to the state's fisc, the indirect costs can only partially be captured by state residents. For example, very few investment banking services are provided in Delaware. The major investment banks that serve Delaware corporations are located almost exclusively in New York. To the extent that Delaware law encourages the use of investment banks, the benefits will largely be captured by people outside the state.

Another example of costs that are shared with people out of state is legal fees. Although the Delaware bar is clearly the most significant in-state recipient of indirect chartering revenues, the bar cannot capture all the legal work on corporate law issues arising under Delaware law. In advisory matters, for example, "Delaware lawyers" are scattered among major law firms across the United States. An attorney does not need to be a member of the Delaware bar in order to provide advice on that state's law. In fact, Delaware lawyers probably operate at a competitive disadvantage vis-à-vis lawyers in other states because of the need in advisory work for extensive and sustained client contact. Our theory, however, predicts that Delaware attempts to maximize the amount of advisory work performed by Delaware counsel. For example, firms chartered in Delaware must obtain Delaware counsel to review their documentation each year. Delaware attorneys also can capture a significant share of revenues from litigation because of legal regulations requiring that court appearances and filings be made by a member of the state bar and because actual court appearances necessarily will take place in Delaware. The filing requirement, however, can be partially avoided through retention of local counsel who adds his or her name to briefs and motions as an accommodation to an out-of-state lawyer who prepares the papers. Delaware lawyers reputedly have developed local counsel services into an art, providing their clients with details and gossip about the inner workings of the Delaware court system, as well as excellent assistance in finding lodging and entertainment while in Wilmington—all for a hefty hourly fee. The requirement that court appearances be by a bar member can easily be surmounted by a motion to appear *pro hac vice,* which is granted as a matter of course. Some lawyers, however, believe that the Delaware judiciary is likely to look with more favor on a case argued by a recognized member of the state's bar; thus many important cases are argued by local counsel. . . .

The power of a political interest group—and therefore the degree to which the equilibrium conditions of legislation will favor its interests—is a function of several different variables. The most important of these variables include the number of members in the group, the amount of their individual stakes in the matter, and the degree to which they are able to cooperate as a single organized unit. The groups that this article identifies as interested in Delaware corporate law differ widely as measured by these variables. The bar is small, discrete, and highly organized. Its members tend to have a large personal stake in the subject matter of the regulation. They also tend to be more wealthy than other groups and to have good political connections. Indeed, many members of the Delaware legislature are themselves members of the bar. Such legislators tend to be represented disproportionately on legislative committees that draft the provisions of the Delaware Corporation Code. As noted above, the bar can be expected to capture much, although not all, of the gains from increasing the amount of legal fees generated by provisions of the Delaware corporate law. Accordingly, the bar is a powerful political force pressing for rules that maximize legal fees but do not necessarily maximize the revenues from corporations chartering in the state.

In contrast to Delaware attorneys, other interest groups within the state are unlikely to be able to galvanize into a coherent force to constitute a potent political threat to the bar. Although large in numbers, the individual stakes of the multitude of competing interests are small, and they are relatively unorganized. The economic theory of regulation predicts that free-rider problems will prevent them from having much success in countering the Delaware bar's drive for increased legal fees, especially because none of these individual interest groups has any assurance that it will be the one to enjoy the fruits of any overall increase to the general corporate treasury. In other words, the competing groups are not organized into an effective political coalition because they lack sufficient incentives to incur the costs of promoting laws that will increase the general revenues of the state. All the costs of pushing for efficient corporate laws must be borne by discrete groups or individual taxpayers. These costs consist of the expenses of making campaign contributions and engaging in other forms of lobbying activities, as well as the search costs of ascertaining what sort of laws are likely to lead to an increase in chartering revenues in the first place. Although the costs of pressing for efficient corporate law in Delaware must be concentrated in specific groups, the benefits of such laws, like the benefits from public interest laws generally, would be spread among all the groups, and perhaps even to the general population. Indeed, should rival interest groups prevail over the Delaware bar in achieving the legislation it wants, these groups would in all likelihood fight a second battle among themselves to determine how the spoils should be divided.

In other words, a rival interest group must win two political battles to achieve a favorable wealth transfer from a change in the Delaware Corporate Code. First it must prevail in its contest with the Delaware bar to obtain the initial revenue increase, and then it must prevail over all other interest groups to obtain the specific legislation that will transfer this additional revenue to it. Perhaps the most important reason the Delaware bar is able to prevail over rival groups in procuring favorable corporate law rules is that, for reasons unrelated to lobbying, the bar has already internalized the significant start-up costs necessary to lobby effectively for the rules it wants. Rival groups must incur these costs in learning what the relevant corporate law rules are, and how they can be written to benefit the state treasury. The state's corporate lawyers, of course, obtain this costly information as a by-product of their normal activities.

Notes and Questions

1. The findings of the event studies summarized in the Romano selection, that shareholders do not experience a wealth loss upon a proposed change in domicile, are consistent with those of a subsequent study. Jeffry Netter and Annette Poulsen investigated Delaware reincorporations from 1986 to 1987, after Delaware adopted a statute enabling shareholders to amend their firms' charters to limit directors' liability for negligence (the impact of this provision is examined in Chapter V, part A, note 10). The reincorporating firms experienced positive abnormal returns in the period surrounding the announcement of the move at a significance level of 10 percent. Netter and Poulsen, "State Corporation Laws and Shareholders: The Recent Experience," 18 *Financial Management* 29 (1989).

Lucian Bebchuk argues that findings of positive or insignificant stock price effects on reincorporations do not provide evidence that state competition benefits shareholders because it is possible that the effect of some desirable provisions obscures or offsets the effect of undesirable laws. Bebchuk, "Federalism and the Corporation: The Desirable Limits on State Competition in Corporate Law," 105 *Harvard Law Review* 1435 (1992). Does the absence of negative abnormal returns for reincorporating firms indicate that more provisions in destination state codes are perceived favorably than adversely by shareholders? Would a negative stock price effect upon enactment of a particular state corporate law lead you to conclude that a federal statute was required or is more necessary to make the case? See the materials on state regulation of takeovers in Chapter VI.

2. Romano's transaction cost explanation for Delaware's predominance in the market for corporate charters builds on the phenomena that she found in her empirical study of reincorporations, Roberta Romano, "Law as a Product: Some Pieces of the Incorporation Puzzle," 1 *Journal of Law, Economics, and Organization* 225 (1985). First, there is a statistically significant positive relation between the responsiveness of a state's corporation code to corporate preferences and the percent of its revenues raised from the franchise tax. Second, firms reincorporate when the legal regime becomes important, that is, when they anticipate engaging in transactions that will be more cheaply undertaken in the destination state (because of code differences and ready availability of case law). In particular, firms reincorporated when they were about to go public, initiate a mergers and acquisitions program, or engage in defensive maneuvering to a takeover (Delaware is the predominant destina-

tion for firms in the first two categories). These findings of dependence on the franchise tax and investments in legal capital are the basis of the hostage explanation of Delaware's success, for they serve as precommitment devices for the state, ensuring that the state will not revamp its code and become unresponsive to corporate needs after a firm reincorporates.

Note that Romano finds that firms that reincorporate in order to lower their tax burden do not migrate to Delaware. In fact, Delaware's franchise tax is higher than that of most states. Thus, firms are paying a premium for a Delaware charter. But the higher fees may be offset by lower costs of doing business under Delaware's code.

3. If Delaware's code is superior to others, why aren't all firms incorporated in Delaware? Romano's transaction cost explanation of reincorporations, discussed in note 2, suggests one reason (not all firms engage in transactions that benefit from a Delaware address). Richard Posner and Kenneth Scott suggest a different explanation involving product differentiation: Delaware specializes in servicing large corporations. Posner and Scott, *Economics of Corporation Law and Securities Regulation* 111 (Boston: Little, Brown, 1980). Their suggestion has limited explanatory power, however, because only half of the largest corporations are incorporated in Delaware.

Barry Baysinger and Henry Butler offer an alternative product differentiation story that incorporates both sides (the Cary and Winter positions) of the state competition debate. They contend that firms locate in states whose laws match their owners' needs, such that firms with diffuse stock ownership choose states with lax codes (i.e., Delaware), while firms with controlling shareholders select states with strict codes (i.e., Cary's preferred regime). Baysinger and Butler, "The Role of Corporate Law in the Theory of the Firm," 28 *Journal of Law and Economics* 179 (1985). They reason as follows: Because controlling shareholders hold large blocks of stock, they would have difficulty selling their shares in the market if they became dissatisfied with firm performance. The high cost of exit makes such owners prefer codes that facilitate shareholder voice and thereby limit management's discretion. Is Baysinger and Butler's thesis affected by the fact that insiders, who retain control when their firms go public, frequently choose to reincorporate in Delaware shortly before the public offering?

4. Delaware has been the leading state for incorporations since shortly after the turn of the century; before then, New Jersey had the commanding position. One reason often cited for New Jersey's replacement by Delaware is political upheaval in New Jersey, where the corporation code was tightened by Governor Woodrow Wilson as a lame duck (when president-elect). But changes in business trust statutes that had contributed to New Jersey's predominance in the chartering market can also be attributed to unfavorable Supreme Court decisions, as well as to economic development in the state that attenuated New Jersey's "hostage" relationship with chartering corporations, rather than just the state's changing political climate. Excellent studies of nineteenth-century competition for charters are Henry N. Butler, "Nine-

teenth-Century Jurisdictional Competition in the Granting of Corporate Privileges," 14 *Journal of Legal Studies* 129 (1985); and Christopher Grandy, "New Jersey Corporate Chartermongering, 1875–1929," 49 *Journal of Economic History* 677 (1989).

5. Elliott Weiss and Lawrence White conclude from their failure to find a significant impact of Delaware court decisions that investors are not concerned with differences in corporate law regimes and, consequently, that there is no state competition to speak of. Weiss and White, "Of Econometrics and Indeterminacy: A Study of Investors' Reactions to 'Changes' in Corporate Law," 75 *California Law Review* 551 (1987). Would you draw the same conclusion? Is a plausible alternative interpretation that investors anticipate Delaware court decisions better than researchers?

Corporate managers typically oppose hostile takeovers by engaging in defensive tactics, which shareholders, desiring the bid's success in order to receive the bid premium, challenge in court. Would you expect a significant market reaction to court decisions on the validity of such tactics? See DoSoung Choi, Sreenivas Kamma, and Joseph Weintrop, "Delaware Courts, Poison Pills, and Shareholder Wealth," 5 *Journal of Law, Economics, and Organization* 375 (1989); Sreenivas Kamma, Joseph Weintrop, and Peggy Wier, "Investors' Perceptions of the Delaware Supreme Court Decision in Unocal v. Mesa," 20 *Journal of Financial Economics* 419 (1988); Michael Ryngaert, "The Effect of Poison Pill Securities on Shareholder Wealth," 20 *Journal of Financial Economics* 377 (1988). Takeover defenses, and their relation to state competition, are examined in Chapter VI.

What effect on stock prices would you expect court decisions to have if the Delaware legislature is known to quickly reverse undesirable decisions? Consider the legislature's reaction to *Smith v. Van Gorkom,* 488 A.2d 858 (Del. 1985), a widely criticized decision holding outside directors had breached their duty of care (acted negligently) for accepting an acquisitive offer too hastily; it enacted a statute permitting shareholders to eliminate outside directors' personal financial liability for negligence. Directors' fiduciary duties and limited liability statutes are examined in Chapter V.

Does the benefit of legal certainty from a Delaware address—knowing how to structure a transaction to avoid liability—suggest another reason for Weiss and White's findings? Namely, if a legal rule defines the rights and obligations of parties who can subsequently transact around the rule without much cost (so that the substantive content of the rule is less important than its existence), would you expect a significant stock price effect upon a judicial rule change? In responding to this question, consider the discussion of corporation codes as enabling regimes and the Coase Theorem in part B of this chapter.

6. Should the dynamics of state competition apply to charters for close corporations? See Ian Ayres, "Judging Close Corporations in the Age of Statutes," 70 *Washington University Law Quarterly* 365 (1992).

7. Should we expect the emergence of a European "Delaware" with the European Community's increasing economic integration? In responding to this question, consider how important to a competitive corporate law market the choice of law rule is. European nations with civil law systems, for example, follow a different corporate choice of law rule than nations like the United States with a common law tradition: the physical presence of corporate headquarters, rather than incorporation state. See Roberta Romano, *The Genius of American Corporate Law* (Washington, D.C.: AEI, 1993). For an interesting study of the Canadian experience (where no one province dominates and there is a competing national corporation statute), see Ronald J. Daniels, "Should Provinces Compete? The Case for a Competitive Corporate Law Market," 36 *McGill Law Journal* 130 (1991).

8. Does Macey and Miller's thesis require a noncompetitive market for corporate lawyers? Or friction in competition across the states? What are the implications of Macey and Miller's thesis for the state competition debate? Consider Judge Ralph Winter's view that "the race to the top applies only when a state legislature is guided by a desire to maximize franchise taxes"; if states have some other political objective, then they will be indifferent to market demand and the resulting laws need not benefit shareholders. Winter, "The 'Race for the Top' Revisited: A Comment on Eisenberg," 89 *Columbia Law Review* 1526 (1989). Would the self-interest of the corporate bar prevent Delaware's code from shifting away from the preferences of corporate charter consumers even if no other states are franchise tax maximizers? This question will be revisited in the discussion of state takeover statutes in Chapter VI.

The Structure of Corporation Laws

The Corporate Contract

FRANK H. EASTERBROOK AND DANIEL R. FISCHEL

The corporate code in almost every state is an "enabling" statute. An enabling statute allows managers and investors to write their own tickets, to establish systems of governance without substantive scrutiny from a regulator and without effective restraint on the permissible methods of corporate governance. The handiwork of managers is final in all but exceptional or trivial instances. Courts apply the "business judgment doctrine," a hands-off approach that they would never apply to the decisions of administrative agencies or other entities—the officials of which do not stand to profit from their decisions, and therefore, one might think, are not subject to the pressures that cause managers' goals to diverge from those of investors.

Consider the domain of choice. The founders and managers of a firm choose whether to organize as a corporation, trust, partnership, mutual, or cooperative. They choose what the firm will make or do and whether it will operate for profit, not for profit, or hold a middle ground, pursuing profit but not to the exclusion of some other objective (as publishers of newspapers do). They choose whether to allow the public to invest or whether, instead, the firm will be closely held. They choose what kinds of claims (debt, equity, warrants) to issue, in what ratios, for what price, with what entitlements, including not only the right to receive payments (how often, in what amounts), but also whether these investments allow their holders to vote—and if to vote, how many votes, and on what subjects. They choose where to incorporate. They choose how the firm will be organized (as a pyramidal hierarchy or as a loose, multidivisional collective), whether central leadership will be strong or weak, and whether the firm grows (internally or by merger) or shrinks (sells existing assets or spins off divisions). Investors select the members of the board of directors, who may be "inside" or "outside," and the board decides who exercises which powers on the firm's behalf. As a practical matter boards are self-perpetuating until investors become dissatisfied and a majority decides to redo everything to a new taste. With trivial exceptions, all business decisions—including the managers' pay, bonuses, stock options, pensions, and perquisites—are taken by or under the supervision of this board, with no substantial inquiry by anyone else. Anyone who asks a court to inquire will be brushed off with a reference to the business judgment rule.

Some things are off limits. States almost uniformly forbid perpetual directorships; they set quorum rules, which typically require a third of the board and sometimes half of the investors to participate on critical decisions; they require "major" transactions to be presented to the board (occasionally shareholders too) rather than stand approved by managers or a committee; they forbid the sale of votes divorced from the investment interest and the accumulation of votes in a corporate treasury; they require managers to live up to a duty of loyalty to investors. Federal law requires firms to reveal certain things when they issue securities, and public firms must make annual disclosures. Determined investors and managers can get 'round many of these rules, but accommodation is a sidelight. Any theory of corporate law must account for the mandatory as well as the enabling features of the law, and must account for the pattern of regulation—one that leaves managers effectively free to set their own salaries yet forbids them to delegate certain questions to subcommittees, that gives shareholders no entitlement to dividends or distributions of any kind but specifies a quorum of one-third of the board for certain decisions. . . .

Why does corporate law allow managers to set the terms under which they will govern corporate assets? Why do courts grant more discretion to self-interested managers than to disinterested regulators? Why do investors entrust such stupendous sums to managers whose acts are essentially unconstrained by legal rules? The answers lie in, and help explain, the economic structure of corporate law. The corporation is a complex set of explicit and implicit contracts, and corporate law enables the participants to select the optimal arrangement for the many different sets of risks and opportunities that are available in a large economy. No one set of terms will be best for all; hence the "enabling" structure of corporate law.

Although managers are self-interested, this interest can be aligned with that of investors through automatic devices, devices that are useless when those in control are "disinterested"; hence the apparent contradiction that self-interested managers have more freedom than disinterested regulators. Of course controls are not free, and much of corporate law is designed to reduce the costs of aligning the interests of managers and investors. . . . The corporate structure is a set of contracts through which managers and certain other participants exercise a great deal of discretion that is "reviewed" by interactions with other self-interested actors. This interaction often occurs in markets, and we shall sometimes call the pressures these interactions produce "market forces." . . .

Managers and investors . . . assume their roles with knowledge of the consequences. Investors part with their money willingly, putting dollars in equities instead of bonds or banks or land or gold because they believe the returns of equities more attractive. Managers obtain their positions after much trouble and toil, competing against others who wanted them. All interested persons participate. Firms are born small and grow. They must attract customers and investors by promising *and delivering* what those people value. Corporations that do not do so will not survive. When people observe that firms are very large in relation to single investors, they observe the product of success in satisfying investors and customers.

How is it that managers came to control such resources? It is not exactly secret that scattered shareholders can't control managers directly. If the investors know that the managers have lots of discretion, why did they give their money to these managers in the first place? If managers promise to return but a pittance, the investors will not put up very much money. The investors simply pay less for the paper the firms issue. There is therefore a limit on managers' efforts to enrich themselves at investors' expense. Managers may do their best to take advantage of their investors, but they find that the dynamics of the market drive them to act as if they had investors' interests at heart. It is almost as if there were an invisible hand. . . .

Corporate Contracts

The arrangements among the actors constituting the corporation usually depend on contracts and on positive law, not on corporate law or the status of the corporation as an entity. More often than not a reference to the corporation as an entity will hide the essence of the transaction. So we often speak . . . of the corporation as a "nexus of contracts" or a set of implicit and explicit contracts. This reference, too, is just a shorthand for the complex arrangements of many sorts that those who associate voluntarily in the corporation will work out among themselves. The form of reference is a reminder that the corporation is a voluntary adventure, and that we must always examine the terms on which real people have agreed to participate. . . .

The way in which corporations run the business, control agency costs, raise money, and reward investors will change from business to business and from time to time within a firm. The structure suited to a dynamic, growing firm such as Xerox in 1965 is quite unsuited to Exxon in 1965 (or to Xerox in 1989). The participants in the venture need to be able to establish the arrangement most conducive to prosperity, and outsiders are unlikely to be able to prescribe a mold for corporations as a whole or even a firm through time. The history of corporations has been that firms failing to adapt their governance structures are ground under by competition. The history of corporate law has been that states attempting to force all firms into a single mold are ground under as well. Corporations flee to find more open-ended statutes that permit adaptations. This is the reason for the drive toward enabling laws that control process but not structure.

To say that a complex relation among many voluntary participants is adaptive is to say that it is contractual. Thus our reference to the corporation as a set of contracts. Voluntary arrangements are contracts. Some may be negotiated over a bargaining table. Some may be a set of terms that are dictated by managers or investors and accepted or not; only the price is negotiated. Some may be fixed and must be accepted at the "going price" (as when people buy investment instruments traded in the market). Some may be implied by courts or legislatures trying to supply the terms that would have been negotiated had people addressed the problem explicitly. Even terms that are invariant—such as the requirement that the board of directors act only by a majority of a quorum—are contractual to the extent that they produce offsetting voluntary arrangements. The result of all of these voluntary arrangements will be contractual. . . .

The corporate venture has many real contracts. The terms present in the articles of incorporation at the time the firm is established or issues stock are real agreements. *Everything* to do with the relation between the firm and the suppliers of labor (employees), goods and services (suppliers and contractors) is contractual. So with the rules in force when the firm raises money— whether by issuing debt, the terms of which often are negotiated at great length over a table, or by issuing equity, the terms of which affect the price of the issue. Many changes in the rules are approved by large investors after negotiation with management. And of course the rules that govern how rules change are also real contracts. The articles of incorporation typically allow changes to be made by bylaw or majority vote; they could as easily prevent changes, or call for supermajority vote, or allow change freely but require nonconsenting investors to be bought out. That the articles allow uncompensated changes through voting is a real contractual choice. And many remaining terms of the corporate arrangement are contractual in the sense that they are "presets" of fallback terms specified by law and not varied by the corporation. These terms become part of the set of contracts just as provisions of the Uniform Commercial Code become part of commercial contracts when not addressed explicitly.

These contracts usually are negotiated by representatives. Indenture trustees negotiate on behalf of bondholders, unions on behalf of employees, investment banks on behalf of equity investors. Sometimes terms are not negotiated directly but are simply promulgated, in the way auto rental companies promulgate the terms of their rental contracts. The entrepreneurs or managers may adopt a set of rules and say "take them or leave them." This is contracting nonetheless. We enforce the terms in auto rental contracts, as we enforce the terms of a trust even though the beneficiaries had no say in their framing. The terms in rental contracts, warranties, and the like are real contracts because their value (or detriment) is reflected in price. . . .

Let us suppose that entrepreneurs simply pick terms out of a hat. They cannot force investors to pay more than what the resulting investment instruments are worth; there are too many other places to put one's money. Unless entrepreneurs can fool the investors, a choice of terms that reduce investors' expected returns will produce a corresponding reduction in price. So the people designing the terms under which the corporation will be run have the right incentives. . . . If the managers make the "wrong" decision—that is, choose the term inferior as investors see things—they must pay for their mistake. . . .

In general, all the terms in corporate governance are contractual in the sense that they are fully priced in transactions among the interested parties. They are thereafter tested for desirable properties; the firms that pick the wrong terms will fail in competition with other firms competing for capital. It is unimportant that terms may not be "negotiated"; the pricing and testing mechanisms are all that matters, as long as there are no effects on third parties. This should come as no shock to anyone familiar with the Coase Theorem [defined in note 1 following the readings—ED.]

Are the terms of corporate governance priced? The provisions in articles of incorporation and bylaws often are picky and obscure. Many are not listed in the prospectus of the firm's stock. Buyers of the original issue and in the aftermarket alike may know nothing of the terms in use, let alone whether a staggered board of directors or the existence of cumulative voting will make them better off. Yet it is unimportant whether knowledge about the nature or effect of the terms is widespread, at least for public corporations. The mechanism by which stocks are valued ensures that the price reflects the terms of governance and operation, just as it reflects the identity of the managers and the products the firm produces. . . .

A great deal of data, including evidence that most professional investors are unable to "beat the market," supports the position that prices quickly and accurately reflect public information about firms. Amateur investors then trade at the same price the professionals obtain. These amateurs do not need to know anything about corporate governance and other provisions; the value of these mysterious things is wrapped up in the price established by the professionals. The price reflects the effects, good or bad, of corporate law and contracts, just as it reflects the effects of good and bad products. This is yet another example of the way in which markets transmit the value of information through price, which is more "informed" than any single participant in the market.

To say that the price of a stock reflects the value of the firm's governance and related rules is not necessarily to say that the price does so perfectly. There may be surprises in store, for a firm or for all firms, that make estimates about the effects of governance provisions inaccurate. But these problems of information and assessment also affect any other way of evaluating the effects of governance devices. That is, if professional investors with their fortunes on the line are unable to anticipate the true effects of nonvoting stock or some other wrinkle in a concrete case, how are members of state legislatures or other alternative rule givers to do better? To put this differently, it does not matter if markets are not perfectly efficient, unless some other social institution does better at evaluating the likely effects of corporate governance devices. The prices will be more informative than the next best alternative, which is all anyone can demand of any device. . . .

If it is possible to demonstrate that the terms chosen by firms are both (a) unpriced and (b) systematically perverse from investors' standards, then it might be possible to justify the prescription of a mandatory term by law. This makes sense, however, only when one is sure that the selected term will increase the joint wealth of the participants—that is, that it is the term that the parties would have selected with full information and costless contracting. But this, too, is a contractual way of looking at the corporation. This formula is the one courts use to fill the gaps in explicit contracts that inevitably arise because it is impossible to cover every contingency. . . .

The Latecomer Term

Much of the discussion has proceeded as if all parts of the corporate contract were established at the beginning. "The beginning" for any participant is when he enters the venture—when he becomes an employee, invests, and so on. This is the critical time for most purposes because the time of entry is when the costs and benefits of governance arrangements are priced. If a term is good or bad at the beginning, adjustments in the prices even everything up. But of course many things change after the beginning. The firm may reincorporate in Nevada. It may adopt staggered terms for members of the board of directors or a "fair price amendment." It may abolish the executive committee of the board, get rid of all the independent directors, or create a board with a majority of independent directors. What are we to make of these changes?

Changes of this sort have some things in common: they are proposed by the existing managers (unless approved by the board of directors, no change in an ongoing firm's rules will be adopted), the proposals are accepted by voting among the equity investors, and the winning side in the vote does not compensate the losing side. If the changes are adverse to existing participants in the venture, there will be price adjustments, but these adjustments do not compensate the participants. If an amendment reduces the expected profitability of the firm by an amount worth one dollar per share, the price will fall and existing investors will experience a capital loss of one dollar per share. They can sell, but they can't avoid the loss. The buyers will get shares worth what they pay; the investors at the time of the change are out of luck. The mechanism by which entrepreneurs and managers bear the cost of unfavorable terms does not work—not in any direct way, anyway—for latecomer terms. It will work eventually. Latecomer terms that injure investors will reduce the firm's ability to raise money and compete in product markets. But these eventual reactions are not remedies; they explain why firms that choose inferior governance devices do not survive, and they show why widespread, enduring practices are likely to be beneficial, but they do nothing for participants in the ventures that are about to be ground under by the heel of history.

The process of voting controls adverse terms to a degree but not perfectly. Investors are rationally uninterested in votes, not only because no investor's vote will change the outcome of the election but also because the information necessary to cast an informed vote is not readily available. Shareholders' approval of changes is likely to be unreliable as an indicator of their interests, because scattered shareholders in public firms do not have the time, information, or incentive to review all proposed changes. Votes are not sold, at least not without the shares. The difference between governance provisions established at the beginning and provisions added later suggests some caution in treating the two categories alike. Some of the hardest questions in corporate law concern arrangements that are adopted or changed after the firm is under way and the capital has been raised. Thus doctrines of corporate law refusing to allow shareholders to ratify waste (except unanimously) are well founded. Yet the rules for amending the rules are themselves part of the original articles, and it is (or should be) possible to draft limitations on amendment. These most commonly take the form of provisions designating some amendments as transactions from which investors may dissent and demand appraisal. Moreover, amendments to governance structures may spark proxy contests in which investors' attention is focused, and they also may call forth takeover bids. So voting, or at least the opportunity for review set in place by the voting mechanism, is a partial substitute for the pricing mechanism that applies at the beginning.

One candidate for a rule of law that could overcome a problem in the contracting process is a rule that differentiates among terms according to the time of their adoption. It could provide that terms in place at the beginning (at the time the firm is founded, goes public, or issues significant amounts of stock) are always to be honored unless there are demonstrable third-party effects, while terms adopted later that appear to increase the agency costs of management are valid only if adopted by supermajority vote at successive annual meetings or if dissenting investors are bought out. (The dual-meeting rule would allow an intervening proxy or takeover contest to prevent the change from going into effect.) Yet if such a constraint on amendments is beneficial to investors, why are supermajority and dual-meeting requirements so rare in corporate documents? Investors can and do appreciate the risk that latecomer terms will be damaging, yet perhaps rules that slow down the adoption of changes would be more damaging still on balance. It is not our purpose here to draft rules of law. It is important, however, to keep the latecomer term in mind as a potential problem in a contractual approach to corporate law.

Why Is There Corporate Law?

One natural question after all this business of corporation-as-contract is: why law? Why not just abolish corporate law and let people negotiate whatever contracts they please? The short but not entirely satisfactory answer is that corporate law is a set of terms available off-the-rack so that participants in corporate ventures can save the cost of contracting. There are lots of terms, such as rules for voting, establishing quorums, and so on, that almost everyone will want to adopt. Corporate codes and existing judicial decisions supply these terms "for free" to every corporation, enabling the venturers to concentrate on matters that are specific to their undertaking. Even when they work through all the issues they expect will arise, they are apt to miss something. All sorts of complexities will arise later. Corporate law—and in particular the fiduciary principle enforced by courts—fills in the blanks and oversights with the terms that people would have bargained for had they anticipated the problems and been able to transact costlessly in advance. On this view corporate law supplements but never displaces actual bargains—save in situations of third-party effects or latecomer terms. . . .

The story is not complete, however, because it still does not answer the question "why law?" Why don't law firms or corporate service bureaus or investment banks compile sets of terms on which corporations may be constructed? They can peddle these terms and recover the cost of working through all of the problems. Yet it is costly for the parties (or any private supplier of rules) to ponder unusual situations and dicker for the adoption of terms. Parties or their surrogates must identify problems and then transact in sufficient detail to solve them. This may all be wasted effort if the problem does not occur. Because change is the one constant of corporate life, waste is a certainty. Often the type of problem that the firm encounters does not occur to anyone until after the venture is under way. Court systems have a comparative advantage in supplying answers to questions that do not occur in time to be resolved ex ante. Common law systems need not answer questions unless they occur. This is an economizing device; it avoids working through problems that do not arise. The accumulation of cases dealing with unusual problems then supplies a level of detail that is costly to duplicate through private bargaining. To put it differently, "contractual" terms for many kinds of problems turn out to be public goods!

The Mandatory Structure of Corporate Law

JEFFREY N. GORDON

The Role of Mandatory Law in a Contractual System

In thinking about the role that mandatory rules might play in a contractarian-framework, it is important to identify two distinct times at which mandatory law operates. The first, $t = 0$, is the point at which the corporation is formed and the charter is adopted. The second, $t = 1$, is the point at which a charter amendment is contemplated. At both times mandatory corporate law limits the ability of the parties to customize charter terms. Any of at least five

Reprinted by permission from 89 *Columbia Law Review* 1549. © 1989 Jeffrey N. Gordon.

hypotheses might explain the existence of such mandatory rules: the investor protection hypothesis, the uncertainty hypothesis, the public good hypothesis, the innovation hypothesis and the opportunistic amendment hypothesis. All but the first rest on the general argument that any efficiency losses from the rigidity of mandatory rules may be outweighed by the gains such rules generate by addressing defects in the contracting process. I believe the latter three hypotheses—particularly the last—have explanatory weight.

Investor Protection Hypothesis

Most contractarian analyses begin with the assumption that the parties have equivalent access to relevant information and the capacity to evaluate the information in light of their respective interests. This information symmetry assumption is the basis for the assertion by Jensen and Meckling that the promoters of the firm bear all the agency costs associated with the firm's governance arrangements and capital structure. This leads to the claim that in the case of the initial choice of rules at $t = 0$, absolute contractual freedom should apply, since the promoters, not the purchasing shareholders, bear the full costs of features that are undesirable from the investor's point of view. Thus the promoters will have the correct incentives to write charter terms that optimize the joint wealth of shareholders and promoters.

An investor protection argument flows directly from rejection of the contractarian information assumption. Many investors do not read the prospectus or do not understand or fully register the entailments of charter provisions. Promoters may therefore include charter terms that negatively affect shareholder interests without bearing the cost; investors will pay too much for such shares.[1] Thus mandatory law represents the state's setting of nonvariable quality standards to protect investors against the risks of misinformation.

Informed versus uninformed investors. It is a mistake to assume that investors can obtain information only through independent research, that is, to isolate individuals from markets. Well-functioning securities markets aggregate information from all active market participants, embody that information in a single fact—price—and make that fact available for free. . . . The uninformed investors will pay too high a price only if the market is not efficient, that is, only if there are too few sophisticated market participants who choose to become specifically informed. It therefore seems unlikely that investors who buy shares in secondary market trading on the highly efficient national securities markets could be systematically victimized by unexpected charter terms.

The real thrust of the investor protection hypothesis must be that the pricing of novel charter terms occurs in a market that is much less efficient, the initial public offering (IPO) market, where uninformed investors may be

[1]This argument assumes that the promoter has greater knowledge than the investor about the implications of a particular charter term, not that both parties may be ignorant about those implications. If knowledge of these implications were randomly distributed among promoters and investors, then investors would not on average pay too much for the stock.

victimized. There are several problems with this claim, however. First, the IPO market has a heavy institutional component. . . . Since issuers must offer securities on the same terms to all investors, unsophisticated investors can free ride on the efforts of sophisticated investors even in the IPO market.

The role that underwriters play in the IPO market also reduces the investor protection problem. Underwriters are experts in understanding the implications of particular charter provisions on expected investor returns. In explaining to the issuer the trade-off of a particular charter provision against price, the underwriter is in effect the bargaining agent of prospective public shareholders as a group. The underwriter's fidelity to a fair bargain with the issuer on behalf of public shareholders is supported by reputation effects. Underwriters will engage in repetitive dealings with many of the same customers, and news of an underwriter's permitting an unpriced adverse charter term would spread rapidly to future customers as well. Moreover, the underwriter's compensation is a percentage of the offering price, so the underwriter has an independent incentive to discourage the inclusion of charter terms that might reduce the price. Because shares of the same class must have identical terms, the firm cannot offer better terms to sophisticated investors than it does to unsophisticated investors. Thus unsophisticated investors capture the benefits of underwriters' efforts on behalf of sophisticated investors. . . .

The pricing of charter terms. The alternative basis for the investor protection hypothesis is that charter terms, unlike financial terms, will not be priced even in otherwise efficient markets. . . .

Evidence from current practice supports the view that investors do attend to significant variations in charter provisions no less than traditional financial or business information. For example, nonvoting (or limited voting) shares issued through an IPO or through a recapitalization trade at a discount to voting shares. . . .

Another example is the case of senior securities, both debt and preferred stock, which frequently contain complicated contractual provisions relating to the circumstances of voting, representation on the board of directors, conversion into common stock, call protection, redemption exposure, dilution, and other such concerns. These terms emerge through negotiation with the underwriter, acting on behalf of prospective purchasers. The effort the parties put into negotiating these terms strongly suggests that they expect these terms to be priced.[2] . . .

Uncertainty Hypothesis

In a regime of contractual freedom, it is likely that different charter terms will proliferate. Indeed, the corporate form might vary radically among firms, depending on the desires of the promoters and the responses of prospective shareholders. One clear cost imposed under such a regime, as compared to a

[2]For evidence that bond covenants are priced, see Chapter IV [EDITOR'S NOTE].

mandatory regime, is the uncertainty associated with different terms. The uncertainty hypothesis asserts that the desire to eliminate the costs of uncertainty is the basis for mandatory corporate law. . . .

From an ex ante contractarian perspective, however, these accounts of how nonstandard terms produce costs are not a basis for legal intervention, because the costs of uncertainty will be borne by the promoters who author the nonstandard terms. That is, prospective shareholders will foresee the possibility of unpredictable effects on firm payouts because of the customized clauses and will insist on a lower stock price as compensation for this risk. . . .

Public Good Hypothesis

The ex ante contractarian perspective argues that the uncertainty costs of nonmandatory terms will be internalized, that is, borne by the adopting firms, in particular, the promoters. This will be the case only if the problem is analyzed firm by firm. Viewed globally, a regime of complete contractual freedom in corporate law imposes externalities. As charters diverge from the standard form, the uncertainty surrounding even standard form terms begins to grow. Those terms are tested less frequently, either through operation in particular circumstances or through successive judicial interpretation. As a result, costs are imposed on firms with standard charters as well as firms with customized charters whose terms may play against the baseline of the standard form. Over time, the addition of customized terms by more firms will lead to disintegration of the standard form. Thus although firms collectively are better off if the standard form is maintained, individual firms will have incentives to deviate from the standard form in a way that will eventually undermine it. This states the classic free-rider problem that undermines the provision of a public good. In this particular case, maintenance of the public good of a standard corporate form will require a mandatory legal regime. . . .

Innovation Hypothesis

The claim of the innovation hypothesis is that a mandatory regime may aid in the innovation of corporate charter terms. The basic argument is this: Innovation is costly for firms because investors will frequently draw a negative inference from the innovation that will lead them to underpay for the firm's stock. These costs can be avoided by state action that credibly signals that the innovation is desirable from the public shareholder perspective. Thus a regime of mandatory law that permits the state to send such signals through revising or relaxing standard form terms may speed the process of innovation or lead to greater diffusion of innovation. . . .

The legislature's signaling capacity depends on whether shareholders generally believe that the focus of the legislative process is ordinarily shareholder welfare. This is obviously a controversial premise. It is at the heart of the debate over state competition for corporate charters. The literature on rent seeking in the legislative process, and more specifically, the state adoption of antitakeover statutes that presumptively reduce shareholder welfare, give reason for pause. Indeed, mandatory regimes raise the threat of innovations that reduce shareholder welfare. In particular, it is conceivable that managers could occasionally obtain a change from the legislature . . . that could not be obtained from the shareholders. This threat would be reduced if legislative changes that significantly affected management's ability to entrench itself required a shareholder vote to "opt in" or required a later shareholder vote on the question of whether to "opt out." . . .

Opportunistic Amendment Hypothesis

A rather different basis for mandatory corporate law is the claim that mandatory law is a hands-tying mechanism that provides assurance against opportunistic charter amendment. Even if promoters bear the full cost of governance arrangements and capital structure at $t = 0$, this would not be the case subsequently, at $t = 1$, when a charter change is contemplated. Because of the sunk investments of the existing public shareholders, the insiders (managers and controlling shareholders) will not bear the full costs of new features that transfer wealth, that is, cash flows or control, in their favor. This creates an incentive for opportunistic amendment that, if not addressed ex ante, generates a cost that will be borne by the promoters. Mandatory law—which prevents amendment of the charter in certain key respects—may exist as the solution to that problem.

Opportunistic amendment is possible because the corporate contract is inevitably incomplete. The parties cannot specify terms to cover even plausible contingencies, for a number of reasons: the transaction costs of the voluminous drafting, the vagaries of judicial fact finding and interpretation, the limits of human calculation ("bounded rationality"), and the sheer unforeseeability of circumstance, for example. Because the contracts are incomplete and because the corporation endures indefinitely, the parties are likely to create a mechanism to alter their prior arrangements. Responsibility for some alterations can be delegated to a governance structure that responds to both managers and shareholders, such as the board of directors. This accounts for the modern trend to permit directors to amend bylaws, for example. For other matters that touch on the very shape of the governance structure and capital structure, shareholders will insist on the right to consent to changes; a convenient procedure is to put such matters in the corporate charter and to require that a majority of shareholders approve charter amendments.

A formal document and voting rights do not solve the shareholders' problem, however. Proposed charter amendments will be sponsored by a relatively cohesive proponent, the insiders, who will argue that the proposed change will improve the corporation's functioning in a way that will significantly enhance shareholder wealth, that is, is wealth increasing. A diffuse group of public shareholders must evaluate this claim against the possibility that the amendment is merely "wealth neutral," because all or almost all of the gain inures to the insiders, or "wealth reducing," because it will transfer cash flow or control from public shareholders to insiders. In these circumstances, shareholder voting as a means of evaluating and consenting to a proposed charter amendment is fraught with severe problems, in particular, collective action problems in acquiring and disseminating information among shareholders, and strategic behavior by insiders that amounts to economic coercion. Thus insiders can exploit their advantages to obtain approval even for wealth-reducing amendments. From this flows an argument for mandatory corporate law: The gains from eliminating opportunistic amendment through mandatory rules will outweigh the efficiency losses from the resulting rigidities, at least for a set of provisions with the potential for significant entrenchment or distributional effects in favor of insiders. . . .

The Shape of Mandatory Rules

The present system of mandatory rules is constitutive. The rules establish the governance structure and set the standards of conduct to which insiders will be held. . . . There are four sorts of mandatory rules: procedural, power allocating, economic transformative, and fiduciary standard setting. The persistence of these rules may be understood in terms of the public good hypothesis, the innovation hypothesis, and opportunistic amendment hypothesis.

Procedural Rules

[Procedural rules] merely serve the [power or cash flow] allocations that have been made. Although the content of these rules is somewhat arbitrary, their mandatory nature may be justified on two accounts. First, the infrequent litigation regarding such provisions often arises in high-stakes situations, such as heated battles for corporate control. Thus, it is important to build a stock of precedents that construe the same provision (a public good rationale). Moreover, any midstream amendment of such a provision is highly suspect since it is likely to be a tactical maneuver in or anticipating a control struggle (an opportunistic amendment rationale).

Power Allocating

Many mandatory rules of corporate law allocate power throughout the governance structure, affecting, in particular, the balance of power between directors and shareholders. The managerial role of the board, shareholder voting rights in the election of directors, and shareholder removal rights are classic examples. Because the corporate contract is radically incomplete, many future decisions are left to the governance structure. Thus many important economic outcomes will be highly sensitive to changes in the allocation of power.

All three rationales operate in these circumstances. First, the output of a governance structure can be best understood by repeat experience in different settings of identical, or nearly identical, governance forms. As governance structures deviate from the baseline, incidents become too particularized; they lose their experimental usefulness. All profit from maintenance of a standard form. This is a public good rationale. Second, because governance structures are the Archimedean lever, public shareholders will suspect that innovative terms are intended to empower the insiders at their expense. By contrast, a change that comes upon legislative invitation after public deliberation may give assurance of a likely increase in shareholder wealth that eliminates any capital market penalty for the adopting firm. This is the innovation rationale. Finally, because of the key role of the governance structure, shareholders will be highly wary of midstream changes that could significantly shift wealth and control away from them and to the insiders, changes that can be obtained through a voting process known to be riddled with shareholder choice problems. The opportunistic amendment rationale explains why a mandatory legal rule is the indicated solution.

Economic Transformative

Transactions that transform the economic structure of the firm are generally governed by mandatory rules: mergers in which the firm does not survive or the ownership stake of the existing shareholders is significantly diluted, the sale of substantially all assets, or the firm's dissolution. These occur once in the life of a firm. The uniqueness of such an event for any single firm and the potential for widely different outcomes turning on the application of different terms are strong reasons to have a standard form. Thus the public good rationale has weight in explaining why these rules are mandatory. Because of the fear that any midstream change would tilt economic payoffs in a large-scale way, and because of the final period problem that would make such an effort tempting, the opportunistic amendment rationale may be even more important.

Fiduciary Standard Setting

There are two distinct reasons for mandatory fiduciary duties for directors, officers, and controlling shareholders. . . . Since insiders have substantial control over the amendment process, they are continually tempted to relax fiduciary standards that govern their behavior and expose them to liability. A mandatory rule eliminates this threat of opportunism while leaving recourse to the legislative process to modify duties—to innovate—where appropriate. Further, a stable conception of fiduciary duty develops only through applying a single standard across a great range of cases. Such a baseline represents a valuable public good, since the verbal formulas and the standards would vary considerably in the absence of a mandatory rule.

A second reason is that contractarian principles operate differently for fiduciary duties than for statutory corporate law in ways that make opting out of fiduciary duties particularly troublesome and ultimately wrongheaded, especially for elements of the duty of loyalty. Fiduciary duties provide a set of standards to restrain insiders in exercising their discretionary power over the corporation and its shareholders in contingencies not specifically foreseeable and thus over which the parties could not contract. Accordingly, parties do not know the decision rule for matters governed by fiduciary duties at the time that they enter into the corporate contract. . . . It is my argument that parties taking into account the insiders' power and positional advantage would pick a standard of fairness or good faith as measured ex post and that this radically undermines the case for opting out of fiduciary duties.

The Mandatory/Enabling Balance in Corporate Law: An Essay on the Judicial Role

JOHN C. COFFEE, JR.

A half-filled glass of water can be described as either half full or half empty. The structure of American corporate law—partly enabling, partly mandatory in character—can be viewed in much the same way. Some commentators see American corporate law as primarily composed of mandatory rules that the shareholders themselves cannot waive or modify. In their view, this mandatory component compensates both for the absence of true bargaining among the parties and for the inevitable divergence of interests between the principals (the shareholders) and their agents (the managers and directors). Conversely, other commentators, to whom this article will refer as "contractarians," see corporate law as primarily composed of waivable "default rules," which the law provides as a model form contract in order to reduce the transaction costs of contracting. Under this view, the parties are free to "opt out" of these "off-the-rack" rules if they wish to strike a different bargain that is more individually tailored to their specific circumstances.

Contractarians thus view corporate law as simply a modest extension of contract law, while their opponents regard the analogy between contract and corporate law as descriptively inaccurate. Both sides reach their respective positions because of a shared assumption that mandatory legal rules are "anticontractarian"; that is, the description of the corporation as a contract (or as a "nexus of contracts") implies for both sides that the law should permit shareholders to write or amend the corporate contract in virtually any way they see fit. Because the anticontractarians believe that shareholders should not be permitted to opt out from the mandatory core of corporate law, they tend also to resist the contract law analogy. As a result, they have tended to overlook the degree to which modern contract law itself contains important mandatory elements.

Reprinted by permission from 89 *Columbia Law Review* 1618. © 1989 John C. Coffee, Jr.

Once the debate between the contractarians and their opponents has been joined in this fashion, its focus has generally shifted from law to economics and, more specifically, to the question of whether market forces provide an adequate substitute for actual bargaining. Although the issue of how well the market prices corporate governance terms can certainly be sensibly debated, an exclusive focus on economics ignores an important feature common to all forms of long-term relational contracts: namely, that courts have invariably played an active and indispensable role in monitoring and interpreting such agreements. Indeed, the feasibility of such contracting probably depends upon the parties' ability to rely upon the courts to play such a role. In this light, analogizing the corporation to a long-term contract may suggest not that the mandatory features of American corporate law are vestigial remnants of an earlier era that was hostile to private ordering, but rather that these provisions are analogous to similar legal rules that restrict opportunism in other areas of complex, long-term contracting. Put simply, the more closely one looks at long-term contracting, the more one realizes that judicial involvement is not an aberration but an integral part of such contracting.

The intent of this article is not simply to defend the proposition that some mandatory component to corporate law is thus inevitable, but also to understand where the line should be drawn between the mandatory and enabling components. Unfortunately, no [commentator] has undertaken a comparative examination of the mandatory/enabling balance in the corporate law of jurisdictions other than the United States, but a fair generalization is that the law of Great Britain and other Commonwealth countries is far more mandatory in character than is ours. Does this mean that their law is less efficient (as proponents of deregulation would seemingly have to predict)? Such a prediction is oversimple because it misses a key trade-off: to the extent that American courts have permitted greater contractual freedom in corporate law, their relative tolerance has been coupled with greater judicial activism in reading implied terms into the corporate contract and in monitoring for opportunism. Thus, the issue of contractual freedom is inextricably entangled with the issue of institutional competence. Do we rely on prophylactic rules allowing little or no departure from the statutory baseline? Or do we counterbalance contractual freedom with ex post judicial review?

A historical perspective frames the issue similarly. Even the most casual observation reveals that the fiduciary strictures of American corporate law— presumably, corporate law's most mandatory inner core—have changed dramatically over this century. In truth, corporate law has been in flux throughout American history, and the fact that the supposedly mandatory core of corporate law has shrunk significantly presents a problem for those who wish to justify its nonwaivable status. If it has changed in the past, why should it remain static in the future? More importantly, attempts to justify its mandatory core on the ground that any alternative structure would expose shareholders to unconscionable risks must face the fact that alternative configurations to corporate law exist in other countries and have existed in the United States over the last century.

Who then is right—the contractarians or their critics? This article answers that both are right and both are wrong, because both have misstated the problem. In this article's view, contractual innovation can be reconciled with a stable mandatory core of corporate law if we recognize that what is most mandatory in corporate law is not the specific substantive content of any rule, but rather the institution of judicial oversight. Judicial activism is the necessary complement to contractual freedom. In short, because such long-term relational contracting is necessarily incomplete, the court's role becomes that of preventing one party from exercising powers delegated to it for the mutual benefit of all shareholders for purely self-interested ends. Indeed, that courts will at some point intervene is intuitively understood by the bar. In drafting the corporate contract, lawyers rely less on the model form provided by the legislature than on their expectation that courts will prevent either side from taking "opportunistic" advantage of the other. That is, the parties contract in the shadow of the law, knowing that courts will not seek simply to enforce the contract as written, but will to some uncertain extent serve as an arbiter to determine how the powers granted to management by the corporate charter may be exercised under unforeseen circumstances.

Notes and Questions

1. Easterbrook and Fischel introduce two concepts that need further elaboration for readers without an economics background: public goods and the Coase Theorem. A public good has two characteristics that differentiate it from ordinary ("private") commodities and prevent markets from producing an optimal level of supply: nonrivalness and nonexcludability of consumption. Nonrivalness occurs when provision of the good to one individual does not affect or limit the consumption of the good by another (the good provides benefits simultaneously to many consumers); for example, my watching a television show does not prevent you from enjoying it as well, whereas my eating an orange will certainly prevent you from enjoying that same orange. This eliminates the market equilibrium determining output quantity that sets price at marginal cost because nonrivalness means the marginal cost to satisfy an additional user is zero.

Nonexcludability occurs when it is impossible or prohibitively expensive to exclude anyone from consuming (benefiting from) the good. Consider, for example, the light from a street lamp: it would be quite difficult, and costly, to single out particular passersby, such as nontaxpayers, on a summer's night and prevent them from using (benefiting from) the light. This creates a free-rider problem, in which individuals can avoid paying their share of the good's production because nonpayment will not keep them from obtaining the benefit of the good. Markets operate by excluding individuals from consumption who are unwilling to pay the going price. But this is not possible here because the good can be consumed for free.

Public goods thus provide an economic rationale for government: government must provide public goods (directly or by financing private production) because markets do not operate effectively for them. A review of the theory of public goods can be found in any public finance textbook, such as Robin W. Boadway and David E. Wildasin, *Public Sector Economics,* 2d ed. (Boston: Little, Brown, 1984).

The Coase Theorem is the most famous proposition in law and economics and was put forth in Ronald Coase, "The Problem of Social Cost," 3 *Journal of Law and Economics* 1 (1960). In the absence of transaction costs, no matter which party in an exchange transaction bears the burden of a legal rule, the parties will reach the same outcome in the exchange, and it will be efficient.

Consider Coase's example of a factory located next to a laundry. When the factory produces widgets, it emits smoke into the air that blackens the clothing the laundry is trying to clean. According to Coase, whether pollution is banned or permitted, whether the factory has the right to pollute or the laundry has the right to be free from pollution, the same level of pollution will be achieved and it will be efficient. The idea is that so long as there are gains from trade, the parties will achieve them (remember Coase has assumed that bargaining costs are low). For example, suppose the laundry's loss from the smoke is $2,000, but the factory can install equipment that will eliminate the pollution at a cost of $1,000. In this case, it is inefficient for the factory to pollute, regardless of who has the pollution rights, because the cost of avoiding the harm is less than the harm itself. If the laundry has the right to be free from pollution, the factory will purchase the equipment. If the factory has the right to pollute, one might expect a different outcome, in which it pollutes and the laundry suffers the $2,000 loss. However, according to Coase, there will be no difference in outcome with the change in legal rule: the laundry will pay the factory $1,000 to purchase the antipollution equipment and perhaps an additional sum, say $500, to give up its right to pollute. The device will then be installed and both parties are better off (the laundry sustains a smaller loss than $2,000 and the factory gains $500).

While the outcome is still efficient in the latter case, the choice of legal rule does have a distributional impact, affecting the wealth of the parties. In the former case, the factory pays for the equipment whereas in the latter case, the laundry bears the cost. There is a massive literature on the Coase Theorem. A good place to begin is with an introductory textbook, such as Robert Cooter and Thomas Ulen, *Law and Economics* (Glenview, Ill.: Scott, Foresman, 1988); A. Mitchell Polinsky, *An Introduction to Law and Economics,* 2d ed. (Boston: Little, Brown, 1989); Richard A. Posner, *Economic Analysis of Law,* 4th ed. (Boston: Little, Brown, 1992).

2. Easterbrook and Fischel use the Coase Theorem to make two fundamental points concerning the structure of corporation codes. First, since transaction costs are low in capital markets, the parties to the corporate contract

(the corporation's charter) will reach the same agreement regardless of the provisions in the governing statute; and second, courts construing incomplete corporate contracts (such as courts applying fiduciary duty law) should choose the outcome the parties would have selected had they bargained over the issue (i.e., if contracting were costless and information complete). For a comprehensive synthesis of their influential contribution to corporate law, which builds from a contractual view of the corporation, see Easterbrook and Fischel, *The Economic Structure of Corporate Law* (Cambridge, Mass.: Harvard University Press, 1991).

Easterbrook and Fischel's analysis need not be robust in situations where one party to a contract has important private information, which creates what is referred to in the economics literature as a problem of adverse selection. The term adverse selection, like moral hazard (see Chapter I, part A, note 1), originated in the insurance field and refers to the tendency that people who buy insurance are not a random sample of the population, but rather, are those who expect to have the highest-valued claims. As in the moral hazard problem, private information creates a problem for the insurer: only the buyer knows her claim type. The difference is that with adverse selection, opportunism occurs before entering into the contract, whereas with moral hazard it occurs after purchase. If the insurer charges a premium based on the average expected claim, it will lose money because the policy purchasers will come disproportionately from the upper tail of the claim distribution, but if it charges a premium above average it simply exacerbates the problem, as individuals with low expected claims are priced out of the market. Just as the insurer would like to be able to sort buyers by expected claim type, the low expected claim types would like to be able to signal their type and, in some situations, costly self-selection can occur.

Consider the archetypal, simple signaling model of this problem in a corporate context: There are two types of firms, firms with good projects and firms with bad projects. Firms know their project type, but investors do not. If investors know the distribution of firm types (what proportion have good projects), then they will provide capital at the average firm's cost (i.e., the market pools the value of investing in good and bad project firms). To obtain better project financing terms, firms with good projects must expend effort to credibly signal their type to investors. In particular, they must adopt a sufficiently expensive signal so that firms with bad projects will not want to duplicate it (e.g., they could guarantee investors large payments if the firm fails), which produces a socially inefficient outcome compared to the pooling outcome. Ian Ayres provides a detailed and accessible analysis of this example, which synthesizes the signaling literature with the incomplete contract problem. Ayres, "The Possibility of Inefficient Contracts," 60 *University of Cincinnati Law Review* 387 (1991). He concludes that the economic theory implies that courts applying Easterbrook and Fischel's hypothetical contract rule in an asymmetric information setting will not reach efficient outcomes (as the contract the parties would have reached would be inefficient). The example is drawn from a formal model by Philippe Aghion and Benjamin Hermalin,

"Legal Restrictions on Private Contracts Can Enhance Efficiency," 6 *Journal of Law, Economics, and Organization* 381 (1990). What empirical evidence would be useful for determining if the economic theory is apposite in the corporate law setting? Does the signaling game seem more apt for closely held firms than publicly traded ones? Are corporate charter provisions likely to involve excessively costly signaling efforts?

3. Is a rule mandatory in a meaningful sense if its constraint is not binding on the parties, that is, if the parties would voluntarily adopt the rule on their own, or if corporations can easily, and legally, sidestep it? As an example of the former, nonbinding constraints, consider a provision that supporters of mandatory rules typically emphasize, the duty of loyalty (preventing managerial self-dealing). How much do you think investors would pay to invest in a new firm whose charter contains a provision eliminating the duty of loyalty? Do you think it plausible that shareholders will vote to amend their corporate charters to permit managers to steal corporate assets by rescinding the duty of loyalty? If investors are so poorly informed as to vote to transfer their wealth to managers without compensation, can we have confidence that they will make the fundamental allocative investment decisions required in a capitalist economy? As an example of the readily avoidable provision, consider another commonly cited mandatory rule, the requirement that shareholders vote on major corporate changes. If this rule is mandatory, should management be able to restructure a transaction to avoid a shareholder vote that it expected to lose? The Delaware Supreme Court has no objection. See *Paramount Communications v. Time, Inc.*, 571 A.2d 1140 (Del. 1989). Would it be justifiable to conclude from these examples that a mandatory/enabling dichotomy of corporate laws misses the mark, as these are laws without bite?

4. How persuasive is Gordon's public good explanation of mandatory laws? Suppose that Clause x is standard under a state corporate code, but firms routinely choose Clause y over x. Would you infer from such behavior—firms' revealed choice of y when they could have chosen x and did not—that Clause x is suboptimal? Wouldn't it be inconsistent to maintain on efficiency grounds (the public goods hypothesis) that deviations (Clause y) should be prohibited to perserve the utility of the standard form (Clause x)? Is the innovation hypothesis any more persuasive? Why should shareholders assume that an innovative charter term is at variance with their interests? Is the purchase of common stock uniquely different from other consumer purchases, such as computers or stereophonic equipment, for which the state does not authorize innovation? If innovations routinely occur in ordinary product markets, why would they fare so poorly in the capital market, which is far more informationally efficient? In other markets, new products are often accompanied with lengthy warranties or generous return provisions to ensure product quality. Can the firm's promoter make an analogous promise? For example, could the proportion of shares retained by insiders in a public offering be a credible signal of an issue's quality? Would state-mandated innovation reduce the value of such a signal?

5. One solution to the problem of latecomer terms in corporate charters is to extend appraisal rights to shareholders voting against a charter amendment. If an amendment caused the stock to decline in value, dissenters could recoup the loss in an appraisal proceeding and would thus not be harmed by the yes votes of rationally apathetic shareholders. Some states, but not Delaware, provide appraisal rights for charter amendments. Should such appraisal rights be mandatory? If latecomer terms were considered a serious problem by investors, would appraisal rights for charter amendments be more prevalent (i.e., be the default rule in Delaware's code)? How would a court determine what dissenters should receive for their shares? Would an event study (see Chapter I, part B, note 4) of the charter amendment help?

6. Another mechanism for avoiding the difficulties of opportunistic amendments is the use of lock-in clauses, which require a supermajority vote to modify a charter provision. When antitakeover provisions are placed in corporate charters, for example, they are frequently accompanied by lock-ins, see Ronald J. Gilson, "The Case Against Shark Repellent Amendments: Structural Limitations on the Enabling Concept," 34 *Stanford Law Review* 775 (1980). Lock-in provisions otherwise appear to be an unusual occurrence. Is this evidence that the benefits of deterring opportunistic latecomer terms are outweighed by the costs of lost flexibility in charter amendment? Or is this evidence of the free-rider problem at its worst? See Chapter VI for a discussion of defensive tactics to takeovers.

7. How serious a problem do you think the rationally ignorant voter is? In most public corporations, a majority of the shares are held by sophisticated institutional investors. Might it not be cheaper than Easterbrook and Fischel and Gordon think for diversified investors to become informed since what they learn about a governance issue for one vote can be used numerous times across their many investments? See Roberta Romano, "Answering the Wrong Question: The Tenuous Case for Mandatory Corporate Laws," 89 *Columbia Law Review* 1599 (1989); Bernard S. Black, "Shareholder Passivity Reexamined," 89 *Michigan Law Review* 520 (1990). For a detailed analysis of shareholders' collective action problems and whether institutional investors are the solution, see Edward B. Rock, "The Logic and (Uncertain) Significance of Institutional Shareholder Activism," 79 *Georgetown Law Journal* 445 (1991). The question of institutional investors and shareholder activism in corporate governance will be revisited in Chapter V.

8. How important is product differentiation as an explanation of state competition if corporation codes are largely enabling statutes? More generally, what is the relation between the structure of corporation laws and state competition? Do proposals for mandatory corporation laws require a national corporation code? Would restrictions (or heavy taxes) on reincorporation suffice? Would such penalties be constitutional under the commerce clause? See the discussion in Chapter VI on state regulation of takeovers.

IV

Financing the Corporation

There are two mechanisms by which corporations raise capital: issuing debt and issuing equity. While Chapter I emphasized the agency costs of outside equity, this chapter introduces the agency costs of debt. For the sole-owner firm, issuing debt avoids the agency costs of the separation of ownership and control because the owner remains the sole equity holder, and thus, the trade-off between pecuniary and nonpecuniary benefits, discussed by Jensen and Meckling (Chapter I), is unchanged. The use of debt in the capital structure, however, generates its own principal-agent problem: creditors are also concerned about manager-agent opportunism, but their problem takes a different form from that of the outside equity investors. By altering the risk of a firm's investments after credit is obtained, manager-shareholders can transfer wealth from bondholders to themselves.

A simple example will illustrate the agency problem of debt. Suppose a firm has two potential projects, each of which requires a $500,000 investment. There are two equally likely states of the world, a lucky and an unlucky state. Project L returns $600,000 in the lucky state and $500,000 in the unlucky state, so its expected value is $.5(\$600,000) + .5(\$500,000) = \$550,000$. Project H returns $800,000 in the lucky state and 0 in the unlucky, for an expected value of $.5(\$800,000) + .5(0) = \$400,000$. The expected value of Project L is greater than that of Project H, and L is less risky than H. L's expected value is also greater than its cost (it is a positive net present value project). The firm's investment decision is therefore clear: choose Project L. This maximizes the value of the firm.

Suppose the firm decides to finance the project with debt. Project L will be preferred to Project H by the bondholders. Under Project L, the bondholders are paid off no matter which state occurs, as the realized value is always at least \$500,000, and so the expected value of L to them is \$500,000. The expected value of Project H to the bondholders is, however, only .5(\$500,000) + .5(0) = \$250,000. It seems that the bondholders' and firm's interests are aligned: both prefer Project L to H.

But the introduction of debt in the capital structure (which is referred to as leverage) alters the shareholders' calculation of the benefits of the projects. The shareholders receive whatever is left after they repay the \$500,000 debt. The expected value of Project L to the shareholders is .5(\$100,000) + .5(0) = \$50,000. The expected value of Project H is .5(\$300,000) + .5(0) = \$150,000. Thus, shareholders prefer Project H over L, even though it has the lower expected value. Remember that limited liability, as discussed in Chapter II, shifts the risk of failure to the debtholders: if the firm does not earn \$500,000, shareholders do not have to ante up any more money to repay the debt.

The shareholders prefer Project H over L for the following reason. If they are unlucky, they get nothing in either case—the bondholders own the firm— but if they are lucky, they get more from Project L than from H (\$150,000 > \$50,000). When debt is included in the capital structure, limited liability truncates the distribution of returns of concern to shareholders: they will be concerned only with the upside returns and not the downside. (This effect is explained in greater detail in note 1 following the readings). Thus, as shareholders control the firm, they will choose the project with a lower expected value, which is not the bondholders' choice and would not be the choice of a 100 percent equity firm.

The perverse incentive for risk taking by the equity holders due to limited liability is the agency cost of debt. This is one reason why, despite the tax advantages of debt over equity (interest is deductible but dividends are not), we do not see 100 percent debt-financed firms. Because shareholders' risk-shifting behavior is foreseeable, bondholders will charge for the greater risk. In our example, the bondholders could require a personal guarantee of the debt, lend only \$250,000 (their claim's expected value under shareholder opportunism), require a higher interest rate, and so forth.

The cost of debt capital can be reduced if governance structures are devised that constrain shareholder opportunism. Clifford Smith and Jerold Warner detail the common contractual provisions in bond indentures that serve such a function. Such contracts are a key protective device for bondholders because they are not owed any fiduciary duty by managers. Another source of protection is their priority over equity holders in bankruptcy (including the right to force the firm into bankruptcy upon a default).

In the 1980s, a new risk was created for corporate bondholders from leveraged buyouts (LBOs), transactions in which corporations restructured by drastically increasing their debt. Kenneth Lehn and Annette Poulsen document the response to this development, contractual provisions known as event-risk clauses. Michael Jensen emphasizes that the high level of debt in LBOs provides incentives to work out financial difficulties without initiating formal bankruptcy proceedings, a benefit to investors that he terms the privatization of bankruptcy, which reduces the cost of foreclosure rights as a form of bondholder protection. LBOs are discussed further as a unique form of corporate governance in Chapter V and in relation to control changes in Chapter VI.

New high-risk firms are typically not able to raise capital by conventional debt and equity offerings. The selection by William Sahlman examines venture capital, a principal financing technique for such firms. Sahlman analyzes the unique agency problems involved in the use of venture capital and the organizational means by which these costs are reduced.

Mature firms have a third source of funding besides debt and equity for investment projects: retained earnings. Internally generated cash flows are, in fact, the most popular source of long-term corporate financing (over 80 percent for U.S. industrial firms during most of the 1980s). The tax code favors this practice because dividends are taxable to investors upon receipt, while the appreciation in stock value from retained earnings is not taxed until the shares are sold. Paradoxically, despite the tax advantage, most firms also pay out dividends to their investors. The selection by Frank Easterbrook relates the decision of a firm to pay dividends, rather than retain cash, to efforts at mitigating the agency problem created by the separation of ownership and control.

On Financial Contracting: An Analysis of Bond Covenants

CLIFFORD W. SMITH, JR., AND JEROLD B. WARNER

In this article, we examine how debt contracts are written to control the bondholder–stockholder conflict. We investigate the various kinds of bond covenants which are included in actual debt contracts. A bond covenant is a provision, such as a limitation on the payment of dividends, which restricts the firm from engaging in specified actions after the bonds are sold. . . .

Sources of the Bondholder–Stockholder Conflict

With risky bonds outstanding, management, acting in the stockholders' interest, has incentives to design the firm's operating characteristics and financial structure in ways which benefit stockholders to the detriment of bondholders. Because investment, financing, and dividend policies are endogenous, there are four major sources of conflict which arise between bondholders and stockholders:

Dividend Payment. If a firm issues bonds and the bonds are priced assuming the firm will maintain its dividend policy, the value of the bonds is reduced by raising the dividend rate and financing the increase by reducing investment. At the limit, if the firm sells all its assets and pays a liquidating dividend to the stockholders, the bondholders are left with worthless claims.

Claim Dilution. If the firm sells bonds, and the bonds are priced assuming that no additional debt will be issued, the value of the bondholders' claims is reduced by issuing additional debt of the same or higher priority.

Asset Substitution. If a firm sells bonds for the stated purpose of engaging in low-variance projects and the bonds are valued at prices commensurate with that low risk, the value of the stockholders' equity rises and the value of the bondholders' claim is reduced by substituting projects which increase the firm's variance rate.

Underinvestment. . . . [A] substantial portion of the value of the firm is composed of intangible assets in the form of future investment opportunities. A firm with outstanding bonds can have incentives to reject projects which have a positive net present value if the benefit from accepting the project accrues to the bondholders.

Reprinted by permission from 7 *Journal of Financial Economics* 117 (Amsterdam: Elsevier Science Pub., 1979).

The bondholder–stockholder conflict is of course recognized by capital market participants. Rational bondholders recognize the incentives faced by the stockholders. They understand that after the bonds are issued, any action which increases the wealth of the stockholders will be taken. In pricing the bond issue, bondholders make estimates of the behavior of the stockholders, given the investment, financing, and dividend policies available to the stockholders. The price which bondholders pay for the issue will be lower to reflect the possibility of subsequent wealth transfers to stockholders. . . .

Financial contracting is assumed to be costly. However, bond covenants, even if they involve costs, can increase the value of the firm at the time bonds are issued by reducing the opportunity loss which results when stockholders of a levered firm follow a policy which does not maximize the value of the firm. Furthermore, in the case of the claim dilution problem (which involves only a wealth transfer), if covenants lower the costs which bondholders incur in monitoring stockholders, the cost-reducing benefits of the covenants accrue to the firm's owners. With such covenants, the firm is worth more at the time the bonds are issued. . . . We selected a random sample of eighty-seven public issues of debt which were registered with the Securities and Exchange Commission between January 1974 and December 1975. . . . Standardized provisions . . . are used frequently: 90.8 percent of the bonds contain restrictions on the issuance of additional debt, 23.0 percent have restrictions on dividend payments, 39.1 percent restrict merger activities, and 35.6 percent constrain the firm's disposition of assets. Furthermore, we found that when a particular provision is included, a boilerplate . . . is used almost exclusively. . . .

Restrictions on the Firm's Production/Investment Policy

The stockholders' production/investment decisions could be directly constrained by explicitly specifying the projects which the firm is allowed to undertake. Alternatively, if it were costless to enforce, the debt contract could simply require the shareholders to accept all projects (and engage in only those actions) with positive net present values. Although certain covenants directly restrict the firm's investment policy, debt contracts . . . do not generally contain extensive restrictions of either form.

Restrictions on Investments

Bond covenants frequently restrict the extent to which the firm can become a claimholder in another business enterprise. That restriction, known as the "investment" restriction, applies to common stock investments, loans, extensions of credit, and advances. . . . We suggest that stockholders contractually restrict their ability to acquire financial assets in order to limit their ability to engage in asset substitution after the bonds are issued. . . .

Secured Debt

Securing debt gives the bondholders title to pledged assets until the bonds are paid in full. Thus, when secured debt is issued the firm cannot dispose of the pledged assets without first obtaining permission of the bondholders.

We suggest that the issuance of secured debt lowers the total costs of borrowing by controlling the incentives for stockholders to take projects which reduce the value of the firm; since bondholders hold title to the assets, secured debt limits asset substitution. Secured debt also lowers administrative costs and enforcement costs by ensuring that the lender has clear title to the assets and by preventing the lender's claim from being jeopardized if the borrower subsequently issues additional debt. In addition, collateralization reduces expected foreclosure expenses because it is less expensive to take possession of property to which the lender already has established title.

However, secured debt involves out-of-pocket costs (e.g., required reports to the debtholders, filing fees, and other administrative expenses). Securing debt also involves opportunity costs by restricting the firm from potentially profitable dispositions of collateral. . . .

Restrictions on Mergers

Some indenture agreements contain a flat prohibition on mergers. Others permit the acquisition of other firms provided that certain conditions are met. . . .

With no contractual constraints against mergers, the value of the bondholders' claims can be reduced due to the effect of a difference in variance rates or a difference in capital structures. Our analysis implies, then, that merger restrictions limit the stockholders' ability to use mergers to increase either the firm's variance rate or the debt to asset ratio to the detriment of the bondholders. Note that to the extent that synergistic mergers are prevented by this covenant, the firm suffers an opportunity loss. . . .

Bond Covenants Restricting the Payment of Dividends

Cash dividend payments to stockholders, if financed by a reduction in investment, reduce the value of the firm's bonds by decreasing the expected value of the firm's assets at the maturity date of the bonds, making default more likely. Thus, it is not surprising that bond covenants frequently[1] restrict the payment of cash dividends to shareholders. Since the payment of dividends *in cash* is just one form which distributions to stockholders can take, actual dividend covenants reflect alternative possibilities. For example, if the firm enters the market

---

[1]Kalay reports that in a sample of 150 randomly selected industrial firms, every firm had a dividend restriction in at least one of its debt instruments.

and repurchases its own stock the coverage on the debt decreases in exactly the same way as it would if a cash dividend were paid. The constraints . . . relate not only to cash dividends, but to "all distributions on account of or in respect of capital stock . . . whether they be dividends, redemptions, purchases, retirements, partial liquidations or capital reductions and whether in cash, in kind, or in the form of debt obligations of the company." . . .

This form of dividend covenant has several interesting features. The dividend restriction is not an outright prohibition on the payment of dividends. In fact, the stockholders are permitted to have any level of dividends they choose, so long as the payment of those dividends is financed out of new earnings or through the sale of new equity claims. The dividend covenant acts as a restriction not on dividends per se, but on the payment of dividends financed by issuing debt or by the sale of the firm's existing assets, either of which would reduce the coverage on, and thus the value of, the debt. . . .

By placing a maximum on distributions, the dividend covenant effectively places a minimum on investment expenditures by the owners of the firm. . . . This reduces the underinvestment problem . . . since so long as the firm *has* to invest, profitable projects are less likely to be turned down. . . .

Bond Covenants Restricting Subsequent Financing Policy

Limitations on Debt and Priority

Covenants . . . limit stockholders' actions in this area [issuance of additional debt of higher priority to reduce the value of outstanding debt by diluting its claim on the firm's assets—ED.] in one of two ways: either through a simple prohibition against issuing claims with a higher priority, or through a restriction on the creation of a claim with higher priority unless the existing bonds are upgraded to have equal priority. The latter restriction requires, for example, that if secured debt is sold after the issuance of the bonds, the existing bondholders must have their priority upgraded and be given an equal claim on the collateral with the secured debtholders. . . .

It is important to note the scope of the restrictions imposed through the covenants limiting the issuance of additional debt. In addition to money borrowed, the covenants also apply to other liabilities incurred by the firm. . . .

Bond Covenants Modifying the Pattern of Payoffs to Bondholders

There are several provisions which specify a particular pattern of payoffs to bondholders in a way which controls various sources of stockholder-bondholder conflict of interest.

Sinking Funds

A sinking fund is simply a means of amortizing part or all of an indebtedness prior to its maturity. . . . In the case of a public bond issue, the periodic payments can be invested either in the bonds which are to be retired by the fund or in some other securities. The sinking fund payments can be fixed, variable, or contingent. For the years 1963–65, 82 percent of all publicly offered issues included sinking fund provisions.

A sinking fund affects the firm's production/investment policy through the dividend constraint. . . . [It] reduces the possibility that the dividend constraint will require investment when no profitable projects are available. One potential cost associated with the dividend constraint is thus reduced.

Myers has suggested that sinking funds are a device to reduce creditors' exposure in parallel with the expected decline in the value of the assets supporting the debt. Myers' analysis implies that sinking funds would be more likely to be included in debt issues (1) the higher the fraction of debt in the capital structure, (2) the greater the anticipated future discretionary investment by the firm, and (3) the higher the probability that the project will have a limited lifetime. One industry which illustrates an extreme of the last of these characteristics is the gas pipeline industry. The sinking fund payments required in some gas pipeline debentures are related to the remaining available gas in the field. . . .

Convertibility Provisions

A convertible debenture is one which gives the holder the right to exchange the debentures for other securities of the company, usually shares of common stock and usually without payment of further compensation. . . . With nonconvertible debt outstanding, the stockholders have the incentive to take projects which raise the variability of the firm's cash flows. The stockholders can increase the value of the equity by adding a new project with a negative net present value if the firm's cash flow variability rises sufficiently. The inclusion of a convertibility provision in the debt reduces this incentive. The conversion privilege is like a call option written by the stockholders and attached to the debt contract. It reduces the stockholders' incentive to increase the variability of the firm's cash flows, because with a higher variance rate, the attached call option becomes more valuable. Therefore the stockholders' gain from increasing the variance rate is smaller with the convertible debt outstanding than with nonconvertible debt.

However, not all debt contracts include a convertibility provision since it is costly to do so. For example, the underinvestment problem is exacerbated with convertible debt outstanding.

Mikkelson presents cross-sectional evidence that the probability of the inclusion of the conversion privilege is positively related to (1) the firm's debt/equity ratio, (2) the firm's level of discretionary investment expenditure, and

(3) the time to maturity of the debt. Each of these relationships is consistent with the . . . hypothesis that the benefits of convertible debt are related to a reduction in the bondholder–stockholder conflict. . . .

Covenants Specifying Bonding Activities by the Firm

Potential bondholders estimate the costs associated with monitoring the firm to assure that the bond covenants have not been violated, and the estimate is reflected in the price when the bonds are sold. Since the value of the firm at the time the bonds are issued is influenced by anticipated monitoring costs, it is in the interests of the firm's owners to include contractual provisions which lower the costs of monitoring. For example, observed provisions often include the requirement that the firm supply audited annual financial statements to the bondholders. . . .

Required Reports

Our analysis suggests that bondholders find financial statements to be useful in ascertaining whether the provisions of the contract have been (or are about to be) violated. If the firm can produce this information at a lower cost than the bondholders (perhaps because much of the information is already being collected for internal decision-making purposes), it pays the firm's stockholders to contract to provide this information to the bondholders. The market value of the firm increases by the reduction in agency costs. . . .

The Required Purchase of Insurance

Indenture agreements frequently include provisions requiring the firm to purchase insurance. . . . Our analysis suggests that the corporate purchase of insurance is a bonding activity engaged in by firms to reduce agency costs between bondholders and stockholders (as well as between the managers and the owners of a corporation). If insurance firms have a comparative advantage in monitoring aspects of the firm's activities, then a firm which purchases insurance will engage in a different set of activities from a firm which does not. . . .

The Role of the Trust Indenture and the Trustee

If the firm's debt is not held by a single borrower, then a number of problems related to enforcement of the debt contract arise. For example, any individual's holdings of the firm's debt may be so small that no single bondholder has much incentive to expend resources in covenant enforcement. But it is not the case that individual bondholders necessarily expend "too few" resources in covenant enforcement. If the number of bondholders is small, then there can actually be overinvestment in enforcement in the sense that there is either a duplication of effort, or that creditors expend resources which simply result in change in the distribution of the proceeds. Our analysis implies that the firm's owners offer a contract which appoints a trustee to help assure that the optimal amount of covenant enforcement will take place.

Having the firm pay the trustees directly solves the "free-rider" problem which would be inherent in making individual bondholders pay the trustee for enforcing the covenants. However after the bonds have been sold, the stockholders have an incentive to bribe the trustee so that they can violate the debt covenants. There are several factors which prevent such bribery from taking place.

Bribing the trustee is expensive if the trustee's reputation has significant value in the marketplace. Ex ante, it is in the interests of the firm's owners to choose an "honest" trustee—that is, one who is expensive to bribe. This is because the value of the firm at the time it issues the debt contract reflects the probability of covenant enforcement. To the extent that enforcement by an "honest" trustee reduces the problems of adverse borrower behavior induced by risky debt, the value of the firm is higher. Our analysis therefore implies that those chosen as trustees stand to lose much if they are caught accepting bribes. In fact, the indenture trustee is "generally a large banking institution," which has significant revenues from activities unrelated to being a trustee and which also depend on the market's perception of its trustworthiness. Furthermore, the behavior of the trustee is restricted by both trust and contract law.

Contractual Resolution of Bondholder–Stockholder Conflicts in Leveraged Buyouts

KENNETH LEHN AND ANNETTE POULSEN

The conflict between bondholders and stockholders in leveraged buyouts has focused attention on the costs of contracting in the bond market. Advocates of statutory or regulatory protection of bondholders in these transactions often cite the costs associated with negotiating and enforcing protective bond covenants as a "market failure" that requires corrective action. In addition, they argue that restrictive bond covenants may inhibit managers from undertaking value-maximizing investment opportunities that become available after the bond contract has been written. These costs, they argue, diminish the

Reprinted by permission from 34 *Journal of Law and Economics* 645. © 1991 by The University of Chicago.

efficacy of relying on private contracts to mitigate bondholder–shareholder conflicts. McDaniel has argued for two policy changes: a fiduciary duty requiring directors of corporations to treat bondholders "fairly" in leveraged buyouts, and a disclosure rule requiring public companies to state whether each class of security holders is treated fairly in leveraged buyouts.

Notwithstanding the costs associated with writing and enforcing protective bond covenants, it is reasonable to expect bondholders to seek contractual means of mitigating the risk associated with leveraged buyouts. Anecdotal evidence suggests that many large industrial companies have issued debt during the past few years that contains either "poison put" provisions (that is, provisions that give bondholders a right to sell their bonds back to the issuer at par if a change of control occurs)[1] or provisions that require issuers to increase the coupon payments on their bonds (in order to maintain the pretakeover value of the bonds) if there is a change in control. . . .

In addition to contracting for explicit protection against the risk of leveraged buyouts, bondholders can mitigate this risk in at least two other ways. First, convertible bonds provide bondholders with the opportunity to participate in the gains associated with leveraged buyouts (if, of course, the conversion price is less than the price offered in the buyout). Bondholders also can hedge the risks associated with leveraged buyouts by simultaneously owning both the bonds and stocks of the same issuers. . . .

At present, little is known about the extent to which bondholders systematically have chosen to mitigate the risk of leveraged buyouts through protective covenants, convertible bonds, and cross-ownership of bonds and stocks of the same issuer. We empirically examine these choices and find the following results.

1. In a sample of 327 nonfinancial issuances of nonconvertible debt during 1989, 32.1 percent had explicit covenants protecting bondholders from leveraged buyouts (that is, event-risk covenants). For the same issuers in 1986, only three firms issued debt with event-risk covenants. Hence, the dramatic increase in the frequency of such protective covenants occurred after 1986 (several years after the advent of large leveraged buyouts), suggesting that event-risk covenants have evolved rather sluggishly.

2. Event-risk covenants were more common in nonsecured debt issues than secured debt issues (42.7 percent versus 8.8 percent). For the sample of 225 nonsecured debt issues, 50.5 percent of the securities issued by companies that were the targets of takeover attempts or takeover rumors included event-risk covenants. Only 36.1 percent of the securities issued by other companies included these covenants. Other proxies for the likelihood of leveraged buyouts reveal similar patterns in the incidence of event-risk covenants. Hence, these covenants are more than simply boilerplate provisions appended to all debt issues—they appear where the risks of leveraged buyouts are the greatest.

[1]Poison puts may also be triggered by a substantial downgrading of the debt by Moody's or Standard & Poor's [EDITOR'S NOTE].

3. Event-risk covenants are observed more frequently in bond contracts that contain covenants restricting dividends and the issuance of additional debt than they are in bond contracts that do not contain these other covenants. This suggests that dividend restrictions and restrictions on additional debt do not provide sufficient protection against the risk of leveraged buyouts.

4. Using a sample of more than 3,000 debt issues during the 1980s, we find that the ratio of the value of convertible-debt issues to straight-debt issues actually fell in the latter part of the 1980s. Contrary to our expectations, the decline in this ratio was largest for firms operating in "high-takeover-intensity" industries. Firms in "low-takeover-intensity" industries experienced little change in this ratio. This suggests that convertible debt has not been used to alleviate bondholders' concern about event risk.

5. For a sample of twenty-four mutual funds that held stocks and non-convertible bonds in the 1980s, we find no significant change in the cross-ownership of stocks and bonds of the same issuer. In 1979–80, 7.3 percent of the bondholdings of the mutual funds consisted of bonds issued by companies in which these funds also owned stock. By 1987–88, this percentage had risen to 13.4 percent, a statistically insignificant change. Notwithstanding the small sample size, these data suggest that cross-ownership of stocks and bonds is not widely used by bondholders to mitigate the risk of leveraged buyouts.

Active Investors, LBOs, and the Privatization of Bankruptcy

MICHAEL C. JENSEN

High Leverage and the Privatization of Bankruptcy

One important and interesting characteristic of the LBO organization is its intensive use of debt. The debt-to-value ratio in the business units of these organizations averages close to 90 percent on a book value basis. LBOs, however, are not the only organizations that are making use of high debt ratios. Public corporations are also following suit as witnessed by recapitalizations, highly leveraged mergers, and stock repurchases.

There has been much concern in the press and in public policy circles about the dangers of high debt ratios in these new organizations. What is not generally recognized, however, is that high debt has benefits as a monitoring and incentive device, especially in slow-growing or shrinking firms. Even less

Reprinted by permission from 27 *The Continental Bank Journal of Applied Corporate Finance* 35 (Spring 1989). © 1989 Stern Stewart Management Services, Inc.

well-known, the costs for a firm in insolvency—the situation in which a firm cannot meet its contractual obligations to make payments—are likely to be much smaller in the new world of high leverage ratios than they have been historically. . . .

In a world of 20 percent debt-to-value ratios (with value based on the going concern value of a healthy company), the liquidation or salvage value is much closer to the face value of the debt than in the same company with an 85 percent debt/value ratio. [Consider] a $100 million company under these two leverage ratios, and [assume] that the salvage or liquidation value of the assets is 10 percent of the going concern value of $10 million. Thus, if the company experiences such a decline in value during bad times that it cannot meet its payments on $20 million of debt, it is also likely that its value is below its liquidation value.

An identical company with an 85 percent debt ratio, however, is nowhere near liquidation when it experiences times sufficiently difficult to cause it to be unable to meet the payments on its $85 million of debt. That situation could occur when the company still has total value in excess of $80 million. In this case there is $70 million in value that can be preserved by resolving the insolvency problem in a fashion that minimizes the value lost through the bankruptcy process. In the former case, when the firm is worth less than $20 million, there may be so little value left that the economically sensible action is liquidation, with all its attendant conflicts and dislocation.

The incentives to preserve value in the new leverage model imply that a very different set of institutional arrangements and practices will arise to substitute for the usual bankruptcy process. In effect, bankruptcy will be taken out of the courts and "privatized." This institutional innovation will take place to recognize the large economic value that can be preserved by privately resolving the conflicts of interest among claimants to the firm. When the going concern value of the firm is vastly greater than the liquidation value, it is likely to be more costly to trigger the cumbersome court-supervised bankruptcy process that diverts management time and attention away from managing the enterprise to focus on the abrogation of contracts that the bankruptcy process is set up to accomplish.

These large potential losses provide incentives for the parties to accomplish reorganization of the claims more efficiently outside the courtroom. This fact is reflected in the strip financing practices commonly observed in LBOs whereby claimants hold approximately proportional strips of all securities and thereby reduce the conflicts of interest among classes of claimants. Incentives to manage the insolvency process better are also reflected in the extremely low frequency with which these new organizations actually enter bankruptcy. The recent Revco case is both the largest such bankruptcy of an LBO and one of the handful that have occurred.

LBOs frequently get in trouble, but they seldom enter formal bankruptcy. Instead they are reorganized in a short period of time (several months is common), often under new management, and at apparently lower cost than would occur in the courts. . . .

There has been much concern about the ability of LBO firms to withstand sharp increases in interest rates, given that the bank debt which frequently amounts to 50 percent of the total debt is primarily at floating rates. This problem is mitigated by the fact that most LBOs now protect themselves against sharp increases in interest rates by purchasing caps that limit any increase or by using swaps that convert the floating-rate debt to fixed rates. Indeed it has become common for banks to require such protection for the buyout firm as a condition for lending. These new financial techniques are another means whereby some of the risks can be hedged away in the market, and therefore the total risks to the buyout firm are less than they would have been in past years at equivalent debt levels.

It will undoubtedly take time for the institutional innovation in reorganization practices to mature and for participants in the process to understand that insolvency will be a more frequent and less costly event than it has been historically.

The Structure and Governance of Venture Capital Organizations

WILLIAM A. SAHLMAN

The venture capital industry has evolved operating procedures and contracting practices that are well adapted to environments characterized by uncertainty and information asymmetries between principals and agents. By venture capital I mean a professionally managed pool of capital that is invested in equity-linked securities of private ventures at various stages in their development. Venture capitalists are actively involved in the management of the ventures they fund, typically becoming members of the board of directors and retaining important economic rights in addition to their ownership rights. The prevailing organizational form in the industry is the limited partnership, with the venture capitalists acting as general partners and the outside investors as limited partners.

Venture capital partnerships enter into contracts with both the outside investors who supply their funds and the entrepreneurial ventures in which they invest. The contracts share certain characteristics, notably:

1. staging the commitment of capital and preserving the option to abandon,
2. using compensation systems directly linked to value creation,
3. preserving ways to force management to distribute investment proceeds.

Reprinted by permission from 27 *Journal of Financial Economics* 473 (Amsterdam: Elsevier Science Pub., 1990).

These elements of the contracts address three fundamental problems:

1. the sorting problem: how to select the best venture capital organizations and the best entrepreneurial ventures,
2. the agency problem: how to minimize the present value of agency costs,
3. the operating-cost problem: how to minimize the present value of operating costs, including taxes.

From one perspective, venture capital can be viewed as an alternative model for organizing capital investments. Like corporations, venture capital firms raise money to invest in projects. Many projects funded by venture capitalists (for example, the development of a new computer hardware peripheral) are similar to projects funded within traditional corporations. But the governance systems used by venture capital organizations and traditional corporations are very different. . . .

General Industry Background

In 1988 an estimated 658 venture capital firms in the United States managed slightly over $31 billion in capital and employed 2,500 professionals. . . . Industry resources were concentrated: the largest eighty-nine firms controlled approximately 58 percent of the total capital. . . . Although a typical large venture capital firm receives up to 1,000 proposals each year, it invests in only a dozen or so new companies. . . .

Approximately 15 percent of the capital disbursed in each of the last three years went to ventures in early stages, whereas 65 percent was invested in later-stage companies, typically still privately held. The remaining 20 percent was invested in leveraged buyout or acquisition deals. In recent years venture capitalists have channeled roughly two-thirds of the capital invested each year into companies already in their portfolios, and one-third into new investments. Venture capitalists often participate in several rounds of financing with the same portfolio company. . . .

Venture capital investing plays a small role in overall new-business formation. . . . Venture capital investing is also modest in comparison with the level of capital investment in the domestic corporate sector. . . .

Although comprehensive data are difficult to obtain, the overall rate of return on venture capital seems to have been high from the mid-1960s through the mid-1980s, the only period for which reliable data are currently available. Between 1965 and 1984, for example, the median realized compound rate of return on twenty-nine venture capital partnerships over the life of each partnership (an average of 8.6 years) exceeded 26 percent per year. . . .

A more recent and comprehensive study [by Venture Economics—ED.] suggests that funds started before 1981 experienced generally positive returns through 1987. . . . This study also reveals that rates of return have declined since 1983, particularly for funds started later in the period. It is extremely

difficult to estimate the extent to which returns have declined, however, because accounting practices in the industry typically reflect a downward
bias. . . .

Returns on individual investments in a venture capital portfolio vary
widely. . . . According to Venture Economics more than one-third of 383
investments made by thirteen firms between 1969 and 1985 resulted in an
absolute loss. . . . Nevertheless, the returns on a few investments have more
than offset these disappointments. Venture Economics reports, for example,
that 6.8 percent of the investments resulted in payoffs greater than ten times
cost and yielded 49.4 percent of the ending value of the aggregate portfolio
(61.4 percent of the profits). . . .

The Most Common Structure of Venture Capital Firms

By 1988 the typical venture capital firm was organized as a limited partnership, with the venture capitalists serving as general partners and the investors
as limited partners. . . .

Venture capital firms tend to specialize by industry or stage of investment.
Some firms focus on computer-related companies, others on biotechnology or
specialty retailers. Some will invest only in early-stage deals, whereas others
concentrate on later-stage financings. Many firms also limit their geographic
scope. . . .

The relationship between investors and managers of the venture funds is
governed by a partnership agreement that spells out the rights and obligations
of each group. . . .

Analyzing the Relationship Between External Investors and Venture Capitalists

Venture capitalists act as agents for the limited partners, who choose to invest
in entrepreneurial ventures through an intermediary rather than directly. In
such situations, conflicts arise between the agent and the principal, which
must be addressed in the contracts and other mechanisms that govern their
relationship.

In the venture capital industry, the agency problem is likely to be particularly difficult. There is inevitably a high degree of information asymmetry
between the venture capitalists, who play an active role in the portfolio companies, and the limited partners, who cannot monitor the prospects of each
individual investment as closely. . . .

Contracts are designed with several key provisions to protect the limited
partners from the possibility that the venture capitalists will make decisions
against their interests. First, the life of a venture capital fund is limited; the
venture capitalist cannot keep the money forever. Organizational models like

mutual funds or corporations, in contrast, have indefinite life spans. Implicitly, the investors also preserve the right not to invest in any later fund managed by the same venture capitalists.

Second, the limited partners preserve the right to withdraw from funding the partnership by reneging on their commitments to invest beyond the initial capital infusion . . . Third, the compensation system is structured to give the venture capitalists the appropriate incentives. The fund managers are typically entitled to receive 20 percent of the profits generated by the fund. For reasons which will be explored more fully below, the profit participation and other aspects of the contract encourage the venture capitalist to allocate the management fee to activities that will increase the total value of the portfolio.

Fourth, the mandatory distribution policy defuses potential differences of opinion about what to do with the proceeds from the sale of assets in the portfolio. The general partners cannot choose to invest in securities that serve their own private interests at the expense of the limited partners.

Finally, the contract addresses obvious areas of conflict between the venture capitalist and the limited partner. Thus, the venture capitalist is often explicitly prohibited from self-dealing (for example, being able to buy stock in the portfolio on preferential terms or receiving distributions different from those given to the limited partner). Also, the venture capitalists are contractually required to commit a certain percentage of their effort to the activities of the fund. Although this requirement is difficult to monitor, egregious violations can be the subject of litigation if fund performance is poor. . . .

The compensation system plays a critical role in aligning the interests of the venture capitalists and the limited partners. . . . The carried interest component of compensation is large in relation to the other components.[1] The implication is that the venture capitalists have incentives to engage in activities that increase the value of the carried interest, which is precisely what benefits the limited partners. . . .

The Sorting Problem

The governance structure also helps potential investors distinguish between good venture capitalists and weak ones. The basic argument is simple: good venture capitalists are more likely than weak venture capitalists to accept a finite life for each new partnership and a compensation system heavily dependent on investment returns. By doing so, they agree explicitly to have their performance reviewed at least every few years: if they engage in opportunistic acts or are incompetent, they will be denied access to funds. In addition, most of their expected compensation comes from a share in the fund's profits. If they perform well, they will participate handsomely in the fund's success. They will also be rewarded by being able to raise additional capital and, most

[1]Carried interest refers to the venture capitalist's (general partner's) share of the profits of the investment, which is a fixed payment received at the end of the fund's duration (hence, it is a "carried" interest). The other components of compensation are management fees and salaries [EDITOR'S NOTE].

likely, benefit from the various economies characteristic of the business. If they are not confident of performing well, or if they intend to neglect the interests of the limited partners, they will probably not agree to the basic terms of the contract. . . .

The Relationship Between the Venture Capitalists and the Entrepreneurial Ventures

Each year venture capitalists screen hundreds of investment proposals before deciding which ideas and teams to support. The success or failure of any given venture depends on the effort and skill of the people involved as well as on certain factors outside their control (for example, the economy), but the capabilities of the individuals involved are difficult to gauge up front.

Once investment decisions are made and deals consummated, it is difficult to monitor progress. The probability of failure is high. . . . The venture capitalist and the entrepreneur are also likely to have different information. Even with the same information, they are likely to disagree on certain issues, including if and when to abandon a venture and how and when to cash in on investments.

Venture capitalists attack these problems in several ways. First, they structure their investments so they can keep firm control. The most important mechanism for controlling the venture is staging the infusion of capital. Second, they devise compensation schemes that provide venture managers with appropriate incentives. Third, they become actively involved in managing the companies they fund, in effect functioning as consultants. Finally, venture capitalists preserve mechanisms to make their investments liquid.

Staging the Commitment of Capital and Other Control Mechanisms

Venture capitalists rarely, if ever, invest all the external capital that a company will require to accomplish its business plan: instead, they invest in companies at distinct stages in their development. As a result, each company begins life knowing that it has only enough capital to reach the next stage. By staging capital the venture capitalists preserve the right to abandon a project whose prospects look dim. The right to abandon is essential because an entrepreneur will almost never stop investing in a failing project as long as others are providing capital.

Staging the capital also provides incentives to the entrepreneurial team. Capital is a scarce and expensive resource for individual ventures. Misuse of capital is very costly to venture capitalists but not necessarily to management. To encourage managers to conserve capital, venture capital firms apply strong sanctions if it is misused. These sanctions ordinarily take two basic forms. First, increased capital requirements invariably dilute management's equity share at an increasingly punitive rate. . . . Second, the staged investment process enables venture capital firms to shut down operations completely. The credible threat to abandon a venture, even when the firm might be economically viable, is the key to the relationship between the entrepreneur and the venture capitalist. . . . By denying capital, the venture capitalist also signals other capital suppliers that the company in question is a bad investment risk.

Short of denying the company capital, venture capitalists can discipline wayward managers by firing or demoting them. . . .

Entrepreneurs accept the staged capital process because they usually have great confidence in their own abilities to meet targets. They understand that if they meet those goals, they will end up owning a significantly larger share of the company than if they had insisted on receiving all of the capital up front. . . .

The Compensation Scheme

Entrepreneurs who accept venture capital typically take smaller cash salaries than they could earn in the labor market. The shortfall in current income is offset by stock ownership in the ventures they start. Common stock and any subsequent stock options received will not pay off, however, unless the company creates value and affords an opportunity to convert illiquid holdings to cash. In this regard, the interests of the venture capital investor and entrepreneur are aligned.

This compensation system penalizes poor performance by an employee. If the employee is terminated, all unvested shares or options are returned to the company. In almost all cases, the company retains the right to repurchase shares from the employee at predetermined prices.

Without sanctions, entrepreneurs might sometimes have an incentive to increase risk without an adequate increase in return. An entrepreneur's compensation package can be viewed as a contingent claim, whose value increases with volatility. The sanctions, combined with the venture capitalists' active role in the management of the venture, help to mitigate the incentive to increase risk.

Active Involvement of Venture Capitalists in Portfolio Companies

No contract between an entrepreneur and venture capitalist can anticipate every possible disagreement or conflict. Partly for this reason, the venture capitalist typically plays a role in the operation of the company.

Venture capitalists sit on boards of directors, help recruit and compensate key individuals, work with suppliers and customers, help establish tactics and strategy, play a major role in raising capital, and help structure transactions such as mergers and acquisitions. They often assume more direct control by changing management and are sometimes willing to take over day-to-day operations themselves. All of these activities are designed to increase the likelihood of success and improve return on investment: they also protect the interests of the venture capitalist and ameliorate the information asymmetry. . . .

The adverse-selection problem is a difficult one in venture capital. Venture capitalists argue that by playing a positive role in the venture, they can increase total value by enough to offset the high cost of the capital they provide. To the extent that venture capitalists make good on this claim, the adverse-selection issue is effectively mitigated. In addition, the due diligence conducted before an investment is made is intended partly to make sure the entrepreneurs are qualified.

Although it seems that venture capitalists retain much of the power in the relationship with entrepreneurial ventures, there are checks and balances in the system. Venture capitalists who abuse their power will find it hard to attract the best entrepreneurs, who have the option of approaching other venture capitalists or sources other than venture capital. In this regard, the decision to accept money from a venture capitalist can be seen as a conscious present-value-maximizing choice by the entrepreneur.

Two Agency-Cost Explanations of Dividends

FRANK H. EASTERBROOK

Economists find dividends mysterious. The celebrated articles by Merton Miller and Franco Modigliani declared them irrelevant because investors could home brew their own dividends by selling from or borrowing against their portfolios. Meanwhile the firms that issued the dividends would also incur costs to float new securities to maintain their optimal investment policies. Dividends are, moreover, taxable to many investors, while firms can reduce taxes by holding and reinvesting their profits. Although dividends might make sense in connection with a change in investment policy—when, for example, the firms are disinvesting because they are liquidating or, for other reasons, shareholders can make better use of the money than managers—they are all cost and no benefit in the remaining cases of invariant investment policies. . . .

One form of agency cost is the cost of monitoring of managers. This is costly for shareholders, and the problem of collective action ensures that shareholders undertake too little of it. Although a monitor-shareholder would incur the full costs of monitoring, he would reap gains only in proportion to his holdings. Because shares are widely held, no one shareholder can capture even a little of the gain. Shareholders would be wealthier if there were some

Reprinted by permission from 74 *American Economic Review* 650. © 1984 American Economic Association.

person, comparable to the bondholders' indenture trustee, who monitored managers on shareholders' behalf.

A second source of agency costs is risk aversion on the part of managers. The investors, with diversified portfolios of stocks, will be concerned only about any nondiversifiable risk with respect to any one firm's ventures. Managers, though, have a substantial part of their personal wealth tied up in their firms. If the firms do poorly or, worse, go bankrupt, the managers will lose their jobs and any wealth tied up in their firms' stock. Managers therefore will be concerned about total risk, and their personal risk aversion will magnify this concern.

The risk-averse managers may choose projects that are safe but have a lower expected return than riskier ventures. Shareholders have the opposite preference. Riskier ventures enrich shareholders at the expense of creditors (because shareholders do not pay any of the gains to bondholders, yet bondholders bear part of the risk of failure), and shareholders would want managers to behave as risk preferrers. . . .

Both the monitoring problem and the risk-aversion problem are less serious if the firm is constantly in the market for new capital. When it issues new securities, the firm's affairs will be reviewed by an investment banker or some similar intermediary acting as a monitor for the collective interest of shareholders, and by the purchasers of the new instruments. . . .

New investors do not suffer under the collective choice disabilities of existing investors. They can examine managers' behavior before investing, and they will not buy new stock unless they are offered compensation (in the form of reduced prices) for any remediable agency costs of management. Managers who are in the capital market thus have incentives to reduce those agency costs in order to collect the highest possible price for their new instruments. New investors are better than old ones at chiseling down agency costs.

Of course, new investors need information, and that may be hard to come by. Neither auditors nor the managers themselves are perfectly reliable unless there is a foolproof legal remedy for fraud. Other forms of information gathering, such as shareholders' inquiries and stockbrokers' studies, suffer from the problem that none of the persons making inquiry can capture very much of the gain of this endeavor, and thus there will be too little information gathered. There would be savings if some information gatherers had larger proportionate stakes, and if the verification of information could be accomplished at lower cost. Underwriters of stock and large lenders may supply the lower-cost verification. These firms put their own money on the line, and any information inferred from this risk-taking behavior by third parties may be very valuable to other investors. This form of verification by acceptance of risk is one of the savings that arise when dividends keep firms in the capital market. . . .

Expected, continuing dividends compel firms to raise new money in order to carry out their activities. They therefore precipitate the monitoring and debt-equity adjustments that benefit stockholders. Moreover, even when dividends are not accompanied by the raising of new capital, they at least increase the debt-equity ratio so that shareholders are not giving (as much) wealth away to bondholders. In other words, dividends set in motion mechanisms that reduce the agency costs of management. . . .

[T]he explanations I have offered are not unique explanations *of dividends*. Nothing here suggests that repurchases of shares would not do as well as or better than dividends. The issuance of debt instruments in series, so that payments and refinancings are continuous, serves the same function as dividends. I have "explained" only mechanisms that keep firms in the capital market in ways that instigate consistent monitoring and consistent readjustment of the risk among investors.

Notes and Questions

1. The discussion of risk shifting in this chapter's introduction and the various selections is based on the option pricing theory (OPT) of modern corporate finance, a mathematically complex model derived by Fisher Black and Myron Scholes and widely used in the investment community. Black and Scholes, "The Pricing of Options and Corporate Liabilities," 81 *Journal of Political Economy* 637 (1973). This note briefly introduces the intuition underlying OPT in order to clarify the connection between capital structure and shareholder opportunism.

An option is a contract to buy (or sell) an asset at a fixed price, called the exercise price, on or by a fixed date in the future, the maturity date. The technical term for an option to buy an asset is a call option and for the right to sell, a put; the word "option" is, however, frequently used without a modifier, and in this usage it means a call option. A specially valuable feature of an option is its contingent nature: the option owner has the right, but no obligation, to purchase the asset. The owner buys the asset only if it is to her advantage. Depending upon the difference between the exercise price and the price of the underlying security at the maturity date, the option will either be exercised or it will expire unexercised. If the price is not right (i.e., exercise will not produce a profit), the owner of the option can simply walk away from the contract. For example, if the exercise price exceeds the price of the asset, it is cheaper for the option holder to purchase the asset directly than to exercise the option.

Options are derivative instruments: they derive their value from something else, the underlying or primitive asset. While options can be written on any type of asset, the most typical type of option is written on common stock. An option's value depends on many factors, including the exercise price and the stock price. The key variable in option pricing is, however, the volatility of the underlying stock. The greater the volatility (the greater the variance) of the return of the stock, the greater the value of the option.

Consider two stocks, S and V, whose expected return is the same but whose volatility differs (the variance of V is greater than that of S). In CAPM terms, the stocks have the same beta but different firm-specific risk. Assuming that the distribution of stock returns is normal (symmetrical around the mean), then there is a 50 percent chance the stock will be above the expected value and a 50 percent chance it will be below it. Suppose now that an option is written on each stock, and the exercise price equals the expected value. This means that there is a 50 percent chance that the options will expire worthless (the stock price will be below the exercise price) and a 50 percent chance that they will be valuable (the stock will rise above the exercise price).

If the stock prices rise, then V, the stock with the greater variance, has a greater probability of rising more than S, the stock with the lower variance; that is what variance measures (the dispersion of returns from the mean). There is, then, a larger chance of a big payoff from the more volatile stock. But the chance of a zero payoff is the same for both (50 percent). There is no negative payoff probability because the option does not have to be exercised if the stock price falls below the exercise price. This is the key to the value of volatility. There is a downside limit on an option's value, it cannot be worth less than zero (because if it is worth a negative amount, it is not exercised). Therefore when the stock's variance increases, the option holder's only concern is with the greater magnitude of favorable outcomes that greater dispersion brings; the upside potential increases with the higher variance, while the increase in high negative payoffs is of no consequence.

Is it peculiar, given our assumptions of investor behavior in portfolio theory, to find greater variance a desirable investment characteristic? The answer is no. When an investor owns stock, the entire probability distribution of returns, good and bad, affects her, and being risk averse, she dislikes the larger variance. When she owns an option, however, the distribution of returns is truncated, only the upside tail matters; the contingent nature of the claim eliminates the exposure from downside risk and, hence, greater spread is better.

Stock in a leveraged firm is equivalent to a call option. The exercise price is the face value of the debt; the underlying asset is the firm. When the debt matures, the shareholders decide whether or not to exercise their option (pay off the debt) and buy the firm. If, when the bond matures, the corporation's assets are worth less than the debt that is due, then the shareholders do not repay the debt (they do not exercise their option). Because of limited liability, zero is the worst result for them, and that is what their stock is worth if they default. If the value of the assets exceeds that of the bonds, then the shareholders exercise their option (they pay off the bondholders), and they buy the firm.

The analogy makes evident that the variables that create value in options will also increase stock value, and this is the key to the risk-shifting game, the agency costs of debt. Because the value of an option increases with the risk of the underlying security, increasing the risk of the firm's assets (choosing Project H over Project L in the example in the introductory note to this chapter) enhances the value of the stock.

Note that investors buy options not only to speculate in price movements (the optimist buys calls and the pessimist buys puts) but also to hedge against price changes. This is because a judicious combination of stocks, options, and borrowing will create a portfolio with a riskless return (for example, the gain or loss on call option sales will offset the gain or loss on a combination of stock and put purchases). Any good finance text will provide an introduction to option pricing, such as Stephen A. Ross, Randolph W. Westerfield, and Jeffrey F. Jaffe, *Corporate Finance,* 3d ed. (Homewood, Ill.: Irwin, 1993).

2. Consistent with Smith and Warner's thesis that debt contract terms are priced, Paul Asquith and Thierry Wizman provide evidence of the value of bond covenants: in an event study of the effect of LBOs on bond prices, they found that debt with strong covenant protection (secured debt or debt with covenants restricting net worth of the surviving firm of a merger and limiting total funded debt) experienced positive abnormal returns (+2.6 percent), whereas those with no protection experienced negative abnormal returns (−5.2 percent). Asquith and Wizman, "Event Risk, Covenants, and Bondholder Returns in Leveraged Buyouts," 27 *Journal of Financial Economics* 195 (1990). The increase in value of the former is, presumably, a function of the covenants, as the LBO will constitute a default requiring the bonds to be retired or the terms improved. This explanation is supported by the additional finding that most of the protected bonds are no longer outstanding postbuyout (i.e., they were retired), whereas unprotected bonds are much more likely to remain outstanding. Asquith and Wizman conclude that covenants "make a difference."

3. Private organizations, such as Standard & Poor's and Moody's Investor Service, rate corporate debt according to credit worthiness (the likelihood of default and the protections afforded by the bond indenture should a default occur). The higher the rating, the lower the interest rate firms must pay investors. Bonds with ratings below investment grade are referred to as junk bonds. Equivalent terms are high-yield or low-grade bonds. The terms are used interchangeably in the investment community and are simply intended to refer to the security's riskiness.

Junk bond financing increased dramatically during the 1980s, rising from 5 percent of total U.S. public bond issues in 1980 to 20 percent in 1986 (in absolute dollars, an increase from $2 to $46 billion). Kevin J. Perry and Robert A. Taggart, Jr., "The Growing Role of Junk Bonds in Corporate Finance," 1 *The Continental Bank Journal of Applied Corporate Finance* 37 (Spring 1988). As Glenn Yago remarks, only 5 percent of over 20,000 U.S. businesses with sales over $35 million qualify for investment-grade quality debt ratings. Prior to the development of the junk bond market, the thousands of companies that did not so qualify were excluded from the public debt market and had to pay the higher cost of capital required by commercial bank loans or equity offerings to undertake new projects. Glenn Yago, *Junk Bonds: How High Yield Securities Restructured Corporate America* (New York: Oxford University Press, 1991). Moreover, venture capital funds are not available for most such firms, for as the Sahlman selection indicates, these funds are used for start-up firm financing up through the firm's initial public offering, rather than for firms' long-term growth and expansion. The depth of the junk bond market is also much greater than the venture capital market; for example, in 1986, the total venture capital pool was $24 billion, whereas close to $46 billion in junk bonds were issued in that year alone.

In recent years junk bonds have been a source of controversy because of their use in corporate takeovers and restructurings. They enable small firms to finance the acquisition of much larger entities, whose assets will generate earnings to repay the debt. Chapter VI will examine in detail whether concern over corporate takeovers is reasonable, and, correspondingly, answer the question whether proposals to restrict junk bonds are good public policy. Contrary to the perception behind the controversy, the vast majority of junk bond financing has not been used in corporate restructurings and LBOs. Most high-yield debt offerings fund the internal growth and expansion of small- and medium-sized companies, many of which are in the more innovative sectors of the economy (e.g., cable television, cellular phones, personal computers). The proportion of mergers and other control changes financed by high-yield issues is insignificant. See Perry and Taggart, supra.

4. Consider Lehn and Poulsen's discussion of event-risk provisions in light of the finance theory introduced in the Malkiel selection in Chapter I. Would diversification of an investor's debt portfolio be an adequate alternative to protection by event-risk provisions?

5. The privatization of bankruptcy that Jensen sees as a benefit of the LBO organizational form, through which firms in financial distress voluntarily restructure their debt without using the formal legal procedure (called workouts), enables the participants to realize substantial cost savings. The best estimates indicate that direct costs of a voluntary debt exchange (.65 percent of the book value of the assets measured just prior to the exchange) are substantially less than direct bankruptcy costs (between 2.8 percent and 7.5 percent of total assets measured one year before filing). Compare Stuart C. Gilson, Kose John, and Larry H. P. Lang, "Troubled Debt Restructurings: An Empirical Study of Private Reorganization of Firms in Default," 27 *Journal of Financial Economics* 315 (1990), with Jerold Warner, "Bankruptcy Costs: Some Evidence," 32 *Journal of Finance* 337 (1977); James Ang, Jess Chua, and John McConnell, "The Administrative Costs of Corporate Bankruptcy: A Note," 37 *Journal of Finance* 219 (1982); and Lawrence A. Weiss, "Bankruptcy Resolution: Direct Costs and Violation of Priority of Claims," 27 *Journal of Financial Economics* 285 (1990).

The number of LBO firms filing bankruptcy petitions increased in the early 1990s, despite Jensen's prediction that the LBO organization offered the ability to restructure in times of financial distress without paying the costs of formal bankruptcy. In addition, Stuart Gilson finds that a majority of these firms did not try to privately restructure their debt before filing, whereas in the 1980s a majority of filers did. Gilson, "Managing Default: Some Evidence on How Firms Choose between Workouts and Chapter 11," 4 *The Continental Bank Journal of Applied Corporate Finance* 62, 69 (Summer 1988). In subsequent work, Jensen explains: "[M]y predictions about the continuing privatization of bankruptcy could not have been more wrong. What I failed to anticipate were major new regulatory initiatives, a critical change in the tax code, and a misguided bankruptcy court decision that together are forcing many troubled companies into Chapter 11 [Chapter 11 is the federal bankruptcy statute—ED.]." Michael C. Jensen, "Corporate Control and the Politics of Finance," 4 *The Continental Bank Journal of Applied Corporate Finance* 13, 25 (Summer 1991).

There is, however, a method by which firms can still obtain some of the cost advantages of a workout without giving up the benefits accompanying a bankruptcy filing: the "prepackaged" bankruptcy plan. In such an arrangement, at the same time a firm files for bankruptcy it files a reorganization plan, which has already been approved by most creditors. The time (and money) spent in bankruptcy court will be substantially reduced under these circumstances. For a discussion of prepackaged plans, which have become increasingly popular, see John J. McConnell, "The Economics of Prepackaged Bankruptcy," 4 *The Continental Bank Journal of Applied Corporate Finance* 93 (Summer 1991).

6. Sahlman indicates that the return on venture capital funds has been approximately 26 percent per year. Compare the historical return on the stock market and on small stocks presented in Table I.2 of the Malkiel selection in Chapter I. For what kinds of investors is a venture capital fund an appropriate investment? Do you think the contractual arrangements that Sahlman describes are adequate to resolve the agency problems of such investments? Do venture capital contractual arrangements provide empirical support for Jensen and Meckling's principal-agent theory of firm organization (Chapter I) as Sahlman maintains, or are there other explanations for such arrangements?

7. Is corporate law's placement of dividend policy within the business judgment rule, which entrusts it to management's discretion, consistent with Easterbrook's thesis? With the finance position that he mentions of the irrelevance of dividends? What if firms appeal to different clienteles of investors, such that high-payout firms are owned by low-tax bracket investors and low-payout firms by high-tax bracket individuals?

The financial economist's dividend puzzle, to which Easterbrook alludes, is actually a revisiting of a famous debate over capital structure. Franco Modigliani and Merton Miller (who are often referred to as "M & M") proposed an irrelevance theorem: the value of the firm is independent of its mix of debt and equity claims. Modigliani and Miller, "The Cost of Capital, Corporation Finance and the Theory of Investment," 48 *American Economic Review* 261 (1958). The proof consists of an arbitrage argument, similar to the proof of CAPM in the Malkiel selection and discussed in Chapter I, part B, note 3. In brief, M & M showed that equity investors in a leveraged firm can obtain the same returns by borrowing on their own account to invest in an unleveraged firm, and consequently, the stock of the leveraged firm cannot be worth more than the unleveraged equity.

The irrelevance theorem of firm capital structure does not hold when corporate taxes are introduced because of the tax code's differential treatment of the returns to debt and equity capital—interest on debt is deductible but dividends, the return on equity capital, are not. In the world with taxes, M & M offered a second proposition that to maximize value, firms should be financed 100 percent with debt. Modigliani and Miller, "Corporate Income Taxes and the Cost of Capital: A Correction," 53 *American Economic Review* 433 (1963). Although LBO firms are highly leveraged, no firms go as far as having 100 percent debt. This chapter's principal theme, the agency costs of debt, which was introduced by Jensen and Meckling in their classic article excerpted in Chapter I, explains why M & M's second theorem is not confirmed by real-world practice. Easterbrook's insight parallels Jensen and Meckling's, combining the theory of the firm with finance to explain otherwise seemingly irrational (non-value-maximizing) behavior.

8. The incentive for firms to retain earnings rather than distribute them was even greater prior to the Tax Reform Act of 1986 because appreciation was taxed at the then lower capital gains rate (60 percent of the ordinary income tax rate). From Easterbrook's perspective, the 1986 reform that eliminated the differential in tax rates is in shareholders' interests: it better aligns corporations' tax incentives with behavior that is desirable from the point of view of agency theory. Note that the distribution policy preference of corporate shareholders differs from individuals because corporations can exclude 70 percent of dividends they receive on equity investments from the corporate tax; thus, for shareholder-corporations there is less conflict between tax code and agency cost incentives. For further instances where tax incentives complement agency incentives and where the two are at odds, see Saul Levmore and Hideki Kanda, "Taxes, Agency Costs and the Price of Incorporation," 77 *Virginia Law Review* 211 (1991); and Myron S. Scholes and Mark A. Wolfson, *Taxes and Business Strategy* 134 (Englewood Cliffs, N.J.: Prentice-Hall, 1992).

9. Easterbrook's monitoring explanation of dividends relies, as he states, on a corporation's need to raise new capital. Why should new equity investors be more perspicacious monitors than the firm's current shareholders? Does Easterbrook's analysis lose force if shareholders are not the rationally apathetic investors depicted in the Easterbrook and Fischel and the Gordon selections in Chapter III?

To what extent does Easterbrook's thesis depend upon an active role for investment bankers? In this regard, consider Rule 415, an innovation by the Securities and Exchange Commission in the 1980s that permits "shelf registration" of new equity issues by established firms. In a shelf registration, new stock is registered with the SEC but not issued (it remains on the shelf). The shares are instead issued at later dates, chosen by the firm, without further action, that is, without a new registration statement and with less time for the underwriter's due diligence check. The firm undertakes, however, at the time of the shelf registration to file amendments to the registration, upon the shares' subsequent issuance, if there are any material changes in circumstances.

Rule 415 is intended to save the transaction costs of equity financing, including investment bankers' fees, by opening up competition. See Securities and Exchange Commission, Securities Act Release No. 6499 (Nov. 17, 1983); and compare Barbara A. Banoff, "Regulatory Subsidies, Efficient Markets and Shelf Registration: An Analysis of Rule 415," 70 *Virginia Law Review* 135 (1984), which views the rule favorably, with Merritt Fox, "Shelf Registration, Integrated Disclosure, and Underwriter Due Diligence," 70 *Virginia Law Review* 1005 (1984), which does not. From Easterbrook's perspective, is there an objection to reducing these costs? Would competition among underwriters, which resulted in lower fees, reduce or enhance the value of their reputational capital?

Rule 415 was promulgated under the Securities Act of 1933, the federal statute regulating new equity issues. For an assessment of the effect of the 1933 Act on firm value, see the Simon selection in Chapter VII.

Internal Governance Structures

V

The readings in this chapter analyze internal governance structures that mitigate the agency problem. The core mechanisms are the board of directors, stockholder voting rights to elect directors and approve fundamental corporate changes, and fiduciary duties, which are policed by shareholder suits. Two additional important internal control devices that are not embedded in corporation codes are executive compensation and outside shareholders with sizeable blocks of stock.

The board of directors plays a key role in corporate governance. It monitors managers to ensure that they maximize equity share prices. Oliver Williamson provides an explanation of why the board's function is best defined in this way, as a monitor for equity investments, as opposed to a representative assembly calibrating the interests of all firm participants. The selection by Clifford Smith and Ross Watts focuses on one of the board's more important tasks, the setting of executive compensation. In describing the features of executive compensation plans, they are, in effect, detailing how well the board performs its monitoring function by motivating management. Compensation can be used as a crucial incentive-alignment device, and as Smith and Watts discuss, incentive pay is a key component of large corporations' executive compensation plans. Smith and Watts also demonstrate that the plans boards adopt are best explained by incentive (reducing agency costs), rather than tax, reasons.

The readings on the board do not address the thorny issue of who monitors the monitors? Fiduciary duty law, by which officers and directors are person-ally liable for acts of negligence or undue self-interest, is corporate law's principal response to this problem. Shareholders may bring lawsuits for breach of fiduciary duty, either in their own right (individually or aggregated as a class), or derivatively on behalf of the corporation. Roberta Romano examines how well shareholder suits perform as a governance device by inves-tigating whether they achieve compensatory or deterrent goals, and what interaction, if any, there is between litigation and the other institutions of corporate governance discussed in this chapter. Shareholder suits are not, however, the only potential legal liability that directors and officers face. As analyzed and critiqued in the selection by John Coffee, in recent years, direc-tors and officers have been criminally prosecuted for actions that would not even result in civil liability.

Shareholder voting is yet another monitoring device: shareholders exer-cise ultimate control over the board, and hence management, through annual elections. This control often seems purely nominal because corporate elec-tions are usually uncontested and the nominees are selected by management. Shareholders also exert oversight in additional contexts, as corporation codes specify voting rights for major corporate changes, such as mergers, sales of substantially all assets, and charter amendments. Even if shareholder voting is beset with collective action problems, as discussed in Chapter III, voting rights are still valuable and an important disciplining device, for votes can be concentrated in one shareholder's hands, in a takeover, and then exercised effectively.

Frank Easterbrook and Daniel Fischel develop the thesis that corporate voting rules are a governance device to reduce agency costs. They also make the case for the conventional voting rule of one share, one vote. This rule ensures that control belongs to the residual claimant (votes follow cash flows), and the rationale follows straightforwardly from the incentive problem expli-cated in Jensen and Meckling's changing benefit trade-off as ownership is separated from control (see Chapter I): The party with the incentive to maxi-mize firm value is the one who should control firm decision making. Jeffrey Gordon suggests that stock exchange listing requirements, and in particular, the New York Stock Exchange's historic requirement of one share–one vote common stock, serve as bonding devices that enable managers to credibly commit not to propose opportunistic charter amendments, such as restructur-ing voting rights to the outside shareholders' detriment. Proxy fights are the rare instance of contested elections, when outside shareholders vie against incumbent management for board or issue control. John Pound maintains that even in this setting where investors' attention is focused and information cheaply available, management is able to pursue its interest at the sharehold-ers' expense because the rules governing the proxy process advantage insiders.

Outside blockholders can serve as an important governance structure even without engaging in proxy fights by monitoring management. Such shareholders are, at least in theory, more willing to undertake monitoring efforts because the cost is spread over a larger investment and thus less likely to outweigh the pro rata benefits. Mark Roe provides a political account of why such a governance structure is less important in the United States than other countries: management's success in obtaining federal legislation restricting active stock ownership of financial institutions, which are the ideal outside blockholders. Bernard Black emphasizes the desirability of institutional investors' monitoring of management by exercising voice—initiating proposals or opposing management at shareholder meetings—rather than by obtaining control (assuming the restrictions noted by Roe are relaxed). Michael Jensen contends that an alternative institution has developed to fill the void: the LBO organization. He emphasizes the beneficial properties of debt, the key feature of the LBO firm, for mitigating the agency problem and what he considers to be a "failed model" of corporate governance, the conventional model discussed in the readings in part A of this chapter, in which the board of directors monitors management.

The materials in this chapter indicate that boards and outside shareholders, while occasionally successful, are comparatively weak constraints on managers. In particular, the data on executive compensation and shareholder litigation are not especially encouraging. The conclusion to draw from these data is not that managers can ignore shareholders with impunity. Rather, it is that we must look elsewhere for potent disciplining devices. For instance, the market for corporate control's threat that voting rights will be concentrated and management replaced is one such mechanism, and it is the subject of the next chapter.

Boards of Directors and Fiduciary Duties

Corporate Governance
OLIVER E. WILLIAMSON

The premises of this article are: First, the relation between each constituency and the firm needs to be evaluated in contractual terms. Second, special-purpose governance structures (of which the board of directors is one) arise in response to the needs of an exchange relation for contractual integrity. And third, lest the design benefits be dissipated, the special purpose character of each governance structure must be respected. . . . [R]epresentation on the board is never warranted for constituencies located at node *A*, but may be warranted for constituencies located at node *B*.[1] The question of why a node *B* constituency has not successfully forged a bilateral governance structure (thereby moving to node *C*) is germane. Finally, constituencies that have forged a viable governance structure at node *C* do not require voting representation on the board but sometimes should be included for informational purposes.

Representation is unwarranted for constituencies at node *A* because of the negligible exposure of their transaction-specific assets. Moreover, their legitimate interests are adequately safeguarded through neoclassical market contracting. Such constituencies have neither informational nor decisional needs to be served through board membership. Constituencies located at node *B*, however, have exposed assets and will charge a higher price ($\bar{p}$) unless safeguards can be devised. If parties cannot devise effective bilateral safeguards, generalized safeguards through voting board membership may be warranted.

Constituencies located at node *C* do not normally need membership on the board of directors to safeguard their interests. Such constituencies have devised a structure of bilateral governance to safeguard their interests. Such specialized structures will normally be better attuned to the adaptive needs and dispute settlement requirements of a constituency than will access to a generalized instrument. If board membership is warranted at all, participation should normally be limited to informational access only. . . .

---

[1]The figure illustrating nodes *A, B,* and *C* is identical to Figure I.1 in Williamson, Chapter I, and is therefore not reproduced here [EDITOR'S NOTE].

Reprinted by permission of The Yale Law Journal Company and Fred B. Rothman & Company from *The Yale Law Journal,* Vol. 93, pp. 1197–1230 (1984).

Labor

Enthusiasts of codetermination regard participation for informational purposes as inadequate. They maintain that codetermination should extend the
influence of workers to include "general issues of investments, market planning, decisions about output, and so forth."

This argument is clearly mistaken as applied to workers with general purpose skills and knowledge (node A). Such workers can quit and be replaced
without productive loss to either worker or firm.[2] Consider, therefore, workers that make firm-specific investments and are located at nodes B or C.
Ordinarily, it can be presumed that workers and firms will recognize the
benefits of creating specialized structures of governance to safeguard firm-
specific assets. Failure to provide such safeguards will cause demands for
higher wages. . . .

Efficient governance, however, requires more than realignment of incentives. The institutions of contract also matter. Machinery for settling disputes
and for adapting to changed circumstances is needed if continuing relations
between the firm and workers located at node C are to operate smoothly. The
grievance machinery and associated job structures—ports of entry, promotion ladders, bumping, and so forth—are thus important parts of efficient
governance.

Another important factor is asymmetric information. A chronic difficulty
with long-term labor agreements is that misallocation will result if wages are
set first and employment levels are unilaterally determined by management
later. . . . Even if wages and employment are both established at the outset,
the agreement may drift out of alignment during the contract's execution to
the disadvantage of the less-informed member of the contracting pair. Such a
result might be avoided by imparting more information to labor. Labor membership on the board of directors for informational purposes is one means of
achieving this result. . . .

Owners

The term "owners" is usually reserved for stockholders, but debtholders may
also assume this status. However described, suppliers of finance bear a unique
relation to the firm: The whole of their investment in the firm is potentially
placed at hazard. By contrast, the productive assets (plant and equipment;
human capital) of suppliers of raw material, labor, intermediate products,
electric power, and the like normally remain in the suppliers' possession. If
located at node A, therefore, these suppliers of finance must secure repay-

---

[2]This is an oversimplification. It assumes easy reemployment and ignores transitional costs,
including the impact on the family. Unemployment insurance may provide a necessary buffer. We
may want to create some barriers to deter termination without cause and reduce transition costs.
The basic point, however, is comparative. Workers located at node A have the least concern over
expropriation.

ment or otherwise repossess their investments to effect redeployment. Accordingly, the suppliers of finance are, in effect, always located on the $k > 0$ branch. The only question is whether their investments are protected well (node C) or poorly (node B).

Stockholders

Although a well-developed market in shares permits individual stockholders to terminate ownership easily by selling their shares, it does not follow that stockholders as a group have a limited stake in the firm. What is available to individual stockholders may be unavailable to stockholders in the aggregate. Although some students of governance see only an attenuated relation between stockholders and the corporation, this view is based on a fallacy of composition. Stockholders as a group bear a unique relation to the firm. They are the only voluntary constituency whose relation with the corporation does not come up for periodic renewal. Labor, suppliers in the intermediate product market, debt holders, and consumers all have opportunities to renegotiate terms when contracts are renewed. Stockholders, by contrast, invest for the life of the firm and their claims are located at the end of the queue should liquidation occur.

Stockholders are also unique in that their investments are not associated with particular assets. The diffuse character of their investments puts shareholders at an enormous disadvantage in crafting the kind of bilateral safeguards normally associated with node C. Unless, therefore, a governance structure of broad scope is somehow devised, stockholders are unavoidably located at node B.

Recall that the critical attributes of suppliers located at node B are their investments in specific assets and the premium they require for their services because of the hazard of expropriation. This premium can be regarded as a penalty imposed on the firm for its failure to craft node C safeguards. The incentive for the firm to secure relief from this penalty is clear. This article considers the board of directors to be a governance structure whose principal purpose is to safeguard those who face a diffuse but significant risk of expropriation because the assets in question are numerous and ill defined, and cannot be protected in a well-focused, transaction-specific way. Thus regarded, the board of directors should be seen as a governance instrument of the stockholders. Whether other constituencies also qualify depends upon their contracting relation with the firm.

Such protection for stockholders can be and often is supplemented by other measures. Corporate charter restrictions and informational disclosure requirements are examples. Firms recognize stockholders' needs for controls and many attempt responsibly to provide them. Some managements, however, play "end games" (undisclosed strategic decisions to cut and run before corrective measures can be taken) and individual managers commonly disclose information selectively or distort data. Additional checks against such concealment and distortion can be devised to give shareholders greater confidence. Arguably, an audit committee composed of outside directors and the certification of financial reports by an accredited accounting firm promote these purposes. Another possibility is the required disclosure of financial reports to a public agency with powers of investigation. The efficacy of these devices is difficult to gauge.

Lenders

In certain atypical circumstances, lenders may deserve board representation. Unlike stockholders, lenders commonly make short-term loans for general business purposes or longer-term loans against earmarked assets. Proof that the firm is currently financially sound, coupled with short maturity, places short-term lenders at node C. Lenders who make longer-term loans commonly place preemptive claims against durable assets. If the assets cannot be easily redeployed, lenders usually require partial financing through equity collateral. Thus, long-term lenders usually carefully align incentives and protect themselves with safeguards of the sort associated with node C. . . . As the exposure to risk increases, these debt holders become more concerned with the details of the firm's operating decisions and strategic plans: With high debt-equity ratios the creditors become more like shareholders and greater consultation between the management and its major creditors results. A banking presence in a voting capacity on the board of directors may be warranted in these circumstances. More generally, a banking presence may be appropriate for firms experiencing adversity, but this should change as evidence of recovery accumulates.

Suppliers

Whether suppliers have a stake in a firm depends upon whether they have made substantial investments in durable assets that could not be redeployed without sacrificing productive value if the relationship with the firm were to be terminated prematurely. The mere fact that one firm does a considerable amount of business with another, however, does not establish that specific assets have thereby been exposed. At worst, suppliers located at node A experience modest transitional expenses if the relation is terminated. Neither specialized bilateral governance nor membership on the board of directors is needed to safeguard their interests. The protection afforded by the market suffices.

Suppliers who make substantial firm-specific investments in support of an exchange will demand either a price premium (as at node B, where the projected break-even price is $\bar{p}$) or special governance safeguards (as at node C). Progress payments and the use of hostages (i.e., credible commitments) to support exchange are illustrations of node C safeguards. An agreement to settle disputes through arbitration, rather than through litigation, is also in the spirit of node C governance.

Considering the variety of widely applicable governance devices to which firms and their suppliers have access, there is little need to accord suppliers additional protection through membership on the board of directors. There could be exceptions, of course, where a large volume of business is at stake and a common information base is needed to coordinate investment planning. Ordinarily, however, the governance structure that firm and supplier devise at the time of contract (and help to support through a web of interfirm relationships) will afford adequate protection. Membership on the board, if it occurs at all, should be restricted to informational participation.

Customers

The main protection for customers located at node A is generally the option to take their trade elsewhere. Products that have delayed health effects are an exception, and consumer durables can also pose special problems. Membership on the board of directors is not, however, clearly indicated for either reason.

Health hazards pose problems if consumers are poorly organized in relation to the firm and lack the relevant information. If consumers can organize only with difficulty because they are unknown to one another, or because of the ease of free riding, then a bilateral governance structure between firm and consumers may fail to materialize. Protection by third parties may be warranted instead. A regulatory agency equipped to receive complaints and screen products for health hazards could serve to infuse confidence in such markets.

Whether consumer membership on the board would afford additional protection is problematic. Who are representative consumers? How do they communicate with their constituency? Token representation may create only unwarranted confidence.

Similar problems of consumer organization and ignorance arise in conjunction with consumer durables. This is true whether the consumer durable requires no follow-up service or a great deal of such service. Among the available types of consumer protection are brand names, warranties, and arbitration panels. Shoppers who choose node B are presumably looking for bargains. They will spurn these additional protections in favor of a lower price. Such customers implicitly accept a higher risk and should accept occasional disappointments. There are other consumers, however, who value protections at node C. Some are prepared to pay a premium for a brand name item. Brand names effectively extend a firm's planning horizon and create incentives for the firm to behave "more responsibly." (To be sure, these assumptions merit qualification. Firms sometimes build up a reputation and thereafter expend it by taking advantage of lagging consumer perceptions.) Warranties are explicit forms of follow-on protection, and many are available on optional terms. The recent introduction of consumer arbitration panels is likewise responsive to concerns over consumer protection. Consumers concerned about fair play in the postsales service period will presumably concentrate their purchases on brands for which arbitration is available.

Further innovations to offer consumer protection on a discriminating basis may be needed. With the possible exception of large customers with special informational needs, however, a general case for inclusion of consumers on the board of directors is not compelling.

The Community

Community interest in the corporation is a very large subject. I consider two concerns here—externalities and the hazards of expropriation.

Externalities commonly arise where the parties in question do not bear a contracting relation to each other. Pollution is one example. Corrections can be interpreted as an effort by the community to impose a contract where none existed. For example, the community may place a pollution tax (price) on the firm, or it may stipulate that pollution abatement regulations must be satisfied as a condition for doing business.

A chronic problem in this area is to secure the knowledge on which to base an informed pollution control policy. Firms often possess the necessary knowledge and may disclose it only in a selective or distorted manner. Public membership on the board of directors could conceivably reduce misinformation. But the remedy would come at a high cost if the corporation were thereby politicized or deflected from its chief purpose of serving as an economizing instrument. Penalties against misinformation coupled with moral suasion may be more effective. This is an area in which there may simply be no unambiguously good choices.

The hazards of expropriation are even less of a justification for public membership on boards. Communities often construct durable infrastructures to support a new plant or renewal investments by old firms. Expropriation is possible if the firm is able to capitalize these public investments and realize a gain upon selling off the facility. Such concerns are much greater if the firm makes general purpose rather than special investments. Communities that make investments in support of a firm should therefore scrutinize the character of the investments that the firm itself makes.

As elsewhere, expropriation hazards will be mitigated if the parties can locate themselves at node C. Insistence that the firm make specialized investments is akin to the use of hostages to support exchange. In general, specially crafted node C protection, rather than public membership on the board of directors, has much to commend it as the main basis for safeguarding community investments.

Management as a Constituency

Management Contracting

Since no firm-specific human assets are exposed by managers located at node A, no specialized governance is needed. Like any other constituency with attributes of node A, such managers look to the market for basic protection. Managers who develop a firm-specific asset relationship with the firm, however, are located at nodes B or C.

Those managers who contract with the firm in a node B manner will receive higher current compensation than those who are accorded internal governance protection of a node C kind. This is the familiar $\bar{p} > \hat{p}$ result. To what types of governance protection do managers located at node C have access? . . .

Compensation Schemes

Both the firm and its managers should recognize the merits of drafting compensation packages that deter both hasty dismissals and unwanted departures. Requiring firms to make severance payments upon dismissal and managers to sacrifice nonvested rights should they quit would help safeguard specific assets. . . .

Board Membership

Suppose that the appropriate alignments of incentive have been worked out. Can the firm realize additional improvements by including the management on the board of directors? Putting the issue this way presumes that the central function of the board is to safeguard the interests of the stockholders. Such a conception of the board has been described by others as the "monitoring model." . . . An alternative conception of the board . . . is . . . the "participative board." The outside board members are invited to join with the management to enhance the quality of strategic decisions. Such involvement can come at a high cost, however, if objectivity is thereby sacrificed.

Since managers enjoy huge informational advantages because of their full-time status and inside knowledge, the participating board easily becomes an instrument of the management. . . .

Rejection of the participating model in favor of a control model of the decision-ratification and monitoring kind does not, however, imply that the management should be excluded altogether. So long as the basic control relation of the board to the corporation is not upset, management's participation on the board affords three benefits. First, it permits the board to observe and evaluate the process of decision making as well as the outcomes. The board thereby gains superior knowledge of management's competence which can help it to avoid appointment errors or to correct them more quickly. Second, the board must make choices among competing investment proposals. Management's participation may elicit more and better information than a formal presentation would permit. Finally, management's participation may help safeguard the employment relation between management and the firm—an important function in view of the inadequacy of formal grievance procedures.

According to the contractual conception advanced here, however, these are supplemental purposes. To the extent that management participation permits reviews on the merits to be done more responsibly and serves to safeguard an employment relationship that would otherwise be exposed to excessive risk, management may be added to the core membership. But the principal function of the board remains that of providing governance structure protection for the stockholders. Management participation should not become so extensive as to upset this basic board purpose.

Incentive and Tax Effects of Executive Compensation Plans

CLIFFORD W. SMITH, JR. AND ROSS L. WATTS

Components of Compensation Plans

Salary

The most common component of executive compensation plans is a prespecified salary. All firms use salary as a means of executive compensation. It typically accounts for more dollars of compensation than any other form. While salary is typically the greatest proportion of compensation it is usually not the only form of compensation. In 1979 each one of the largest 100 firms in the United States had at least one type of incentive provision. . . .

Bonus Plans

While the managerial salaries vary with past performance, they are not formally tied at the beginning of a compensation period to the firm's performance in that period. But incentive plans do formally tie compensation ex ante (before the fact) to performance. The most common type of incentive plan is the bonus plan. Over 75 percent of medium and larger-size manufacturing firms have bonus plans. Typically under a bonus plan the executive is rewarded at year-end on the basis of that year's performance, measured as a function of accounting earnings. . . . The bonus plan is administered by the compensation committee (a committee of the board of directors composed of nonmanagement directors). . . . There are often maximums for the award to any individual. The most common such maximum across all salary levels is 50 percent of base salary. . . . The proportion of bonus to salary increases both with the size of the firm and the manager's rank within the firm. . . .

The bonus plan is administered so that top executives typically receive an award. Even in 1975 when the after-tax profits of the Fortune 500 firms fell

Reprinted by permission from 7 *Australian Journal of Management* 139 (1982).

13 percent, only 16 percent of those firms with bonus plans did not pay a bonus. . . .

Stock Options

Many companies compensate their top executives by providing them with options to purchase a given number of the firm's shares at any time within a given period (exercise period) at a prescribed price (the exercise price). (Technically, these options are warrants since they are issued by the firm itself.) In 1980, 83 percent of the 100 largest companies in the United States had option plans. The popularity of the plans declines with the size of the company. . . .

Stock Appreciation Rights

Stock appreciation rights are offered by companies along with options. In 1980, 68 of the largest 100 U.S. corporations had stock appreciation rights. Under a stock appreciation right executives may choose to give up their option and receive the difference between the stock price and the exercise price (the appreciation). Providing managers alternatives of options and rights enables them to reduce transactions costs associated with exercising options and selling shares should they want cash. If they want shares, options will minimize transactions costs. . . .

Performance Units

Under these plans performance goals are established in terms of accounting numbers (earnings per share, growth in earnings per share, accounting rate of return on assets, etc.) at the beginning of the award period which usually ranges from four to five years. . . . Each executive in the plan is allocated a given number of units of fixed dollar value at the start of the award period. At the end of the period the executive's compensation is the number of units earned out times the fixed value per unit. The proportion of the number of allocated units which are earned out depends on the extent to which the performance goal is achieved over the award period. . . .

The use of performance units has grown rapidly over the last ten years. In 1972 none of the largest 100 firms used these plans; by 1980, 29 out of the 100 used them.

Performance Shares

Performance shares are similar to performance units in that performance goals are established in terms of accounting numbers over award periods of four to five years. However, instead of being allocated units of fixed value at the beginning of the award period the executive is allocated a number of shares. The proportion of the allocated shares earned out over the period depends on the extent to which the goals are met. The executive's compensation is the number of shares earned out times the market value of the shares at the end of the award period. Compensation under performance share plans as under performance unit plans depends on the extent to which the goals are met. The executive's compensation is the number of shares earned out times the market value of the shares at the end of the award period. Compensation under performance share plans as under performance unit plans depends on performance measured in terms of accounting numbers. However, unlike performance units, performance shares cause compensation to be affected by the change in the stock price over the award period.

Like performance units, performance shares have become much more popular in recent years. However, performance shares are not as popular as performance units (29 of the top 100 firms in 1980 had units, 11 had shares). The Conference Board reports that together performance shares and units are less popular with smaller firms. . . .

Control of the Horizon Problem

Conditional payments. Incentive plans explicitly tie the manager's compensation to a measure of the firm's value or change in value. The manager's compensation is *conditional* on the measure. . . .

Instead of compensation depending on performance after the fact (ex post), under an incentive plan it is tied before the fact (ex ante) to some measure of performance. This formal tie to performance reduces the horizon problem for the sixty-four-year-old manager; the final year's compensation depends on that year's performance (e.g., the bonus depends on that year's reported profits).

The horizon problem is also reduced in incentive plans by deferring compensation to the retirement period. Ninety-eight of the 100 largest industrial firms in the United States have bonus plans with 63 of those 98 containing provisions which allow the compensation committee to defer payment of the bonus (either cash or stock) paying it in future installments. . . . Those plans make provisions for forfeiture of any installments not yet paid if and when the compensation committee finds that the manager committed "any act of omission or commission prejudicial or detrimental to the interests" of the firm. Such a provision will affect the incentives of a manager facing retirement.

Managers whose horizons are short because they are considering leaving the firm for other employment will also be influenced by the deferment provisions included in compensation plans. Deferred payments under bonus plans as well as performance unit and share plans are forfeited if the manager leaves the firm or is fired. Similarly, option plans [and] stock appreciation right plans . . . carry the threat of forfeiture if the executive leaves the firm before the date of the exercise of the option or right, or the date of removal of the restrictions on the stock. Deferral of compensation with the threat of forfeiture reduces the probability that the manager will cheat or steal from the firm and increases the incentive to be efficient. . . .

On the other hand, it might be thought that the forfeiture provision would enable the firm to cheat its managers out of their deferred compensation. However, such a policy would impose substantial costs on the firm. Capricious use of the forfeiture provision would cause current and future managers to discount the value of the deferred compensation for potential expropriation. This discounting would force the firm to increase its current and future managerial compensation to retain a management team of a given quality. In addition, discounting of the deferred compensation would reduce the incentive benefits of the plan. The incentive loss alone should be sufficient to ensure that the firm will not use the forfeiture provision to renege on its contracts. . . .

Deferred compensation via incentive plans has other effects in addition to the provision of incentives to be efficient and not to cheat. For example, a manager accumulates a lot of knowledge specific to the firm's industry. Part of that industry-specific human capital is provided (at some cost) by the firm. However, the manager who leaves the firm to work for another firm in the industry will be rewarded for that industry-specific capital in the form of higher compensation. In effect the firm bears the cost and the manager receives the benefit. Without some mechanism to capture the return on that investment the firm would invest less in providing its managers with industry-specific capital (e.g., via training programs). The problem can be controlled by deferred compensation subject to forfeiture if the manager leaves the firm. The deferred compensation means the manager is less likely to leave the firm and the firm is therefore more likely to provide the training. Both the firm and the manager benefit. . . .

Control of Manager's Risk Aversion

Incentive plans (and to a lesser extent renegotiated salaries) control an incentive problem which is endemic to compensation via a fixed salary, invariant to the value of the firm. If the top executive of a corporation were paid a fixed claim on the firm, the manager would have incentives to take some investment projects which actually reduced the value of the firm. If an investment project reduces the volatility of the firm's cash flows, it can increase the value of the manager's fixed claim (and other fixed claims, e.g., debt) while reducing the firm's value. The reason the value of the fixed claim increases is that the probability of the firm's cash flows covering the fixed claim increases. However, the expected cash flows to the firm decrease, so the value of the firm falls. . . .

To control the manager's risk aversion, the compensation plan must include provisions with positive incentives to increase volatility. The expected payoff to a stock option increases with the volatility of the stock price. Thus, options or stock appreciation rights provide the manager with incentives to invest in projects which increase the volatility of the firm's cash flows. As a result, it is possible by augmenting a manager's fixed salary with the right amount and type of stock options or stock appreciation rights to offset the manager's incentives to take volatility-reducing negative value projects or turn down volatility-increasing positive value projects (i.e., to be too risk averse). . . .

Attributes Which Can Be Explained by Taxes

Tax effects can also explain why incentive plans have deferral provisions. If the deferral is in the form of options, stock appreciation rights, and so forth, capital gains taxes are reduced and if the deferral involves investment by the firm to provide for the deferred compensation, taxes are reduced to the extent the corporate tax rate is less than the manager's tax rate. . . .

Many aspects of the design of compensation plans are consistent with providing managers with incentives to maximize the value of the firm; some design aspects are also consistent with the objective of tax reduction. For example, deferral of compensation can be explained as a means of inducing greater efficiency, or as a means of reducing taxes if the corporation's tax rate is lower than the manager's tax rate. The two explanations are not mutually exclusive; both are likely important. But to establish a convincing case for the existence of the incentive function, it is necessary to present observed attributes of incentive plans which can be explained only by the incentive function, and by the tax reduction function.

There are serious questions about the explanatory power of the tax reduction hypothesis. First, the very examples used to demonstrate the advantages of incentive provisions show that the same advantages can be obtained without establishing a formal plan tying compensation to performance (i.e., an incentive plan). Salary compensation plus a pension plan achieves the same tax effects as options, and so forth. Salary compensation with some of the salary deferred achieves the tax advantages of deferral with investment. Hence, why should firms bother with incentive plans? A tax explanation is that the tax code restrictions on pension contributions are binding. However, if tax reduction is the sole motivation for incentive plans one would expect, given the manager's risk aversion, for the payments to be conditional on the variable with low dispersion. For example, bonus plans would use earnings rather than what is actually used, the excess of earnings over a return on equity or assets.

Second, . . . most executives to whom deferral provisions apply probably have marginal tax rates below the corporate tax rate, so that deferral with investment would be disadvantageous in tax terms.

The conclusion is that tax effects are at best a partial explanation for the nature, existence, and growing popularity of incentive plans and are not likely to be the primary explanation. Incentive effects would appear to be more important given their ability to explain the nature of, and cross-sectional variations in, the plans.

Empirical Regularities Which Can Be Explained by Incentive Effects But Not by Tax Effects

Disaggregation of performance. If incentive plans are designed to provide individual managers with incentives to make decisions in the shareholders' interests one would expect the manager's performance and rewards under the plan to be sensitive to the decisions made. Measurements of overall firm performance such as share values, total profits, and so forth, are appropriate for the president or chairman of the board who is virtually responsible for all the firm's cash flows. But, such measurements and compensation based on them may not be well suited for a divisional manager. The divisional manager may do an excellent job, but be responsible for only 5 percent of the firm's cash flows. When an accounting is made for the impact of the managers of the other 95 percent, rewards based solely on overall firm performance may provide the manager with little incentive to maximize the value of the firm. The effect of personal decisions and efforts on compensation is swamped by the decisions and efforts of others. With an incentive plan based on such measures the manager has incentives to "free ride" on the efforts of other managers.

Thus, it would be expected that incentive plans which measure performance on a companywide basis would be less likely to apply to managers other than top management than incentive plans in which performance is measured on a divisional or more disaggregated level. This expectation is generally consistent with the evidence. Performance units and share plans use overall corporate performance and are usually restricted to top management. On the other hand, bonus plans do measure the performance of individual managers and they typically include divisional and lower-level managers. . . .

Taxes cannot explain the differences in the nature of performance measures across bonus plans. Indeed, the tax motivation would not even lead to the prediction that incentive plans isolate individual performance. But some of those plans do isolate individual performance, particularly the most popular bonus plans. . . .

Comparison of performance across the industry. Given the manager's risk aversion, it would be beneficial if the performance measure used in a compensation plan varied only with factors subject to the manager's control. Variability in the performance measure due to factors beyond the manager's control would increase the manager's exposure to risk without providing incentives to maximize the value of the firm. Thus, performance plans explicitly try to remove variability beyond the manager's control, often attempting to remove uncontrollable market and industry fluctuations in performance by comparing the firm's performance to that of other firms in the same industry. For example, Toro Co.'s plan compares its growth in returns on assets to that of its competitors, while Champion International Corp.'s plan compares its growth in earnings per share to that of its competitors.

If tax reduction were the primary motivation for incentive plans one would expect the plans to try to reduce the manager's exposure to risk. In fact, it is apparent that from this dimension, salary plus a pension plan (which did *not* invest in the firm) would be superior to incentive plans which tie compensation to the firm's performance. By diversifying the pension plan could eliminate risk specific to the firm and thereby substantially reduce the manager's risk exposure. The performance plans which measure performance relative to other firms in the industry do not do this. They remove sources of market and industry volatility, but leave the volatility specific to the firm which, while diversifiable, is more controllable by the manager. This suggests that the incentive advantage of making the manager responsible for the firm-specific variability exceeds the cost of the increased compensation for the increased exposure to risk. . . .

Conclusion

It is apparent that tax motivations cannot explain the existence of, and variations in, U.S. firms' compensation plans. A combination of salary and pension plan is as efficient as the incentive plans in reducing taxes and does not increase the manager's exposure to risk as much as incentive plans. Further, taxes cannot explain the extent to which bonus plans go to isolate individual performance or why incentive plans are less frequent in regulated industries. On the other hand, the incentive effects of compensation plans can explain these phenomena. This suggests that the stated objective of incentive plans (i.e., to align the manager's incentives with firm value maximization) is principally responsible for the popularity of incentive plans in the United States.

The Shareholder Suit: Litigation Without Foundation?
ROBERTA ROMANO

Shareholder litigation is accorded an important stopgap role in corporate law. Liability rules are thought to be called into play when the primary governance mechanisms . . . fail in their monitoring efforts but the misconduct is not of sufficient magnitude to make a control change worthwhile. By imposing personal liability on corporate officers and directors for breach of the duties of care (negligence) and loyalty (conflict of interest), litigation is thought to align managers' incentives with shareholders' interests.

The efficacy of shareholder litigation as a governance mechanism is hampered by collective action problems because the cost of bringing a lawsuit, while less than the shareholders' aggregate gain, is typically greater than a shareholder-plaintiff's pro rata benefit. To mitigate this difficulty, successful plaintiffs are awarded counsel fees, providing a financial incentive to the plaintiff's attorney to police management. There is, however, a principal-agent problem with such an arrangement: the attorney's incentives need not coincide with the shareholders' interest. For instance, settlement recoveries in shareholder litigation may provide only for payment of attorneys' fees. Critics of the shareholder suit assert that most of the suits are frivolous and that the plaintiff's bar is the true beneficiary of the litigation. . . .

There are two categories of suits enforcing directors' and officers' duties: derivative actions, brought by shareholders on the corporation's behalf, and direct actions, brought by shareholders in their own right, either individually or as a class. Procedural requirements differ, with more hurdles placed before plaintiffs in derivative suits. . . . This distinction also has important financial ramifications that bear on the incentives for parties to settle even frivolous claims. In particular, in most states, payments in settlement or judgment of a derivative suit cannot be reimbursed, and indemnification is instead limited to legal expenses. However, all states permit corporations to purchase D & O [Directors' and Officers'] liability insurance for their executives, and policies can cover losses that cannot be indemnified. Policies routinely exempt losses from adjudication of dishonesty, but if a claim is settled, courts prohibit insurers from seeking an adjudication of guilt and thereby avoiding the claim's payment. Similarly, while all policies exclude losses involving personal profit, if a suit alleging breach of both the duty of care and loyalty is settled, the insurer is required to cover the entire claim (although it is able to limit reimbursement of defense costs if it can prove how much was spent solely on defending the excluded claim).

The combination of differential indemnification rights, insurance policy exclusions, and plaintiffs' counsel as the real party-in-interest creates powerful incentives for settlement. For an individual defendant, a settlement entails no personal expenditures, while if the claim is litigated, there is some probability, however small, of being held liable with no reimbursement.[1] The plaintiff's attorney's calculus nets similar results. With a settlement, attorneys' fees will be recovered, as defendants routinely agree not to oppose petitions for fees, and, in any event, the benefit the plaintiff has conferred on the firm will be recognized in the settlement. If a claim is litigated, however, there is some probability that the plaintiff will lose. The plaintiff's attorney thus also has a powerful incentive to settle. Because D & O insurers reimburse both sides' expenses in a settlement, unlike other civil litigation, in shareholder suits neither party internalizes litigation costs. A corporation's insurance premium may well rise following a lawsuit, but this cost is borne by all of the shareholders, rather than the litigating parties. Besides insurers, courts also have a monitoring role: they must approve settlement proposals for class and derivative actions for the settlement to estop subsequent claimants. But they rarely scrutinize settlements and, consequently, attorneys' incentives are the key factor in shareholder litigation. . . .

---

[1]There is, of course, an obvious impetus for defendants to reach a settlement quickly apart from these financial incentives: court approval estops other shareholders from bringing similar claims.

Direct Benefits from Shareholder Suits

This study consists of all shareholder suits brought from the late 1960s through 1987 (the easing of the insurance crisis) against a sample of 535 public corporations, randomly selected from firms currently traded on the New York Stock Exchange (NYSE) and over-the-counter (OTC) in the National Association of Securities Dealers Automated Quotation National Market System, and from firms that have ceased trading in those markets. . . .

Ninety-nine firms (19 percent of the sample) experienced a shareholder suit. There were 139 suits.[2] The litigation trends are consistent with [earlier] studies: shareholder suits are rare, and they cluster (i.e., the distribution of litigation is skewed). Litigation frequency is one shareholder suit every forty-eight years, and 29 percent of firms with lawsuits account for approximately half of all suits. Although the likelihood of an individual director being sued is considerably higher than that of a firm if, as is common, the director serves on several boards, shareholder litigation is nonetheless an infrequent experience. . . .

Sample lawsuits can be grouped into five categories (of roughly the same size): (1) acquisitions, including challenges to friendly mergers, and proxy fights; (2) challenges to takeover defensive tactics; (3) challenges to executive compensation and other self-interested transactions; (4) misstatements or omissions in financial statements; and (5) a residual category of all other suits. While lawsuit category does not vary significantly by firm subsample, it does over time. In particular, the number of suits involving acquisitions and defensive tactics increased over fivefold after 1980. . . . Most lawsuits (eighty-three of one hundred twenty-eight resolved suits) settled. This aspect of shareholder litigation is unremarkable; most civil suits settle. In fact, with powerful incentives to settle from the structure of indemnification rights and insurance coverage, it is, perhaps, surprising that one-third of the suits did not settle. Shareholder-plaintiffs, however, have abysmal success in court. Only one suit had a judgment for the plaintiff (upholding some of the plaintiff's claims in a ruling on stipulated facts), and in one other suit a state supreme court reinstated a complaint that had been dismissed by the trial court. This is a success rate of 6 percent of adjudicated cases, but plaintiffs actually won no judgments for damages or equitable relief.

The settlements exhibit two striking features. First, only half of all settlements have a monetary recovery (forty-six of eighty-three). Second, awards are paid to attorneys far more frequently than to shareholders (seventy-five of eighty-three). In seven cases (8 percent) the *only* relief was attorneys' fees. Moreover, financial recoveries are highly skewed: the average recovery in thirty-nine settlements that can be valued is $9 million, while the median is $2

---

[2]Unless otherwise stated, the term "lawsuit" is used interchangeably for "disputed transaction." The relevant unit is the disputed transaction and not lawsuit filing, because lawsuit filings are consolidated for disposition. At least 506 separate suits were filed for the 139 disputed transactions. . . .

million.[3] These recoveries are approximately 1.3 percent of firm assets (median: 0.5 percent). One interpretation is that most fiduciary breaches involve minor harm to shareholders. But the settlement pattern is consistent with another, more troubling explanation, that a significant proportion of shareholder suits are without merit.

Settlement funds vary substantially by type of action. The average recovery in derivative suits ($6 million) is about half that in class actions ($11 million). As a percentage of firm assets it is also much less (0.5 percent compared to 1.6 percent). The proportion of derivative suits with a cash payout to shareholders (21 percent) is significantly lower than that of class actions (67 percent). These differences are of interest because the principal debate over the merits of shareholder litigation is focused on derivative, and not class, actions. The quality of a legal claim is presumably positively correlated with the compensation obtained in settlement, as it enhances the party's bargaining position. To the extent that derivative suits consistently return less to shareholders than class actions, there is a greater likelihood that more of these suits are frivolous. It is also possible that the misconduct giving rise to derivative claims is less serious than that of class actions. Proponents of the derivative suit would view these data in a favorable light. Schwartz, for example, contends that derivative suits are intended to deter rather than compensate, with the latter role reserved for class actions. . . . From this perspective, small recoveries in derivative suits indicate that liability rules are deterring egregious misconduct.

The data support the contention that there is a free-rider problem caused by litigation costs overwhelming individual shareholder benefits, as per share recovery is small. Eleven derivative suits with a monetary recovery averaged $0.18 a share ($0.15 net of attorneys' fees). This is approximately 2 percent of the stock price on the day prior to the lawsuits' filing. . . .

Computing per share recoveries in class actions is extremely difficult because information on class size, individual members' losses, and hence their allotted recovery is, for the most part, unavailable. Precise information was obtainable for only seven class actions. Most of these involved acquisitions, where each share in the class is entitled to the same pro rata award. These class actions averaged a considerably higher recovery of $3.28 per share than the derivative suits ($2.83 net of attorneys' fees). This is about 7 percent of the stock price on the day prior to filing. But the small sample makes generalization hazardous. . . . [I]t can plausibly be maintained that the relevant per share valuation of a class action is not recovery by class members, but the impact on firm value. Firm payouts affect future cash flows, so for every class winner there is a shareholder loser; calculating the average gain smooths out a recovery's distributional effect. Moreover, while the vast majority of settlements are paid by D & O insurance, all shareholders bear the cost of the deductible, as well as any future increase in insurance premiums. Under such a calculation, the mean per share value for class actions (thirty-one suits) is $1.66 a share ($1.42 net of

[3]All amounts are in 1988 dollars. The range is from $226,000 to $118 million.

attorneys' fees). This is an average recovery of 5 percent of the stock's market value, which is still considerably higher than derivative suits. . . .

Participating in a settlement fund is not, however, the only source of shareholder benefit from litigation. In twenty-five lawsuits, settlements afforded structural relief. (There was a monetary recovery as well in four of these suits.) While it is impossible to value the benefits from structural settlements with any precision, the gains seem inconsequential.[4] . . .

A likely explanation for cosmetic structural settlements is the need to paper a record to justify an award of attorneys' fees to courts. . . . Successful plaintiffs' attorneys' fees are paid regardless of whether there is a monetary recovery, if the corporation benefits from the suit; a settlement with an apparent structural change provides evidence of a benefit. In only one settlement did a court deny an award, and this settlement did not produce even cosmetic structural reform. The fee awards in structural relief cases, which follow the lodestar formula (hours expended multiplied by a reasonable hourly rate that may be adjusted for complexity, risk, quality of work, or delay in receiving payment), are much lower than the awards in monetary settlements ($287,000 compared to $1.45 million). This difference is consistent with viewing most structural relief as of inconsequential benefit to shareholders, to the extent that an attorney's expenditure of effort is positively correlated with a claim's value. . . .

Settlement Rates and Differential Costs of Legal Rules

Agency costs—the divergence of attorneys' incentives from shareholder interests—provide one explanation of why most suits settle despite the insubstantiality of relief. Economic models of the decision to litigate offer an alternative explanation that involves the effect of legal rules with differential costs on parties' incentives. Namely, when defendants' trial costs are higher than plaintiffs', their bargaining position is weaker and they will settle even nuisance suits for a positive amount.

When settlement rates are examined by lawsuit subject, we find confirmation of the differential cost prediction: there is a heavier burden on defendants in lawsuits involving self-interested transactions than defensive tactics, and the former are settled more frequently than the latter (72 percent compared to 47 percent). A similar, though less compelling, pattern emerges when settlement rates are compared by fiduciary allegation: duty of loyalty suits, in which the defendants' burden is highest, settle more frequently than suits where the defendants' burden is lower, duty of care and entrenchment lawsuits (72, 70, and 41 percent settled, respectively). The difference is at the borderline of significance across the three groups (. . . probability: .058), but the effect is driven by the entrenchment cases. . . . The cases in which the

[4]Such relief includes adding independent directors to the board (nine suits), changing terms in executive compensation plans (three suits), restricting greenmail payments, a defensive tactic to takeovers discussed in Chapter VI (three suits) and specially restricting self-interested transactions by management (five suits) [EDITOR'S NOTE].

defendants' burden is lightest (the duty of care suits) settle almost as frequently as the loyalty cases. In fact, when the care and loyalty cases are separately compared, no statistically significant difference remains. This undercuts the potency of the differential cost explanation of settlement rates.

As a final test of this hypothesis, settlement rates were compared by lawsuit type, as plaintiffs face higher procedural bars in derivative than class actions. While derivative suits do settle less frequently than class actions (66 percent compared to 79 percent), the difference is insignificant. Thus, while there is some support for a differential cost explanation of the disposition pattern, it is not sufficiently compelling to counter the agency cost explanation.

Indirect Benefits of Shareholder Suits

Lawsuit settlements are not the sole measure of litigation benefits. It is possible that when a shareholder suit is filed a corporation reorganizes on its own, leaving little for litigation to achieve, as was acknowledged in three lawsuit settlements. This would be a salutary Coasian outcome: a firm is better able than a court to determine on whom to impose liability, the entity or an individual employee. . . . Plaintiffs then need only be awarded a finders' fee (attorneys' fees) in compensation for monitoring, while the board fine-tunes incentives. In addition, if lawsuits are one among many governance structures aligning managers' incentives with shareholders' interests, their outcomes should vary with the efficacy of the corporation's other governance mechanisms, such as independent board composition . . . and ownership concentration. . . . Firms with insider-dominated boards should be more likely to be litigation targets, as should firms lacking an outside block owner whose investment mitigates the free-rider problem for a shareholder to be an active monitor. Moreover, litigation should be inversely related to inside stock ownership, for as management's investment increases, its incentives will be more aligned with the shareholders' interests, lessening the need for outside monitoring.

To probe these propositions, three key organizational factors, which are the focus in most structural settlements, are examined for changes during litigation—cash compensation and turnover of top management, and board composition—as are data on other governance mechanisms—inside and outside stock ownership. To control for changes that would occur regardless of litigation, a paired sample was constructed that matched lawsuit firms with sample firms that had not experienced a lawsuit and had the same four-digit Standard Industrial Classification (SIC) code. . . .

Does Litigation Produce Coasian Adjustments in Firm Organization?

Changes in top management or their compensation following the commencement of a shareholder suit would be key indicia of incentive realignment induced by litigation. Litigation appears, however, to have no impact on insiders' compensation packages. Cash compensation of top executives of lawsuit firms grew more rapidly than that of their matches, but the difference is not significant. . . . In addition, the CEO's compensation in firms whose suits were dismissed grew more rapidly than that in firms whose suits settled, but again, the difference is insignificant. . . . There is, then, no evidence of financial incentives for top management at the firm level that offset indemnification and insurance practices.

Managers are likely to be motivated more powerfully by the threat of job loss than temporary fluctuations in cash compensation. If firms allocate liability more efficiently than plaintiffs and courts, we would expect to find more instances of loss of employment than compensation decreases after a lawsuit. The data support such a ranking of incentive strategies. Top management of lawsuit firms turned over significantly more frequently than did their matches: 55 percent of lawsuit firms experienced a change in CEO or chair in the years surrounding the litigation while only 31 percent of the matches experienced a change. . . .

A further beneficial organizational change that a lawsuit could produce would be a change in the composition of the board. To minimize a recurrence of litigation, sued firms could enhance the board's monitoring capacity by increasing outsiders on the board. . . . [S]ued firms did increase the proportion of outsiders on their boards over the course of the litigation, and by the end of the lawsuit, they had a slightly higher percentage of outsiders than nonlawsuit firms (48.1 percent compared to 47.9 percent). But the change in outside representation on lawsuit firm boards, while in the predicted direction, is not significantly different from that of matched firms.

It may not be sensible for firms to react and alter board composition upon a lawsuit's filing, as opposed to waiting until the suit is resolved (e.g., if management is exonerated, there was no monitoring lapse and hence no need to change the board). This strategy would be obscured in a consideration of board changes using the aggregate data. If we assume that settlements represent meritorious claims, we would then expect the board's composition to change after settled rather than dismissed lawsuits: such a trend would indicate firms react to successful litigation by beefing up their internal monitoring. The data are consistent with the hypothesis—firms that settled their lawsuits did increase the proportion of outsiders over the course of the lawsuit, while the firms whose suits were dismissed did not . . . but the difference is insignificant (either compared to matches or compared within lawsuit firms across lawsuit outcome). We therefore cannot conclude that firms respond to litigation by increasing the board's independence. . . .

Board composition *does* vary significantly with litigation when changes are calculated by an individual's presence on the board rather than by director status: lawsuit firms' boards turn over more than their matches do. . . .

This difference holds up whether the change in membership is measured by retention rate—the proportion of directors on the board at the lawsuit's end who were on the board at its outset (71 percent for lawsuit firms compared to 81 percent for matches), or by departure rate—the proportion of directors on the board at the start of the litigation who had departed by its end (29 percent for lawsuit firms compared to 21 percent for the matches). Moreover, board turnover is significantly higher for lawsuit firms whose suits settled compared to their matches, but not for those whose suits were dismissed. . . . The difference in board stability between settled and dismissed lawsuit firms, as measured by retention rate, is statistically significant. . . .

The board stability data are difficult to interpret. They appear to suggest an internal monitoring effect in response to litigation: Board members are held accountable for lawsuits, particularly those with greater merit (those that settled), as they depart in greater numbers from sued firms. But turnover need not be a sign of poor monitors being replaced. It could also indicate that more able monitors are leaving for better opportunities because they disliked being sued. One means of distinguishing between these alternative explanations of departures is to determine whether the number of boards on which a director serves declines after litigation. The poor monitor explanation of board departures is more plausible if the number of other corporate boards on which a defendant-director serves decreases upon departure from the target board, because a decrease suggests that being sued has a negative reputational effect in the market for directors.

The data do not support this contention. The board memberships of defendant-directors who departed increased (from 4.0 to 4.2), while those of defendant-directors who remained declined (from 4.4 to 4.3) and the difference in mean change across the groups is not significant. . . .

Although the aggregate data do not reveal much of an independent organizational response to shareholder litigation, in three lawsuits, the defendant corporation made structural changes in reaction to the suit's filing. In these cases, institutional change led to the suits' resolution but was not part of the settlement agreements.[5] . . .

Litigation as a Substitute Monitor of Managers

The three lawsuits that induced voluntary structural change suggest that for block shareholders who are not insiders, litigation can be a valuable mechanism to redirect corporate policy. While strong conclusions cannot be drawn from these few cases, these striking similarities highlight the potential conditions for when shareholder litigation is most useful. They also suggest that litigation is a complement rather than a substitute for outside block ownership as a managerial monitoring device. Block ownership can mitigate the free-rider problem of shareholder litigation, for with a large enough block, the investor's prorated benefit will exceed a lawsuit's cost. This aspect of block ownership may offset the substitution effect that block ownership should make lawsuits less likely because block holders' monitoring deters managerial misconduct. . . . The anecdotal evidence that block owners use lawsuits as a

[5]In all three cases, the plaintiff owned a sizeable block of stock, and in two of these cases, was a former officer or relative of top management [EDITOR'S NOTE].

monitoring device rather than substitute for it is confirmed by the sample data: outside block ownership is significantly higher in lawsuit firms than their nonsued matches. . . .

Many firms, however, do not have outside owners who hold large blocks. The primary monitoring device is, instead, the board of directors. Some commentators consider the board to be most effective as a monitor when the number of independent (i.e., outside) directors is substantial. If litigation is an alternative to board monitoring, then sued firms should have fewer outsiders on their boards than their nonsued matches. Moreover, changes in board membership upon the onset of litigation would suggest that suits function as substitute governance structures. That is, outside representation on the board should increase after litigation, as lawsuits spur the firm to improve its monitoring in order to substitute a cheaper monitor (the board) for the more expensive one (litigation).

Lawsuit firms did have a lower percentage of outside directors than their matches when the suit was filed (47.3 percent compared to 48.3 percent), but the difference is not significant. . . . This datum does not provide much support for the substitute monitoring explanation. However, a related datum . . . provides a glimmer of support. Firms that settled their lawsuits had a lower proportion of outside directors at the start of the litigation than those whose suits were dismissed, and the difference is at the borderline of significance for a one-tailed test. This finding implies a beneficial monitoring effect—outside directors were more numerous in firms that were exposed to the least meritorious suits. But it must be evaluated with caution, because there is an alternative explanation for the difference in board composition that turns on third-party perception rather than board performance. Courts scrutinize more closely decisions by insider-dominated boards, and this could disproportionately encourage firms with such boards to settle. Finally, sued firms increased the number of outsiders on their board over the course of the lawsuit, and, when disaggregated by lawsuit outcome, it is the firms whose suits settled, not those whose suits were dismissed (i.e., the firms whose lawsuit disposition could signal a monitoring failure), that increased the outsiders' representation. But neither difference is significant. The data on board composition therefore leave open the substitute monitor hypothesis.

A final governance characteristic that is important if litigation substitutes as a monitor of management is inside stock ownership. The larger management's stake, the more closely its incentives are aligned with the interest of other shareholders, and, hence, there is less need of monitoring through litigation. This consideration is tempered by the fact that as ownership increases, opportunities to self-deal also increase as the insider consolidates control over decision making. And boards may not adequately police self-interested opportunities because firms with higher levels of inside ownership tend to have fewer outsiders on their boards. Accordingly, we would expect duty of loyalty claims to be correlated with inside ownership, given the greater need for monitoring self-interested transactions by litigation in insider-owned firms, yet we would expect other types of suits to be less likely when there is significant inside ownership.

The data provide limited support for the view that inside ownership substitutes for monitoring by lawsuits. Managers of lawsuit firms have a smaller stock interest than those of matches (19 percent compared to 21 percent), but the difference is not significant. . . . Inside ownership in firms with duty of loyalty claims is much higher than that in firms with duty of care and federal law violation claims (33 percent compared to 9 and 15 percent, respectively). It is also higher in sued firms than in nonsued matches only for the duty of loyalty lawsuit firms. In fact, inside ownership levels in the other lawsuit category firms are significantly lower than those of their matches. The data, therefore, comport nicely with a substitute monitor story of inside ownership that takes into account a nonlinear effect: inside ownership appears to align incentives successfully for negligence claims, but it increases opportunities for self-dealing, which shareholders must monitor through litigation. . . .

A . . . regression was estimated for the effect of firms' governance structures on the likelihood of a lawsuit. The dependent variable is the dichotomous variable of whether or not a firm has been sued. Variables measuring firm size and profitability are included in the regression, as well as the governance variables, to control for other firm characteristics that may affect the probability of litigation. . . .

The most important finding is that outside ownership is significantly positively related to litigation. This is good evidence of shareholder monitoring: firms with shareholders for whom the free-rider problem is least severe are more likely to be sued. . . . The signs of most of the other regressors [are as predicted] but none are significant. . . .

Conclusion

The data support the conclusion that shareholder litigation is a weak, if not ineffective, instrument of corporate governance.

1. Lawsuits are an infrequent occurrence for the public corporation and, while most suits settle, the settlements provide minimal compensation. . . . The principal beneficiaries of the litigation therefore appear to be attorneys, who win fee awards in 90 percent of settled suits.
2. There is little evidence of specific deterrence. . . .
3. A possible interpretation of the data is that managers are so deterred by the prospect of shareholder litigation that suits involve only trivial violations. This proposition cannot be tested against the alternative—that shareholder litigation is by and large an ineffective governance structure—because it is virtually impossible to identify a general deterrent effect as we cannot estimate the number of offenses, and thereby estimate the probability of detection, which is essential for measuring such an effect. Still, the

failure to find much in the way of specific deterrence undercuts this proposition, and suggests that general deterrence is weak. This is because without instances of penalized managers, it is unlikely that anyone else will be deterred.

4. The evidence of indirect benefits from litigation serving as a backup monitor of management is mixed. There is scant evidence that lawsuits function as an alternative governance mechanism to the board. . . . Lawsuits can, however, be useful for outside block owners. . . . Finally, inside ownership seems to serve as a successful alternative monitor for negligence but not for conflicts of interest. . . .

5. One potential social benefit from a shareholder suit that is ancillary to its role as a governance device has not been discussed: legal rules are public goods. All firms benefit from a judicial decision clarifying the scope of permissible conduct. The benefit of clarification is not simply deterrence of future managerial misconduct, but rather, given the contractual setting of the corporation, identification of a rule around which the parties (managers and shareholders) can transact. As few suits produce a legal rule (only two in this sample), this explanation of lawsuit efficacy turns on the need for a large number of lawsuits in order to obtain a ruling. There is no reason to believe that the current level of litigation is optimal in relation to any public good benefits.

Does "Unlawful" Mean "Criminal"?: Reflections on the Disappearing Tort/Crime Distinction in American Law

JOHN C. COFFEE, JR.

My thesis is simple and can be reduced to four assertions. First, the dominant development in substantive federal criminal law over the last decade has been the disappearance of any clearly definable line between civil and criminal law. Second, this blurring of the border between tort and crime predictably will result in injustice, and ultimately will weaken the efficacy of the criminal law as an instrument of social control. Third, to define the proper sphere of the criminal law, one must explain how its purposes and methods differ from those of tort law. Although it is easy to identify distinguishing characteristics of the criminal law—for example, the greater role of intent in the criminal law, the relative unimportance of actual harm to the victim, the special character of incarceration as a sanction, and the criminal law's greater reliance on

Reprinted by permission from 71 *Boston University Law Review* 193. © 1991 John C. Coffee, Jr.

public enforcement—none of these is ultimately decisive. Rather, the factor that most distinguishes the criminal law is its operation as a system of moral education and socialization. The criminal law is obeyed not simply because there is a legal threat underlying it, but because the public perceives its norms to be legitimate and deserving of compliance. Far more than tort law, the criminal law is a system for public communication of values. As a result, the criminal law often and necessarily displays a deliberate disdain for the utility of the criminalized conduct to the defendant. Thus, while tort law seeks to balance private benefits and public costs, criminal law does not (or does so only by way of special affirmative defenses), possibly because balancing would undercut the moral rhetoric of the criminal law. Characteristically, tort law prices, while criminal law prohibits. . . .

Three trends, in particular, stand out. First, the federal law of "white-collar" crime now seems to be judge-made to an unprecedented degree, with courts deciding on a case-by-case, retrospective basis whether conduct falls within often vaguely defined legislative prohibitions.[1] Second, a trend is evident toward the diminution of the mental element (or "*mens rea*") in crime, particularly in many regulatory offenses. Third, although the criminal law has long compromised its adherence to the "method" of the criminal law by also recognizing a special category of subcriminal offenses—often called "public welfare offenses"—in which strict liability could be combined with modest penalties, the last decade has witnessed the unraveling of this uneasy compromise, because the traditional public welfare offenses—now set forth in administrative regulations—have been upgraded to felony status. . . .

The upshot of these trends is that the criminal law seems much closer to being used interchangeably with civil remedies. Sometimes, identically phrased statutes are applicable to the same conduct—one authorizing civil penalties, the other authorizing criminal sanctions. More often, the criminal law is extended to reach behavior previously thought only civilly actionable. Either way, this practice of defining the criminal law to reach all civil law violations in a particular field of law in order to gain additional deterrence may distort the underlying legal standard. What needs to be more clearly recognized is the variety of ways in which such distortion can occur. For example, some civil law standards may be aspirational in character (e.g., the rule that attorneys should avoid any "appearance of impropriety"). Other standards may frame prophylactic rules, which prevent the possibility of misconduct, but involve no element of culpability. . . .

Obviously, new problems may arise for which the criminal law is the most effective instrument, but which involve behavior not historically considered

[1]The leading example of this trend is supplied by recently enacted 18 U.S.C. § 1346 (1988), which invites federal courts to consider any breach of a fiduciary duty or other confidential relationship as a violation of the mail and wire fraud statutes. . . . This new legislative enactment is, however, simply a continuation of a long-standing tradition of case-by-case judicial lawmaking under the mail and wire fraud statutes. . . . The recent evolution of insider trading law is also entirely judge-made, without any legislative statement or administrative rule defining what constitutes insider trading.

blameworthy. Modern technology, the growth of an information-based economy, and the rise of the regulatory state make it increasingly difficult to
maintain that only the common law's traditional crimes merit the criminal
sanction. In fact, historically, the criminal law has never been static or frozen
within a common law mold, but has constantly evolved. This has been especially true within the field of "white-collar" crime. Even the first modern
"white-collar" offenses to be criminally prosecuted—price-fixing, tax fraud,
securities fraud, and, later, foreign bribery—were "regulatory" crimes in the
sense that they had not been traditionally considered blameworthy. In short,
the line between *malum in se* and *malum prohibitum* has been crossed many
times and largely discredited. Today, to rule out worker safety, toxic dumping,
or environmental pollution as necessarily beyond the scope of the criminal law
requires one to defend an antiquarian definition of blameworthiness.

But where does this leave us? . . . if the criminal law is overused, it will
lose its distinctive stigma. While conceding that the criminal law is a system of
socialization, [traditional libertarians] would reply that for precisely that reason it must be used parsimoniously. Once everything wrongful is made criminal, society's ability to reserve special condemnation for some forms of misconduct is either lost or simply reduced to a matter of prosecutorial discretion.
Still, valid as this response is, it does not answer fully the criticism that the
traditional criminal law scholar's focus on blameworthiness is anachronistic
because it freezes the criminal law's necessary evolution. . . .

If so, what alternative is left? What substitute bulwark can prevent the
criminal law from sprawling over the landscape of the civil law? One answer is
to update the notion of blameworthiness, looking not only to historical no-
tions of culpability, but to well-established industry and professional standards
whose violation has been associated with culpability within that narrower
community. Another answer is to focus on the temporal relationship of the
civil and criminal law. At some point, a civil standard can become so deeply
rooted and internalized within an industry or professional community that its
violation becomes blameworthy, even if it was not originally so. . . .

Criminalizing the Civil Law

[F]ew legal categories seem inherently less "criminal" in character than the
civil law applicable to fiduciary duties or to the use of economic duress in
negotiations. Yet, both areas have, to an uncertain extent, been subjected to
the reach of the criminal law. . . .

The federal mail and wire fraud statutes supply the most obvious example
of the criminal law being overlaid on civil law standards. By the mid-1960s,
federal courts had accepted the principle that the term "scheme to defraud"
(which is the critical element in both the mail and wire fraud statutes) required neither that there be any pecuniary or property loss to the victim nor
that the purpose of the scheme be contrary to state or federal law. Rather, it
was sufficient that a victim was defrauded of an "intangible right." . . . [T]he

decisions that had the greatest impact were those that seemed to criminalize any knowing failure to disclose a conflict of interest by a person subject to a fiduciary duty. As late as 1976, the Second Circuit in a decision by Judge Henry Friendly suggested that the "intangible rights" doctrine applied only to public officials and not to private fiduciaries. However, at the beginning of the 1980s, the Second Circuit overrode his thoughtful distinction. In *United States v. Bronston,* it upheld the conviction of a lawyer who secretly represented one client while his law firm represented a rival contender for a public franchise. *Bronston* was a watershed decision, because no bribes or kickbacks were involved, and the evidence did not demonstrate that the defendant had actually used his fiduciary position to injure the firm's client. After *Bronston,* all that seemed necessary to support a mail fraud conviction was a knowing and undisclosed breach of a fiduciary standard. These decisions seemingly turned the mail and wire fraud statutes into mandatory disclosure statutes that required all public officials and private fiduciaries to disclose any conflict of interest to which they were subject.

The high-water mark of this theory of liability came in the mid-1980s, when federal prosecutors successfully used it to reach not only self-dealing conduct by corporate officials against the interests of the corporation, but also actions by corporate officials that were intended to benefit the firm, but had not been adequately disclosed to the board or shareholders. In *United States v. Siegel* and *United States v. Weiss,* corporate officers who created off-book slush funds in order to facilitate questionable corporate payments were convicted of fraud, even though they did not misappropriate any funds. Indeed, in *Weiss,* the subordinate had acted pursuant to direct instructions from his superiors to establish the secret fund. . . . In both cases, the defendants may have violated their duty of care, but they did not engage in self-dealing in any form. As a practical matter, they probably faced relatively little prospect of civil liability, because the duty of care has historically not been strictly enforced. The bottom line then is ironic: the criminal law has been cantilevered out beyond the civil law as defendants have been convicted of a federal felony on facts that would have been unlikely to support civil liability in a derivative suit.

This line of cases came to a screeching, but temporary, halt in 1987 when the Supreme Court decided, in *McNally v. United States,* that the mail fraud statute did not reach schemes to deprive victims of the intangible right of honest services, but only schemes to obtain money or property. For a time, *McNally* seemed a major obstacle to the continuing growth of a judge-made law of white-collar crime. Then, two things happened. First, the Supreme Court announced in *Carpenter v. United States* that confidential business information could amount to a form of intangible property. . . . *Carpenter* threatened to trivialize *McNally* by allowing prosecutors simply to relabel what they had indicted, before *McNally,* as a deprivation of an intangible "right" as a deprivation of intangible "property." To a limited extent, this has in fact happened.

The second post-*McNally* development was even more important: in 1988, Congress enacted a statutory definition of the critical term, "scheme to defraud." New section 1346 defines this term to include any "scheme or artifice to defraud another of the intangible right of honest services." At a stroke, this language may criminalize any violation of fiduciary duties or the law of agency. The expansion of section 1346 then supplies a paradigm of the criminal law being overlaid unthinkingly on top of the civil law, without serious consideration being given to whether the civil law standard in question should be backed by the special threat of the criminal law. . . .

The Diminution of *Mens Rea*

American criminal law scholarship has always placed the issue of *mens rea* at center stage. Its greatest achievement—the Model Penal Code—creates a presumption that *mens rea* applies to every material element in the crime, unless the statute clearly indicates otherwise. In *Morissette v. United States,* the Supreme Court seemed to give such a presumption a quasi-constitutional gloss. . . . More recently, in *Liparota v. United States,* the Court reaffirmed this presumption, at least with respect to those elements in the crime that establish moral blameworthiness. Simultaneously, however, *Liparota* acknowledged that an exception to this generalization existed for "public welfare offenses." . . .

If public safety is the deciding test, the possibility arises that many environmental statutes, which commonly require permits before various conduct (e.g., the disposal of waste, the filling-in of wetlands) may be engaged in, will fall on the strict liability side of the line. Here, the circuit courts of appeal have recently divided. . . .

Vicarious Responsibility

Generally, in American criminal law, individuals are criminally liable only for conduct that: (1) they direct or participate in; (2) they otherwise aid or abet; or (3) with respect to which they conspire. Corporate officers, however, now appear to face an additional form of vicarious liability. In *United States v. Park,* the Supreme Court upheld the imposition of criminal liability upon "corporate employees who have 'a responsible share in the furtherance of the transaction,' " even when the corporate officer took action to prevent the violation. Lower federal courts have extended this principle to apply, even when it has appeared that subordinate employees had purposely failed to follow the superior's orders or that the officer took significant corrective action that could not be implemented in time because of a labor strike. . . .

The "Technicalization" of Crime

[E]nvironmental crime is important, and knowing violations—such as falsification of records or willful endangerment—are serious offenses that do not merit leniency. But, the typical environmental offense involves the mishandling of toxic substances, and recent decisions have reduced or eliminated the role of *mens rea* in these statutes, while also applying *Park*'s doctrine that corporate officers who have a "responsible relation" to the performance of the statutory obligations are liable under them. As a result, the traditional public welfare offense has now been coupled with felony-level penalties. While the defendant in *Park* was only fined, corporate executives in an equivalent position in the future may face years in prison. . . .

In fairness, the federal government's attempt to use criminal sanctions in traditionally civil areas—such as stock parking—has met with some judicial resistance. During the last year, the Second Circuit has reversed several securities fraud convictions in marginal cases, but affirmed others where the evidence of intent was clearer. Still, these decisions lack any clear rationale and tend to depend on specific ad hoc judicial theories that seem in some cases driven by a need to justify reversal on as narrow a ground as possible. Conceivably, the phenomenon of judicial nullification is at work in some of these cases, but such a process is at best an inconsistent, sometime thing.

An Initial Summary: The Uncertain Cost/Benefit Calculus

If the disposal of toxic wastes, securities fraud, the filling-in of wetlands, the failure to conduct aircraft maintenance, and the causing of workplace injuries become crimes that can be regularly indicted on the basis of negligence or less, society as a whole may be made safer, but a substantial population of the American workforce (both at white-collar and blue-collar levels) becomes potentially entangled with the criminal law. Today, most individuals can plan their affairs so as to avoid any realistic risk of coming within a zone where criminal sanctions might apply to their conduct. Few individuals have reason to fear prosecution for murder, robbery, rape, extortion, or any of the other traditional common law crimes. Even the more contemporary, white-collar crimes—price fixing, bribery, insider trading, and so forth—can be easily avoided by those who wish to minimize their risk of criminal liability. At most, these statutes pose problems for individuals who wish to approach the line but who find that no bright line exists. In contrast, modern industrial society inevitably creates toxic wastes that must be disposed of by someone. Similarly, workplace injuries are, to a degree, inevitable. As a result, some individuals must engage in legitimate professional activities that are regulated by criminal sanctions; to this extent, they become unavoidably "entangled" with the criminal law. That is, they cannot plan their affairs so as to be free from the risk that a retrospective evaluation of their conduct, often under the uncertain standard of negligence, will find that they fell short of the legally mandated standard. Ultimately, if the new trend toward greater use of public welfare offenses continues, it will mean a more pervasive use of the criminal sanction, a use that intrudes further into the mainstream of American life and into the everyday life of its citizens than has ever been attempted before.

Several replies are predictable to this claim that there is a social loss in defining the criminal law so that individuals cannot safely avoid its application. Liberals may claim that the traditionally limited use of the criminal sanction was class-biased and that a more pervasive use of it simply corrects that imbalance. Economists may argue that the affected individuals will only demand a "risk premium" in the labor market and, having received one, cannot later complain when the risk for which they were compensated arises. Others may conclude that the anxiety imposed on such employees, while regrettable, is necessary, because it is small in comparison to the lives saved, injuries averted, and other social benefits realized from generating greater deterrence. This may be true, but the cost/benefit calculus is a complex and indeterminate one that depends upon a comparison of marginal gain (in terms of injuries averted) in comparison to other law enforcement strategies (such as greater use of corporate liability or civil penalties) that have not yet been utilized fully. Moreover, on the cost side of the ledger, one must consider not simply the consequences to those actually prosecuted, but the anxiety created within the potential class of criminal defendants. To the extent that liability is imposed for omissions (i.e., failure to detect and correct dangerous conditions), such fear will affect a broad class of employees, most of whom will never be prosecuted or even threatened with prosecution. In addition, there is a cost to civil libertarian values, because statutes that apply broadly can never be enforced evenly. Hence, some instances of "targeting" or selective prosecutions (based on whatever criteria influence the individual prosecutor) become predictable. These costs would be more tolerable if the conduct involved were inherently blameworthy, but negligence, like death and taxes, is inevitable.

Notes and Questions

1. A majority of states have enacted statutes that instruct boards to consider nonshareholder interests in their decision making, although this prescription is often limited to the board's reaction to a takeover bid. Are these statutes, known as other-constituency statutes, plainly at odds with Williamson's analysis? Would you expect an event study of the enactment of these statutes to support William Cary's or Ralph Winter's characterization of state competition? What would be the possible explanations if the event study uncovered no significant stock price effects? See Roberta Romano, "What Is the Value of Other Constituency Statutes?" 43 *University of Toronto Law Journal* (forthcoming 1993). Delaware, the state that Cary identified as leading the race for the bottom, has not adopted an other-constituency statute, and Delaware courts permit such interests to be considered only if in doing so there are rationally related benefits accruing to shareholders. Chapter VI will

reexamine the debate over the benefits of state competition in the context of state takeover laws.

2. A further argument against board representation of nonshareholder groups involves the costs of collective decision making emphasized in the Hansmann selection in Chapter I. Shareholders have homogeneous interests in maximizing the value of the firm, because they can adjust for differences in temporal consumption preferences by borrowing or lending on the value of their shares. This unanimity provides corporate managers with a clean decision rule. When diverse constituencies are added to the board's concerns, the interests will conflict and unanimity is lost, rendering a coherent decision rule impossible. The likely consequence is that management's preferences will determine outcomes. In addition, because value maximization results in the efficient allocation of resources, when corporations pursue other objectives, as they must in the multi-interest board, the market's allocative efficiency will be compromised.

3. The issue of whose interests are represented by the board is closely related to questions concerning the corporation's objective, should it be limited to equity share price maximization or should it include some broader obligation of social responsibility? While the arguments in Williamson and note 2 make the case for the former, narrower goal, the difficulties with the alternative formulation are not uniformly recognized. Consider, for instance the American Law Institute's (A.L.I.) definition of the objective of the business corporation:

> A business corporation should have as its objective the conduct of business activities with a view to enhancing corporate profit and shareholder gain, except that, whether or not corporate profit and shareholder gain are thereby enhanced, the corporation, in the conduct of its business . . . may take into account ethical considerations that are reasonably regarded as appropriate to the responsible conduct of business.

A.L.I., Tent. Draft No. 2, § 2.01 (1984). Can responsible ethical judgments be identified with precision, or does this definition simply create a hopeless muddle, giving managers tremendous discretion that is immunized from shareholder scrutiny? The A.L.I. draft itself provides ample evidence of the difficulty: in the illustrative example to the section, involving an unenforceable contract that if performed will produce a substantial loss to the firm, the drafters state that a manager's decision to honor the contract because of ethical considerations is valid, as is a decision not to honor the contract. (A.L.I. Tent. Draft No. 2, § 2.01, Illustrations 11 and 12). Such vague decision rules exacerbate the agency problem; instructing managers to maximize profits minimizes it.

There is at least one further important difference in the justification for corporate as opposed to individual altruism that should be noted, as it is tied to concerns over agency costs. David Engel notes that individual altruism can always be justified by the pleasure the individual donor obtains from engaging in a charitable act (if all other justifications fail), whereas a consumption motive is unsatisfactory in the corporate context because it would have to involve the managers' gratification, not the shareholders'. Engel, "An Approach to Corporate Social Responsibility," 32 *Stanford Law Review* 1 (1979). This, of course, is a restatement of the principal-agent problem.

Although the contention that shareholders derive consumption benefits from the corporation's charitable giving cannot be dismissed out of hand, it is unlikely that shareholders will be unanimous concerning the appropriate object of that benevolence. Consequently, management, by control of the agenda, will be able to choose its preferred charity. For a fascinating field study of an agenda setter's ability to impose his preferences on others, see Michael Levine and Charles Plott, "Agenda Influence and Its Implications," 63 *Virginia Law Review* 561 (1977).

This analysis implies that shareholders would be better served by management maximizing profits and distributing the cash, which shareholders can then give directly to their preferred charities. But there are some competing considerations. If the corporation's tax bracket is higher than that of individual shareholders, then a charitable contribution is worth more to the firm than to its owners. More important, if there are informational asymmetries between managers and shareholders with respect to charitable giving, for instance, if managers are better (or more cheaply) informed than shareholders about the relative merits of charitable causes, then corporate altruism would be more efficient. How likely do you think such information asymmetries are? Would corporate giving to an executive's alma mater or a charity on whose board she served confirm or refute the information hypothesis? Is the information hypothesis more plausible for some ownership configurations rather than others (consider, for instance, a firm in which a majority of the stock is held by large institutions, or by a few wealthy individuals)?

Do we have to reach the information scenario to justify corporate altruism or can it be cast more directly in a value-maximizing light? Could it function as a marketing device, like advertising?

4. Many commentators believe that corporate boards ought to consist of a majority of outside directors (directors who do not have a significant financial interest in the firm) in order to perform a monitoring function most effectively, for example, Melvin A. Eisenberg, *The Structure of the Corporation* (Boston: Little, Brown, 1976). In fact, the initial draft of the A.L.I.'s corporate governance project, cited in note 3, for which Eisenberg was the reporter, would have required a majority of the boards of large corporations to be outsiders. To evaluate such proposals, Barry Baysinger and Henry Butler examined which directors are better monitors by investigating whether board composition (the proportion of outsiders) affects corporate performance. Baysinger and Butler, "Corporate Governance and the Board of Directors: Performance Effects of Changes in Board Composition," 1 *Journal of Law, Economics, and Organization* 101 (1985). They found that over the period 1970 to 1980, firms with a higher percentage of outside directors at the

start of the decade had superior performance records at the end of the decade. Increasing the number of outsiders on the board over the decade did not, however, result in improved performance. There was also no contemporaneous effect on performance. Finally, the positive effect on performance of board composition was achieved at levels of outside directors much lower than a majority. Baysinger and Butler conclude from these data that proposals like the A.L.I. that mandate independent boards are misguided. Would you reach the same conclusion?

5. An important function of a monitoring board is the hiring and firing of top management. A number of studies have investigated the board's performance in this role. Randall Morck, Andrei Shleifer, and Robert Vishny examined management changes in 454 Fortune 500 corporations during 1981 to 1985. Morck, Shleifer, and Vishny, "Alternative Mechanisms for Corporate Control," 79 *American Economic Review* 842 (1989). They find that firms experiencing complete turnover in top management or a hostile takeover performed more poorly than firms with no management changes. More important, internally precipitated change in top management (i.e., board monitoring) is more likely to occur when the firm is doing poorly compared to the rest of its industry. But when the industry is doing poorly, turnover is accomplished by hostile takeovers and not board action. Morck, Shleifer, and Vishny thus provide evidence that boards and control changes are complementary governance structures. Their interpretation of the data is that boards are reluctant to institute changes unless the firm's poor performance can be attributed to management, rather than exogenous industrywide factors. Boards thus discipline management upon negative relative performance evaluation.

Morck, Shleifer, and Vishny do not, however, explain why boards would have adopted such a practice toward management discipline. Formal principal-agent models provide some theoretical support for the distinction, as they imply that relative performance evaluation is useful. These models suggest that compensation rules should depend on information, that is, performance measures, that filter out uncontrollable factors, so that risk-averse agents' exposure to market (systematic) risk is reduced without simultaneously reducing the agents' incentives. For a test of these models, see Rick Antle and Abbie Smith, "An Empirical Investigation of the Relative Performance Evaluation of Corporate Executives," 24 *Journal of Accounting Research* 1 (1986); Robert Gibbons and Kevin Murphy, "Relative Performance Evaluation for Chief Executive Officers," 43 *Industrial and Labor Relations Review* 305 (1990). The Smith and Watts selection provides some evidence of compensation based on relative performance as well.

Michael Weisbach analyzed the effect of board composition on chief executive officer turnover and also found support for the monitoring thesis. Weisbach, "Outside Directors and CEO Turnover," 20 *Journal of Financial Economics* 431 (1988). In a sample of 495 publicly traded corporations, he found that outside boards (boards with at least 60 percent independent directors) were significantly more likely to fire CEOs, given poor stock performance, than mixed boards. Outside boards were also more likely to fire managers given poor stock performance than inside boards (boards with 40 percent or less outsiders), but this difference was not significant. In addition, for outside boards the probabilities of a CEO's firing ranged from 7 percent for firms in the bottom performance decile to 1.3 percent for firms in the top decile, a difference that is significant, whereas for inside boards, the range in probabilities is insignificant. Weisbach concludes from these data that outside directors do engage in monitoring. His research thus complements Morck, Shleifer, and Vishny's study, by differentiating the kinds of boards that tend to discipline management. But note that while termination of employment increases with poor performance, the probability of discharge is still extremely small.

6. If the purpose of incentive compensation plans is to align managers' and shareholders' interests, as Smith and Watts suggest, then why do firms adopt plans whose performance-based payouts rely on accounting figures rather than stock prices? Would you expect management's reporting of accounting earnings to be influenced by their bonus contracts (i.e., would they select accounting procedures and accruals in order to maximize the value of their awards)? See Paul Healy, "The Effect of Bonus Schemes on Accounting Disclosure," 7 *Journal of Accounting and Economics* 85 (1985). Are shareholders harmed by such behavior?

7. Michael Jensen and Kevin Murphy maintain that executive compensation is a singularly ineffective incentive device because top management's income is only tenuously affected by poor stock performance. In a study of the relation between top management pay and corporate stock performance, they estimated that a CEO's compensation changes by a mere $3.25 for a $1,000 change in stock value. Jensen and Murphy, "Performance Pay and Top-Management Incentives," 98 *Journal of Political Economy* 225 (1990). Of this change, the greatest proportion represents change in the value of management's stock holdings ($2.50, assuming median holding). Salary and bonus plan compensation produced a change of only 30 cents per $1,000, and stock options 15 cents. They interpret the small magnitude of the sensitivity of pay to performance as refuting principal-agent models of optimal contracting. Note that the change in compensation, though small, is statistically significant.

Jensen and Murphy offer a political explanation of the data, which they term an "implicit regulation" thesis: public disclosure of top-management compensation limits the amount that managers can be paid. While disclosure protects shareholders from managerial looting, it also places pressure on boards to agree to lower CEO salaries, due to fear of attacks by unions and the media. Given managerial risk aversion, the enforced truncation at the top of the salary scale must be accompanied by limits on potential losses from poor performance. This results in minimal sensitivity of pay to performance. Jensen and Murphy provide support for the implicit regulation thesis by evidence that the pay–performance relation and the absolute level of CEO compensation have declined since the 1930s. This is one reason why Jensen extols the LBO organization (see the Jensen selection in this chapter): as a private corporation, it can, and does, pay top management extremely high salaries dependent upon performance.

8. Another recent study of shareholder litigation also raises concern about its effectiveness as a governance device. Janet Cooper Alexander examines securities class actions involving claims of fraud in initial public offerings of computer-related companies (nine suits filed in the first half of 1983) and finds that two-thirds of the suits settled in a narrow range of 21 to 27 percent of the stock's market loss. Alexander, "Do the Merits Matter? A Study of Settlements in Securities Class Actions," 43 *Stanford Law Review* 497 (1991). She maintains that the threat of trial does not constrain the bargaining process because litigation costs are so high that the parties would never contemplate going to trial. She thus concludes that the narrow range of recoveries indicates that settlements are unrelated to lawsuit merits. Is another possible explanation that none of the suits in this sample are meritorious? Why don't issuers of directors' and officers' liability insurance check this perverse incentive structure? If the settlement practice is anticipated, insurers will charge corporations higher premiums. What incentives are there for firms to seek lower premiums? Is it possible to precommit to not settle frivolous claims? Would shareholder agreement to eliminate directors' personal liability for damages serve such a function? See Roberta Romano, "Corporate Governance in the Aftermath of the Insurance Crisis," 39 *Emory Law Journal* 1155 (1990).

Can alternative procedures be devised to improve shareholder litigation rather than eliminate it? Consider the following approach permitted under the Michigan Business Corporation Act: judicial appointment of a disinterested party to investigate the shareholder's complaint and instruct the court on the appropriate resolution of a lawsuit, along the lines of special masters. For a first-hand account of the disinterested person procedure and a careful analysis of its comparative strengths, see Joel Seligman, "The Disinterested Person: An Alternative Approach to Shareholder Derivative Litigation," 55 *Law and Contemporary Problems* 357 (1992).

9. Although Romano found no evidence of directors' experiencing reputation effects (a decrease in the number of directorships) after being sued, reputation effects have been identified under other circumstances. Stuart Gilson finds that directors who resign from boards of bankrupt firms thereafter hold significantly fewer directorships with other companies (the mean drops from 2.2 seats to 1.4 seats over the three years after the resignation). Gilson, "Bankruptcy, Boards, Banks and Blockholders: Evidence on Changes in Corporate Ownership and Control When Firms Default," 27 *Journal of Financial Economics* 355 (1990). In addition, Steven Kaplan and David Reishus find that chief executives whose firms cut dividends are less likely to receive new directorships than CEOs who do not cut dividends, although they do not lose any existing outside board positions. Kaplan and Reishus, "Outside Directorships and Corporate Performance," 27 *Journal of Financial Economics* 389 (1990). In view of these other studies' evidence of a market for

directors, is a plausible interpretation of Romano's data that the initiation of shareholder litigation is perceived to have no connection to directors' performance, and thus it has no bearing on directors' reputations?

10. From 1984 through 1987, the market for directors' and officers' liability insurance experienced extreme dislocation—premiums skyrocketed, deductibles rose, limits decreased, substantive coverage contracted, and many commercial carriers left the market. Press reports proliferated of individuals refusing to serve on boards because insurance had lapsed or because they feared inadequate coverage. In addition, the percentage of outside directors declined for the first time in two decades. Finally, many firms resorted to self-insurance through captive insurers or organized their own insurance groups. See Roberta Romano, "What Went Wrong with Directors' and Officers' Liability Insurance?" 14 *Delaware Journal of Corporate Law* 1 (1989).

In response to these trends, state legislatures enacted statutes limiting directors' personal exposure and expanding their indemnification rights. The most popular reform was a limited liability statute, enacted by Delaware in 1986 and adopted by forty-one other states within two years. This statute permits firms to amend their charters to eliminate nonofficer directors' liability for monetary damages for negligence (i.e., the duty of care). No statute permits elimination of liability for breaches of the duty of loyalty or violations of federal laws. The vast majority of corporations in the legislating jurisdictions have opted to eliminate liability, and there has not been shareholder opposition to the changes. Do the materials in Chapter III on the enabling structure of corporation codes shed light on which fiduciary duty rules were reformed?

Several studies have examined the stock price effects of limited liability statutes to evaluate investor perception of their impact on stockholder welfare, for example, Michael Bradley and Cindy Schipani, "The Relevance of the Duty of Care Standard in Corporate Governance," 75 *Iowa Law Review* 1 (1989); Vahan Janjigian and Paul J. Bolster, "The Elimination of Director Liability and Stockholder Returns: An Empirical Study," 13 *Journal of Financial Research* 53 (1990); Roberta Romano, "Corporate Governance in the Aftermath of the Insurance Crisis," 39 *Emory Law Journal* 1155 (1990). If investors perceived duty of care litigation as an important vehicle for compensation or deterrence, then we would expect a negative stock price effect. The studies do not, however, find any significant stock price effect when firms propose charter amendments to opt out of the duty of care. Nor is there any significant effect when the Delaware statute eliminating liability for negligence was proposed. The studies find significant negative abnormal returns on the date the Delaware statute became effective (approximately two weeks after its enactment). Does this finding provide evidence that investors perceive the reform to be harmful to their interests? (Hint: Would any new information be revealed on this later date?)

11. In addition to the focus of the selection by Coffee on criminal prosecutions of corporate directors and officers, the criminal law is used to sanction

the corporation itself. The question whether criminal penalties are more effective than civil penalties when directed at the enterprise rather than individuals is an important one. Jonathan Karpoff and John Lott have sought to answer this question by investigating whether corporations experience reputational damage from criminal activity. In an event study of announcements of allegations, investigations and convictions of corporate fraud, they found that firms experienced significant negative abnormal returns on the announcement dates, ranging between -1.21 and -4.55 percent when the fraud was committed against corporate consumers and suppliers or the government or involved misrepresentation of the firm's financial reporting (i.e., frauds against investors), but not when the fraud involved violations of government regulations. Karpoff and Lott, "The Reputational Penalty Firms Bear from Committing Criminal Fraud," Wharton School, University of Pennsylvania Working Paper (rev. March 1991); discussed in Michael K. Block, "Optimal Penalties, Criminal Law and the Control of Corporate Behavior," 71 *Boston University Law Review* 395 (1991). Because the market penalized firms for engaging in criminal fraud, they conclude that the market, and not criminal sanctions, should be the principal punishment for such misconduct. How would you evaluate the sufficiency of the market penalty for deterring corporate misconduct?

Michael Block, supra, extends Karpoff and Lott's analysis to compare stock price effects of corporate fraud with the effects of certain *malum prohibitum* crimes (crimes that have negligible effect on parties in contractual relations with the corporation, such as tax evasion, money laundering, and currency reporting violations). Block finds a significant negative price effect only for the fraud cases. The reputational costs of crime are therefore fairly specific. As Block puts it, "[s]imple conviction of a criminal act does not generally stigmatize." Id. at 414. Block further examines stock price effects for certain civil fines—federal safety regulation violations by airlines. He finds a significant negative price effect of the same magnitude as that experienced by the criminal fraud firms (-2.2 percent). He concludes that there is little evidence that "criminal enforcement is uniquely suited to impose costs on firms for activities that generate external costs" and that "empirical evidence suggests that civil enforcement of laws regulating corporate behavior is just as effective in imposing reputational penalties" as the criminal law. Id. at 418.

12. The issue of enterprise rather than individual liability, and the flip side, vicarious liability of officers and directors for acts by lower-level employees, also arises in the civil context, in third-party lawsuits against the corporation. Would directors' and officers' insurance and indemnification practices cause a dual liability regime for financial penalties (whether civil or criminal) to collapse into one of enterprise liability only? What is the impact of limited liability on the choice of a liability regime? Which institution, courts or firms, would you expect to be most competent at identifying effective sanctions for misconduct by corporate agents? More generally, will the choice of the target of liability, enterprise or individual, matter in practice? Review the discussion

of the Coase Theorem in Chapter III, part B, note 1, and see Lewis Kornhauser, "An Economic Analysis of the Choice between Enterprise and Personal Liability for Accidents," 70 *California Law Review* 1345 (1982); Reinier H. Kraakman, "Corporate Liability Strategies and the Costs of Legal Controls," 93 *Yale Law Journal* 857 (1984). Should the choice of enterprise or individual liability take into account whether the agent is in the last period of her employment? See Jennifer Arlen and William Carney, "Vicarious Liability for Fraud on Securities Markets: Theory and Evidence," 1992 *University of Illinois Law Review* 691 (1992).

13. An alternative approach to a regime of enterprise and individual liability is to impose liability on third parties ("gatekeepers") who supply specialized expertise to firms, such as accountants and attorneys, and thereby force outsiders to discover and prevent corporate offenses. For example, the Securities Act of 1933 imposes liability for misstatements in registration statements on underwriters and accountants as well as on directors and officers, providing outsiders with a powerful incentive to monitor management. When would a gatekeeper approach be preferable to enterprise or individual liability? See Reinier H. Kraakman, "Corporate Liability Strategies and the Costs of Legal Controls," 93 *Yale Law Journal* 857 (1984); Reinier H. Kraakman, "Gatekeepers: The Anatomy of a Third-Party Enforcement Strategy," 2 *Journal of Law, Economics, and Organization* 53 (1986).

Shareholder Voting Rights and the Exercise of Voice

Voting in Corporate Law

FRANK H. EASTERBROOK AND DANIEL R. FISCHEL

We examine the legal rules and contractual arrangements that determine who votes, on what issues, and using what procedures. We argue that the states' legal rules generally provide investors with the sort of voting arrangements they would find desirable if contracts could be arranged and enforced at low cost. . . .

The code of corporate law is a standard form contract for issues of corporate structure. To the extent they anticipate the desires of the contracting parties, these off-the-rack principles reduce the number of items to be negotiated and the costs of negotiating them. On many occasions the legal rules will not be sufficiently detailed. The standby rule of corporate law, the fiduciary principle, requires actors to behave in the way that they would have agreed to do by contract, if detailed contracts could be reached and enforced at no cost. Yet the structural rules and the fiduciary principle together cover only the outlines of the relations among corporate actors. Something must fill in the details.

Voting serves that function. The right to vote is the right to make all decisions not otherwise provided by contract—whether the contract is express or supplied by legal rule. The right to make the decisions includes the right to delegate them. Thus voters may elect directors and give them discretionary powers over things voters otherwise could control. . . . The collective choice problems that attend voting in corporations with large numbers of contracting parties suggest that voting would rarely have any function except in extremis. When many are entitled to vote, none of the voters expects his votes to decide the contest. Consequently none of the voters has the appropriate incentive at the margin to study the firm's affairs and vote intelligently. . . .

Voters are not fungible. Those who have more shares, such as investment companies, pension trusts, and some insiders, do not face the collective action problem to the same extent. Nonetheless, no shareholder, no matter how large his stake, has the right incentives at the margin unless that stake is 100 percent.

Reprinted by permission from 26 *Journal of Law and Economics* 395. © 1983 by The University of Chicago.

These collective action problems may be overcome by aggregating the shares (and the attached votes) through acquisitions, such as mergers and tender offers. We expect voting to serve its principal role in permitting those who have aggregated equity claims to exercise control. . . .

Voting exists in corporations because someone must have the residual power to act (or delegate) when contracts are not complete. But, on the discussion so far, voting rights could be held by shareholders, bondholders, managers, or other employees in any combination. Given the collective choice problem, one might expect voting rights to be held by a small group with good access to information—the managers themselves. Yet voting rights are universally held by shareholders, to the exclusion of bondholders, managers, and other employees. When a firm's founders take the firm public, they always find it advantageous to sell claims that include votes, and thus ultimately the right to remove the insiders. Why do the insiders sell such claims? Why do investors pay extra for them? (They must pay something, or the insiders would not expose themselves to the risk of removal.)

The reason, we believe, is that shareholders are the residual claimants to the firm's income. Bondholders have fixed claims, and employees generally negotiate compensation schedules in advance of performance. The gains and losses from abnormally good or bad performance are the lot of the shareholders, whose claims stand last in line.

As the residual claimants, the shareholders are the group with the appropriate incentives (collective choice problems to one side) to make discretionary decisions. The firm should invest in new products, plants, and so forth, until the gains and costs are identical at the margin. Yet all of the actors, except the shareholders, lack the appropriate incentives. Those with fixed claims on the income stream may receive only a tiny benefit (in increased security) from the undertaking of a new project. The shareholders receive most of the marginal gains and incur most of the marginal costs. They therefore have the right incentives to exercise discretion. And although the collective choice problem prevents dispersed shareholders from making the decisions day by day, managers' knowledge that they are being monitored by those who have the right incentives, and the further knowledge that the claims could be aggregated and votes exercised at any time, tends to cause managers to act in shareholders' interest in order to advance their own careers and to avoid being ousted. . . .

The right to vote (that is, the right to exercise discretion) follows the residual claim. Owners of common stock have the voting right most of the time. But when the firm undertakes projects that alter its risk, exposing creditors to losses, they too have approval rights. Too, when the firm is in trouble and, for example, omits dividends to preferred stockholders, those stockholders commonly acquire the right to cast controlling votes. When the firm is insolvent, the bondholders and other creditors eventually acquire control, through provisions in bond indentures and other credit agreements or through operation of bankruptcy laws. . . .

The fact that voting rights flow to whichever group holds the residual claim at any given time strongly supports our analysis of the function of voting rights. It also suggests why, ordinarily, only one group holds voting rights at a given time. The inclusion of multiple groups (say employees in addition to shareholders) would be a source of agency costs. People who did not receive the marginal gains would be influencing corporate discretion, and the influence would not be expected to maximize the wealth of the participants as a group. . . .

There is another reason why only one class of participants in the venture commonly holds dispositive voting rights at one time. The voters, and the directors they elect, must determine both the objectives of the firm and the general methods of achieving them. It is well known, however, that when voters hold dissimilar preferences it is not possible to aggregate their preferences into a consistent system of choices. If a firm makes inconsistent choices, it is likely to self-destruct. Consistency is possible, however, when voters commonly hold the same ranking of choices (or when the rankings are at least single peaked).[1]

The preferences of one class of participants are likely to be similar if not identical. This is true of shareholders especially, for people buy and sell in the market so that the shareholders of a given firm at a given time are a reasonably homogeneous group with respect to their desires for the firm. . . .

An Analysis of State Rules Concerning Elections

The Presumption of One Share–One Vote

The most basic statutory rule of voting is the same in every state. It is this: All common shares vote, all votes have the same weight, and no other participant in the venture votes, unless there is some express agreement to the contrary. . . .

The presumptively equal voting right attached to shares is, however, a logical consequence of the function of voting we have discussed above. Voting flows with the residual interest in the firm, and unless each element of the residual interest carries an equal voting right, there will be a needless agency cost of management. Those with disproportionate voting power will not receive shares of the residual gains or losses from new endeavors and arrangements commensurate with their control; as a result they will not make optimal decisions. . . .

It explains, too, why cumulative voting has all but vanished among publicly traded firms and why most state statutes contain a presumption against cumulative voting. Cumulative voting gives disproportionate weight to certain "minority" shares, and the lack of proportion once more creates an agency cost of management. It makes realignments of control blocs very difficult by

---

[1]This is a technical condition on preferences that requires all voters to locate the proposals as lying along the same left–right continuum [EDITOR'S NOTE].

distributing a form of holdup power widely; although every share has the same holdup potential, the aggregate holdup value exceeds the value of the firm and thus makes negotiation very difficult.

Cumulative voting (or any other method of requiring a supermajority consent to certain corporate actions) has the further property of impeding changes of control and thus supporting the position of managers vis-à-vis residual claimants. . . .

The Prohibition of Vote Buying

It is not possible to separate the voting right from the equity interest. Someone who wants to buy a vote must buy the stock too. The restriction on irrevocable proxies, which are possible only when coupled with a pledge of the stock, also ensures that votes go with the equity interest.

These rules are, at first glance, curious limits on the ability of investors to make their own arrangements. Yet they are understandable on much the same basis as the equal-weighting rule. Attaching the vote firmly to the residual equity interest ensures that an unnecessary agency cost will not come into being. Separation of shares from votes introduces a disproportion between expenditure and reward. . . .

The Common Law Rules for the Conduct of Elections

Unlike federal law, . . . state law usually imposes no restrictions on the conduct of elections apart from requiring the incumbents to furnish lists of shareholders to prospective challengers at the challengers' expense. Managers may campaign against shareholders' proposals, and for their own reelection, at corporate expense; the firm may reimburse insurgents' expenses if they win, and incumbents may reimburse insurgents even if they lose (although this is rare).

All of these rules (or, rather, the lack of rules) are consistent with the analysis we have proposed. Because proxy fights may be waged by parties who lack significant ownership of shares, a successful contest could put in office insurgents inferior to the incumbents in managerial skill (but superior in ability to siphon profits). These insurgents gain more in perquisites and side payments than they lose in diminution of the value of their stock. All residual claimants benefit if such insurgents are defeated. Incumbents' use of corporate funds to campaign for reelection, or the election of their nominees, spreads the costs of the election across all of the residual claimants. Like the other principles we have discussed, this reduces the agency costs that would arise if particular directors incurred expenses disproportionate to their shareholdings. The gains and losses of the directors' decisions accrue to all residual claimants; if the costs are not similarly spread, the directors will not equate costs and benefits to the firm at the margin. The same consideration explains why insurgents may reimburse themselves if they prevail and why incumbents may reimburse insurgents if they choose.

It may seem odd, however, that challengers are not reimbursed by the firm as a matter of course. There are substantial free-riding problems in mounting a campaign. The collective choice problem that inhibits voters from learning about the firm in order to cast intelligent ballots applies in spades to waging a fight. The full costs are borne by the challengers in every case, yet they obtain reimbursement only if they prevail, and they obtain the gains (if any) from changes in management only in proportion to their equity interests. The divergence between cost and benefit makes proxy contests rare and drives challengers to the more costly alternative of the tender offer. Because the firm appears to gain whether or not the insurgents prevail, it could be argued that the firm should pick up the expenses of those who seek election to at least the same extent as it picks up the incumbents' expenses.

There is nonetheless a substantial problem with allowing challenges at the firm's expense. The firm's offer to pay for the contest may become an attractive nuisance. There are always publicity seekers willing to stand for office on someone else's money. An offer to pay for the contest is worthwhile only if, in its absence, significant numbers of otherwise-beneficial contests will be stifled, and even then only if there is a good way to distinguish plausible challengers from frivolous ones.

We may put the difficulty of weeding out frivolous candidates to one side. The implausibility that there will be a serious challenge in any given election is a sufficient explanation for the lack of corporate financing in all elections. Challenges are rare not only because of the free-riding problem but also because of the operation of capital markets. . . .

Federal Regulation of the Proxy Machinery

One of our themes is that firms have incentives to locate in states that enable them to adopt voting procedures that maximize the welfare of investors. The practice of firms in allowing shareholders to vote on certain types of issues, and to disclose certain types of information when votes are taken, is good evidence of what constitutes the optimal allocation of resources on voting procedures. . . . The [federal] proxy rules displace private arrangements with respect to both the issues on which shareholders are entitled to vote and the amount of disclosure that must be made when a vote is held. This type of regulation by fiat is not entitled to the same presumption of efficiency as long-standing voluntary arrangements codified by common law rules. The opposite is true. Because federal regulation of voting is not the product of a competition between states to provide rules that maximize the welfare of investors, but rather displaces those rules, the presumption is that federal regulation is welfare decreasing. At the very least, there is no presumption of betterment. . . .

There is an *optimal* amount of monitoring, which firms would facilitate in their own interest. As we pointed out, voting is used only for large events (mergers and the like), when the gains exceed the substantial costs of information and aggregation of blocs. The existence of these gains is no warrant for inferring, as the SEC has done, that if some voting is good, more disclosure and more voting must be better still. Because it is so easy to sell one's shares, and because managers must set attractive terms for new securities (including terms for voting) if they are to maximize their returns, there is no good reason for believing that the voting rules designed by the firms themselves will be inferior to those the SEC can think up.

Ties That Bond: Dual Class Common Stock and the Problem of Shareholder Choice

JEFFREY N. GORDON

Over the past five years, and at an accelerating pace, more than eighty public firms have adopted, or proposed to adopt, capital structures with two classes of common stock. One class, intended principally for public shareholders, carries limited voting rights; the second class, intended principally for management and its associates, carries enhanced, or "super," voting rights. Although proposals for "dual class common stock" vary in their details, their effect would be significantly to unbundle corporate governance from economic participation. Overall, the move toward dual class common stock portends the most important shift in the underlying structure of corporate governance since the rise of institutional stock ownership in the 1960s and 1970s.

Firms capitalized with dual, or even multiple, classes of common stock have been a well-known feature of the corporate landscape. Closely held corporations and public firms with significant dynastic family voices have frequently used the dual class common device. However, the dual class common has typically been part of these firms' capital structure since their initial public offerings (IPOs). It is no secret that the current popularity of dual class common among public firms is a response to the recent wave of hostile takeovers. . . . For if management and its allies hold the voting stock necessary to elect directors, a hostile bid becomes practically impossible. One crucial difference for firms now seeking to adopt the dual class common structure is that the required corporate action is a recapitalization, rather than an initial public offering. . . . [E]xisting public shareholders must be induced to part with their voting stock in order for the scheme to work.

Different stock exchange policies on dual class common stock have complicated matters. The New York Stock Exchange (NYSE), which historically has forbidden dual class common, has the most restrictive policy. The National

Reprinted by permission from 76 *California Lew Review* 2. Copyright © 1988 by Jeffrey N. Gordon.

Association of Securities Dealers (NASD), overseer of the over-the-counter market (OTC), places no limitations on the use of multiple classes of common stock. The American Stock Exchange (Amex) has permitted firms to issue multiple classes of common stock, but lists only those classes that have the right to elect at least 25 percent of the board of directors.

The weakening competitive position of the NYSE in the provision of stock transaction services has put pressure on the exchange to abandon its single class common rule. Formerly, the NYSE could insist on its rule because of the perceived benefits of an NYSE listing. The liquidity and market-making functions provided by the NYSE arguably lowered the firm's cost of capital. A listing also carried prestige that probably entailed pecuniary benefits for the firm and gratification for its principals. In recent years, however, advances in communications technology and the regulatory efforts to create a "national market system" have dramatically enhanced the competitive position of the OTC market. . . . Thus NYSE firms that desire to establish dual class common capital structures are able credibly to threaten a shift from the NYSE to the Amex or the OTC.

This threat has triggered an extraordinary series of actions. Rather than lose listings, listing fees, and commission revenue for its broker-dealer membership, the NYSE Board of Governors proposed to dilute substantially its single class common rule. As overseer of the self-regulatory organizations, the SEC was required to approve the NYSE rule change. Because of great interest in the matter, the SEC held public hearings in December 1986. For several months thereafter the SEC attempted to broker an agreement on a uniform voting rights rule among the NYSE, the Amex, and the NASD. However, these negotiations broke down, largely because of the Amex's insistence on a one share–one vote standard.

In June 1987, after these negotiations failed, the SEC proposed a rule drawn somewhat more narrowly than the voluntary rule nearly agreed upon. Proposed SEC Rule 19c-4 would prohibit the exchanges and the NASD from listing the stock of a firm "that issues any class of a security or takes other corporate action that would have the effect of nullifying, restricting, or disparately reducing the per share voting rights of holders" of stock registered under the 1934 Act. The proposed rule would permit firms on all exchanges, including the NYSE, to issue limited voting common stock, but would prohibit dual class recapitalizations that diminished the power of present shareholders.[1] . . .

This article contends that shareholder approval of a dual class common recapitalization—even by a majority of public shareholders—does not necessarily support a belief that these actions increase shareholder wealth. Indeed, such approval can be elicited even if the recapitalization almost certainly reduces shareholder wealth. This is true because of collective action and strategic choice problems associated with shareholder voting. . . . The proxy statements issued by the firms make relatively candid disclosures that the proposals will tend to entrench the management bloc and, in particular, will

---

[1]The aftermath of rule 19c-4 is discussed in the notes following the readings [EDITOR'S NOTE].

make a hostile takeover bid at a premium price very difficult. Despite this candor, these plans apparently have been adopted whenever proposed. . . .

Collective Action Problems

[In this omitted section, Gordon describes the collective action problems of shareholder voting discussed in chapter III and in this chapter's selection by Easterbrook and Fischel. Gordon notes that the problem is typically exacerbated in the dual class stock context because the corporations making such proposals tend to have lower-than-average institutional ownership (hence fewer informed voters) and higher-than-average stock ownership by insiders (who vote in support of the plan in order to consolidate their control)—ED.]

Strategic Choice Problems

Management control of the structure and timing of a dual class recapitalization proposal permits strategic behavior vis-à-vis public shareholders. First, management can bundle the recapitalization with a "sweetener," an unrelated proposal that shareholders may independently desire. In addition, management can play "chicken" by credibly threatening to pursue less than optimum strategies for the firm if the recapitalization proposal is defeated. . . .

Sweeteners

Management can "sweeten" a proposal that decreases shareholder wealth by bundling it with an unrelated proposal that increases wealth. For example, many firms announce plans to increase cash payouts to shareholders if the recapitalization is adopted but not otherwise. These plans include substantially higher dividends and even open-market repurchases of stock. Exchange offer recapitalizations offer the possibility of a dividend preference upon exchange for limited voting shares. Even if the recapitalization reduces shareholder wealth, these "sweeteners" produce offsetting gains for public shareholders. . . . [A]n objection might be raised: Approval of a sweetened recapitalization proposal means only that management and public shareholders have engaged in a mutually beneficial trade. . . . However, . . . it ignores the impact of a significant insider bloc in a context in which only a simple majority vote is required and in which calculation is difficult.

Ordinarily when public shareholders evaluate a management proposal, there will be a distribution of predictions as to its effect. Shareholders may disagree about the magnitude of the effect, and even about whether it will be positive or negative; disagreement widens if calculation is difficult. If the median point of the distribution is negative—that is, if holders of a majority of shares believe that the proposal decreases shareholder welfare—then the proposal will be defeated even if a substantial number of shareholders "get it wrong." Insider control of a significant block of stock radically changes this scenario. In order to prevail in a simple majority vote regime, the insiders need to obtain the votes of only a minority of public shareholders. Thus even if the median belief of public shareholders is negative, the proposal is likely to pass. The addition of a sweetener to the recapitalization proposal makes a calculation of its effect on shareholder wealth more difficult. This increases the likelihood that a sufficiently large minority will believe the package is wealth increasing, even if the median shareholder belief is otherwise. In this way, a sweetener operates less as a basis for a trade and more as a means for distorting shareholder choice.

Strategic Games: "Chicken" and Its Variants

Management asserts in most cases that the dual class recapitalization proposal stems from a desire to issue equity to pursue profitable projects without diluting management's control. If the projects are pursued, public shareholders benefit, but so does management, because it has large holdings. Conversely, if the projects are not pursued, both managers and public shareholders will lose. This set of outcomes makes recapitalization a variant of the game of "chicken." In the stylized game, two parties face each other on a collision course. If one party yields, the other party is better off, but if neither party yields, both are worse off.

In the recapitalization context, management can employ a combination of incentives, credible threats, and bluffs to increase its chances of winning the game. It may be that the value of the firm increases because of profitable projects pursued upon the issuance of limited voting common. Nevertheless, public shareholders may be worse off in comparison to a scenario in which the recapitalization had not been permitted. Thus even without strong collective action problems, approval of a recapitalization can be driven by strategic considerations that distort shareholder choice rather than by a collective judgment that approval is optimal for public shareholders. . . .

Bonded Non-Renegotiation Rights

Opportunistic managers can exploit [the defects of shareholder voting] to obtain approval of plans that may fail to maximize, or even reduce, shareholder wealth. These problems, however, are foreseeable to a significant extent at the time a firm issues stock. At that time shareholders and managers may make mutually beneficial agreements concerning the possibilities of management opportunism. Shareholders may demand a premium, in the form of a discount on the stock price, for bearing the risk of certain forms of opportunism. To reduce this premium, management may accept certain constraints on its subsequent behavior. The supervisory authority of a board of directors elected by shareholders can be understood as one sort of constraint. An undertaking to maintain a capital structure with a single class of common stock can be understood as another.

The problem for the firm is this: Given the flaws of shareholder voting, how can the firm provide convincing assurances that specific constraints, such as single class common, will have continuing effect? In this context, the NYSE one share–one vote rule may be understood as a way of bonding the firm's promise to maintain the single class capital structure without renegotiation. . . . A stock exchange rule that prohibits dual class common stock permits, but does not require, a firm to select a legal regime that bars recapitalization. The availability of multiple levels of legal regimes allows the firm to decide, first, if it wants dual class common, immediately or as an option, and second, whether it is prepared to bond its choice of single class common against a subsequent renegotiation. This serves the interests of flexibility, in that different firms can organize in different ways, and certainty, in that a particular firm can opt for a legal regime with an absolute prohibition and thus a secure bond.

The virtues of the NYSE rule are contingent, of course, on the inability of firms to migrate to exchanges with less strict rules. Otherwise the NYSE rule will have no more effect than the law of a particular state. Until recently, marketplace forces made such migration highly unlikely. As noted above, an NYSE listing once provided unique liquidity and reputational benefits. Sacrificing these benefits entailed very significant costs for the firm, both in terms of subsequent efforts to raise capital and in the loss of the pecuniary and nonpecuniary rewards of being an NYSE firm. These costs, borne by management both as shareholders and as beneficiaries of the firm's prestige, were sufficiently great to bond the firm's choice of the NYSE corporate governance requirements, and in particular, the rule against dual class common. Indeed, the costs to shareholders of delisting were thought so great as to provide a basis for judicial intervention against action that put the NYSE listing at risk.

The success of NASDAQ's National Market System (NMS) has obviously changed this situation. The costs of losing an NYSE listing have diminished to the point that they are no longer sufficient to bond the choice of single class common. This allows managers to behave opportunistically with regard to public shareholders. The importance of the SEC's backstop for the NYSE rule therefore becomes clear. Market forces have eroded the ability of any exchange to bond a nonrenegotiation right against dual class common. Yet such a bond previously existed and presumably had value. An SEC rule that barred the listing by the Amex or the NASD of a firm delisted by the NYSE because of a dual class recapitalization would simply provide a different mechanism for a bond that previously existed. Put otherwise, a federal rule aimed at migration between exchanges is one effective way that a firm could be given both the opportunity to choose its capital structure and the ability to bond its promise to maintain single class common stock.

Proxy Contests and the Efficiency of Shareholder Oversight

JOHN POUND

Economists and legal scholars alike have traditionally viewed proxy contests as the least efficient approach to disciplining management and, when necessary, transferring corporate control. . . . Data confirm that proxy contests are used less frequently than tender offers to monitor management and transfer control. In the period 1981–84, there were over 250 tender offers for publicly held U.S. corporations, but only about 100 proxy contests, of which fewer than 60 were for control.

Proxy solicitors and other observers have suggested three primary reasons why proxy contests are the less-preferred mechanism for imposing change on corporations. One is that there is inefficiency in the laws governing how managements and dissidents solicit votes from shareholders. These laws are argued to create a chaotic voting process that gives incumbent management a differential vote-getting advantage. A second hypothesis is that some institutional shareholders face conflicts of interest in proxy voting that they do not face in making tendering decisions. As a result, they are alleged to vote with management even when doing so is contrary to their fiduciary interests. A third hypothesis is that because proxy contests involve little resource commitment by dissidents compared with direct acquisition offers, they are regarded with suspicion by other outside shareholders. Practitioners indeed argue that many proxy contests are simply "crank" control bids that should not be taken seriously.

This article explores whether these three hypotheses are valid explanations of the apparent relative disuse of the proxy contest mechanism. . . . The results confirm that each factor—inefficiencies in proxy solicitation, fiduciary voting behavior, and the problem of "crank" proxy initiatives—does indeed deter dissidents from launching and winning proxy contests. . . .

Management's Differential Advantage in Proxy Solicitation

In proxy solicitations, management and dissidents seek from voting shareholders the authority to act as designated voting representatives at shareholder meetings. In purely mechanical terms, the current system imposes cost and uncertainty because the dissident is required to make direct contact with shareholders. A "ballot," in the form of a signed proxy card, must be deliv-

Reprinted by permission from 20 *Journal of Financial Economics* 237 (Amsterdam: Elsevier Science Pub., 1988).

ered to and returned by the shareholder. In tender offers, by contrast, contact must be made in one direction only, from shareholder to bidder. Perversely, the costs imposed by the necessity of direct contact from solicitor to shareholders increase with the liquidity of the market, because higher share turnover makes it more difficult to isolate current voting shareholders. Thus, under current voting rules, the liquid market in shares so desirable from the viewpoint of efficient pricing and informed shareholder decision making is a detriment to efficient proxy voting.

The solicitation problem is compounded by state laws that make it difficult to determine who possesses the right to vote at any given time. Two laws create particular difficulties. The first is the convention of recording the ownership of large proportions of outstanding stock in nominee or "street" name. Stock held in street name is registered to a brokerage house, bank, depository corporation, or other intermediary, which administers the transfer of shares among its clients. The street name convention affords ease of transfer for active traders and anonymity if desired. However, it also means that votes must be solicited through the nominee; and there is no effective way to guarantee the diligence or even the honesty of the nominee holder in contacting actual holders of shares. Further, many states, including Delaware, limit the voting protections accorded to owners listing their stock under nominee name, by presuming in the absence of a formal contract to the contrary that the voting right lies with the nominee rather than the true owner.

A second major problem affecting vote ownership occurs because of the timing of the so-called record date defined in state law. On this date the right to vote is affixed to the known recorded owners of stock. The problem caused by the record date involves vote transfer. Record dates are usually set to fall weeks before the vote occurs and are open to the discretion of management. Generally, after record dates, votes may still be transferred when shares are sold, but only by the affirmative (and costly) process of granting an irrevocable proxy to the new share owner. If this is not done, the voting power remains with the shareholder of record when shares are sold and transferred.

The result is that in a company with high share turnover, often a large proportion of shares is effectively separated from votes during the solicitation period. Holders of record who own votes but no longer hold residual claims face no economic incentive to vote or even to sign over votes after they have sold their shares. Thus the pool of functional voting shares is often reduced substantially by voting time. The reduction in the voting pool creates an additional source of uncertainty for any initiative—either by management or by dissidents—requiring a majority of shares to pass, as it means that a majority of all outstanding shares must be gained from a pool of available votes that is smaller than the full 100 percent of votes nominally outstanding.

In proxy challenges, all of these problems are argued to work to the advantage of incumbent management because of incumbents' long-term experience in maintaining shareholder lists, soliciting votes for annual meetings, and developing relationships with and hence the loyalty of relatively uninformed shareholders. Dissidents indeed often have to sue simply to obtain shareholder lists. They must then start from scratch in unraveling the mystery of who owns voting rights in the corporation, how these parties can be contacted, and who is the best proxy solicitor to be charged with the job. Management's advantage is particularly strong in cases in which shareholders wish to countersolicit against a managerial initiative. In these cases management sets both the agenda and the time that dissidents have available for countersolicitations. The minimum period between meeting announcements and vote dates, set by state law, is often as short as ten days. A management making a proposal that is known to be unpopular will minimize the time available for dissident organization and activity. Dissidents may respond by seeking injunctive delay of the vote through the courts, but this is an expensive and sometimes unsuccessful procedure.

The hypothesis that management enjoys an advantage in proxy solicitation can be tested using several measurable characteristics of proxy challenges. First, if it is indeed more difficult for dissidents to identify shareholders and solicit votes, then dissidents should be less likely to win proxy challenges the more diffuse share ownership is. The larger the number of shareholders that must be contacted to gain control, the higher will be management's solicitation advantage. Second, if dissidents face a more difficult solicitation task, then their chances should be increased the longer the time available for vote solicitation. Time may work to offset management's superior knowledge about shareholder composition, allowing the dissident's proxy solicitor to track down and identify voters. Third, dissidents are likely to be at a disadvantage if the proxy vote is scheduled to occur at a special meeting rather than a regular annual meeting. Special meetings are virtually always called by management, because current regulations make it very cumbersome for an outside shareholder to secure consent for a special meeting. Management can schedule these meetings on short notice, thus exacerbating the disadvantage that dissidents face in soliciting votes. Thus, when voting is set to occur at a special meeting, this should further decrease the dissident's chances for success.

If these relationships are found to be significant, it suggests that proxy contests are deterred to some extent by purely mechanical problems caused by the regulations governing vote ownership, transfer, and solicitation.

The Composition of the Shareholder Pool

The most widely acknowledged flaw in the incentive structure for voting is the problem of collective choice in making informed voting decisions. . . .

It has been argued that given this problem, the most important determinant of efficiency in the proxy process is whether informed shareholders with large holdings have economic incentives to make the voting process efficient at the margin, even in the presence of some uninformed shareholders. Particular emphasis has recently been placed on the informed voting behavior of institutions and large blockholders. These two groups, with large holdings, fiduciary responsibilities, and relatively great expertise in managing investments, have been argued to have the potential to make proxy voting relatively more efficient.

There are alternative hypotheses about the effects of both groups, however. One view is that many large investors simply vote "street rules," meaning that they abstain from voting and sell their holdings rather than vote actively against management. A second possibility is that some large blockholders maintain strategic alliances with incumbent management. This can be observed in acquisition contests, where third-party corporations sometimes act with target management to defeat a bid, and it may also happen in the proxy arena.

A third possibility, which has recently received considerable attention, suggests that institutional investors may vote with management because of conflict-of-interest problems. This behavior is alleged to result from the current regulatory structure, which places no burden on fiduciary managers to vote or to disclose voting behavior to beneficial owners, but concurrently allows incumbent management to know how individual shareholders have voted. In the presence of these rules, fiduciaries are alleged to be frequently swayed in their voting behavior by existing business relationships with incumbent management. For example, an insurance company may hold a significant portion of a corporation's stock and concurrently act as its primary insurer. Voting against management may significantly affect the firm's business relationship with this incumbent management and perhaps others as well, whereas voting with management results in no obvious penalty.

These hypotheses are difficult to test with the available data because of observational equivalence problems. However, a series of tests is possible, which, taken in sum, militate more or less in favor of a particular hypothesis. The conflict-of-interest and strategic alignment hypothesis both suggest that the higher is institutional and block ownership in any given firm, the lower should be the probability of a dissident proxy initiative. These hypotheses also may imply that institutional and block ownership should be lower in contests for full control, because the dissident will be less likely to undertake these high-stake, high-cost initiatives where there is a diminished chance of victory. Finally, the conflict-of-interest and strategic alignment hypotheses suggest that the higher is institutional and block ownership in any proxy contest target, the lower should be the probability of the dissident's winning the proxy initiative.

The "efficient monitoring" hypothesis about institutional ownership has less clear implications for the former two of these relationships and might indeed not suggest very different results from those suggested by the conflict-of-interest theory. If institutional ownership were found to be unusually low among proxy contest targets, this might imply that a lack of efficient monitoring had precipitated the need for a proxy challenge. Similarly, low institutional ownership in targets of full control initiatives might imply that the lack of efficient monitoring of the target firm had precipitated the need for a control challenge.

The third relationship, between institutional ownership and proxy contest outcomes, appears to offer more hope for discriminating between the theories. It would be hard to justify a negative relationship between institutional or block ownership and dissidents' victory chances on the basis of the efficient monitoring theory. The hypothesis would have to be that less credible proxy challenges occur when institutional and block ownership is higher. Efficient monitoring instead appears to suggest that among proxy contests, higher institutional and block ownership should cause a higher chance of dissident success. Proxy contests might be undertaken because of relatively low precontest institutional and block ownership, and hence a lack of efficient precontest monitoring. Within contests, however, if institutions and blockholders increase efficiency, then the more of them there are to vote efficiently, the likelier it should be that dissidents win.

Dissident Commitment and Outside Shareholder Support

Proxy contests can be seen as residing along a continuum of strategies for corporate control. At one end of the continuum is a direct acquisition offer, in which the dissident shareholder simply purchases a controlling interest from other outside shareholders. In a proxy contest, in contrast, the dissident owns less—typically far less—than a controlling interest. (In this article's sample, dissidents averaged only 12 percent of outstanding voting shares.) The dissident borrows the voting rights of other shareholders to depose management or impose particular policy or personnel changes on the firm. After the vote, the borrowed voting rights revert to the original owners, who also maintain their full ex ante residual claims on the firm.

There are two conditions in which a proxy contest may be more efficient than a tender offer as a means for challenging management. One is when the dissident anticipates that a direct acquisition offer will be made difficult or impossible by defensive initiatives undertaken by the target firm. A second is when the dissident's goal is not to gain full control of the company, but rather to overturn a particular managerial initiative or gain minority (monitoring) representation on the board. In the latter case, the proxy contest mechanism allows the dissident to seek substantively different strategic goals than are accomplished by a tender offer, and there is no clear reason why the dissident should seek to acquire controlling interest in the corporation.

Critics argue that these substantive reasons for undertaking a proxy contest are complemented by an adverse selection problem. Dissidents may also choose proxy fights, rather than tender offers, because they have little confidence in their ex post ability to increase share values. A tender offer bonds the dissident to creating and sharing gains and places a lower bound on their magnitude. In contrast, in a proxy contest, which costs the dissident only solicitation costs, the insurgent shareholder may be seeking personal satisfaction only or may be no more than a wealthy crank. For any given proxy contest, moreover, informational asymmetries may mean that only the dissident knows whether the contest is value-increasing. Dissidents and incumbents usually present information to shareholders about the value of their alternative business plans, but this information is difficult for outside shareholders to judge and is also liable to distortion.

There is a natural solution to the adverse selection problem, which is likely to be used by dissidents if the problem is indeed significant. Signaling devices should be adopted by dissidents who are not cranks. These signals should differentiate the dissident as being committed to enacting value-increasing policy changes and thus convince shareholders that the dissident's bid should be taken seriously. Two primary means for signaling commitment suggest themselves. First, the dissident can buy more shares before the proxy vote. Willingness to acquire a larger stake should signal to other outside shareholders that the dissident is relatively more likely to actually increase share values after assuming control. At the limit, of course, acquiring sufficient shares yields control without a proxy fight. A second vehicle for signaling commitment is a promise by the dissident to purchase some fraction of shares contingent on winning control. Such a promise should prove legally binding. The commitment to a postvote share purchase is likely to be substituted for precontest share purchases under two conditions. First, a dissident may prefer an outright acquisition offer, but have been deterred by defensive strategies by incumbent management. Second, the dissident may perceive risks inherent in the proxy contest process and wish to make purchase contingent on actually winning the proxy initiative.

The existence of an adverse selection problem and the need to signal commitment suggest a substantive reason why proxy contests might be relatively less used than tender offers. Available signaling strategies move any given proxy initiative closer to a tender offer in terms of the dissident shareholder's resource commitment and thus diminish whatever special advantages the proxy contest format held over a direct tender offer. Concurrently, because of the solicitation and voting problems outlined in the subsections above, the proxy challenge remains more costly, in terms of administrative mechanisms and the uncertainty associated with seeking and securing proxy ballots. Thus, if the dissident's need to signal passes a certain cost threshold, it will become more profitable to substitute a tender offer for the proxy challenge.

The adverse selection-signaling problem with proxy challenges suggests direct empirical tests. If adverse selection means that outside shareholders are suspicious of proxy challenges and hence that serious dissidents need to differentiate themselves through signaling policies, then dissidents' signaling policies should systematically influence proxy contest outcomes. Higher dissident holdings should increase dissidents' chances, not only because of the additional voting control they confer, but also because the dissident is willing to absorb more of the residual risk associated with his own business plan. Contingent share purchase offers should also increase the probability of dissident victory.

In addition, it is possible that proxy contests may be more successful when they are substantively different than tender offers—in other words, when they are not for full corporate control. In noncontrol contests, the proxy mechanism is used to seek different ends than would be achieved by a tender offer, and shareholders typically face less risk in supporting the dissident initiative. This is particularly true in partial control contests, where the dissident seeks only minority monitoring representation on the board. . . .

To examine proxy contest voting behavior, a data base is used that covers 100 dissident challenges from 1981 to 1985. The 100 proxy contests covered represent 71 percent of the universe of 141 contests on record with the SEC for the sampling period. . . .

Dependent Variable

A first task with the data base involves defining the dependent variable—the proxy contest winner. This is straightforward for those contests that go to a final vote at the shareholder meeting, of which there are 76 in the sample. Twenty-four contests, however, were settled in advance of the meeting, and for these criteria had to be developed to classify the outcomes as management or dissident victories. A strict criterion is used, judging the contests to be management victories unless the settlement gave the dissident the outcome originally sought through the voting challenge. Thus, for example, a dissident seeking a majority of board seats who was awarded a minority of seats in a settlement is judged to have lost the initiative despite having gained minority representation on the board. The strict criterion is used because the purpose of the tests is to judge the direct strategic success of the proxy challenges, rather than make inferences about the potential long-term impact of partial victories.

Even by this strict criterion most settled contests are judged dissident victories. This is not very surprising, given that it is really only the prospect of a management defeat that is likely to engender incentives for a prevote settlement. Before the vote management holds the strategic advantage of being the status quo. The prospect of dissident defeat will not move management to prevote bargaining. Nor is the dissident likely to compromise much in such bargaining, having committed the costs of solicitation to the contest and having observed significant strategic weakness in incumbent management. . . .

Managerial Advantage in Proxy Solicitation

The regression results support the hypothesis that dissidents face a significant disadvantage in soliciting votes in proxy initiatives. The strongest support is provided by the variable measuring the influence of the number of target firm shareholders on dissidents' victory prospects. Throughout the regressions the results show that the higher the number of shareholders the lower the dissident's chance of success. . . .

Although statistically weaker, the coefficient on the variable measuring the time available for the dissident challenge also suggests a managerial advantage in proxy solicitation. The results show that the longer the time available for solicitation, the greater the dissident's chances of victory. The managerial advantage in vote solicitation thus appears to diminish as the dissident's proxy solicitor has additional time available to track down the location of voting shares. The significance of the result varies across subsamples, but the sign and magnitude of the coefficient remain roughly similar to those for the full sample. . . . When the vote is set to occur at a special meeting, rather than a regularly scheduled annual meeting, dissident chances decrease. The result is not statistically significant at conventional levels, but this is not surprising in view of the fact that only six firms in the sample had special meeting votes. . . .

Large Shareholders and Proxy Challenges

[There are] indications that the level of institutional ownership in target firms has systematic effects on proxy contest activity. The data show that proxy contest targets have about 30 percent lower average institutional ownership than the market as a whole, and that in contests that dissidents win, institutional ownership is in turn lower than it is across all target firms. [The] higher levels of institutional ownership are associated with a lower probability of dissident victory in proxy initiatives. This is a consistent result throughout [the] regressions, occurring even when the dissident has made a formal offer to purchase shares conditional on winning the proxy initiative. Finally, the data show that high-stake, full-control proxy initiatives are launched against firms with relatively low institutional holdings, compared with other proxy contest target and market averages.

These results are most easily interpreted as confirming the strategic alignment-conflict-of-interest theories about the role of institutional ownership in proxy contests. . . . In contrast, the efficient voting hypothesis is salvaged only by a convoluted interpretation of the results. . . . The only possible explanation would be that, across proxy contests, those made against firms with higher levels of institutional ownership are more likely to constitute "crank" contests, whereas those made against firms with low institutional ownership are likelier to succeed because they are justified. This is a labored explanation of these results, particularly given the increasing anecdotal evidence in favor of the conflict-of-interest hypothesis.

The results for block ownership are also more consistent with the strategic alignment–conflict-of-interest hypothesis than with efficient voting, although here the results appear somewhat more sensitive to the test specification. Once again, block ownership is lower in successful proxy contests than in unsuccessful contests and lower also in full control bids than in other types of proxy challenges. However, the regression tests relating block ownership to dissident victory are sensitive to the specification of the block ownership variable. . . .

Dissident Commitment and Proxy Contest Outcomes

[The data] consistently confirm that higher dissident commitment translates into an increased chance of winning a proxy challenge. Dissident chances are significantly increased by both higher dissident holdings and the existence of a formal offer to purchase shares contingent on a dissident victory. These results are stable across the different specifications . . . . Only where the sample is restricted to contests involving a contingent share purchase offer are the results different. In these regressions the dissident's relative holdings do not systematically affect contest outcomes. This may indicate that contingent share purchase commitments dominate relative dissident holdings as a positive signal of dissident commitment.

Across the regressions, relative management holdings do not have a signaling effect equivalent to that displayed by dissident purchases. Shareholders apparently do not take increased management holdings to mean that management is more committed to outside shareholders. In part, this is because management holdings are small across all types of proxy challenges and vary little by both type of contest and contest outcome.

Dissident holdings, by contrast, vary significantly both by type of contest and by contest outcome. Dissident holdings [are] higher in full control contests than in other proxy initiatives, which is consistent with the notion that outside shareholders demand higher dissident commitment in these contests because of the increased risk associated with a full control transfer. Moreover, dissident holdings [are] significantly higher in contests won by dissidents than in those won by management—roughly twice as high, on average. Average dissident holdings in successful contests, however, remain below 14 percent of outstanding shares. . . .

Finally, the evidence lends some support to the hypothesis that proxy contests are most successful when they are clearly different, in substantive terms, from tender offers. Full control proxy contests are won by dissidents less frequently than are partial control contests, shareholder proposals, or shareholder opposition to management proposals. The latter result is particularly surprising, because of the strategic disadvantage faced by dissidents in countersoliciting against a management initiative. Overall, the most successful proxy challenges are in partial control contests. This is consistent with intuition, as it shows that outside shareholders tend to award dissidents victory when it involves the lowest ex post risk. . . .

A Political Theory of American Corporate Finance

MARK J. ROE

Why is the public corporation—with its fragmented shareholders buying and selling on the stock exchange—the dominant form of enterprise in the United States? Since Berle and Means, the conventional corporate law story begins with technology dictating large enterprises with capital needs so great that even a few wealthy individuals cannot provide enough. These enterprises consequently must draw capital from many dispersed shareholders. Shareholders diversify their own holdings, further fragmenting ownership. . . .

In the classic story, the large public firm survived because it best balanced the problems of managerial control, risk sharing, and capital needs. In a Darwinian evolution, the large public firm mitigated the managerial agency problems with a board of directors of outsiders, with a managerial headquarters of strategic planners overseeing the operating divisions, and with managerial incentive compensation. Hostile takeovers, proxy contests, and the threat of each further disciplined managers. Fragmented ownership survived because public firms adapted. They solved enough of the governance problems created by the large unwieldy structures needed to meet the huge capital needs of modern technology. In the conventional story, the large public firm evolved as the efficient response to the economics of organization.

I argue here that the public corporation is as much a political adaptation as an economic or technological necessity. The size and technology story fails to completely explain the corporate patterns we observe. There are organizational alternatives to the fragmented ownership of the large public corporation; the most prominent alternative is concentrated institutional ownership, a result prevalent in other countries. But American law and politics deliberately diminished the power of financial institutions to hold the large equity blocks that would foster serious oversight of managers, making the modern American corporation adapt to the political terrain. The modern corporation's origin lies in technology, economics, *and* politics.

Shareholder control of managers arises when the owner holds a large block of stock. Individuals rarely have enough money to buy big blocks. Institutional investors do. But law creates barriers to the institutions' taking big blocks. Banks, the institution with the most money, cannot own stock. Mutual funds generally cannot own control blocks of stock. Insurance companies can put only a fragment of their investment portfolio into the stock of any one company. Pension funds own stock, but they also face restrictions. More importantly, corporate managers control private pension funds, not the other way around.

And we have just exhausted the major financial institutions in America; none can readily and without legal restraint control an industrial company. That is the first step of my argument: law prohibits or raises the cost of institutional influence in industrial companies.

The second step is to examine the politics of corporate financial structure. Many legal restraints had public-spirited backers; some rules would be those that wise regulators, unburdened by politics, would reach. But many important rules do not fit into this public-spirited mold. . . . Main Street America did not want a powerful Wall Street. Laws discouraging and prohibiting control resulted. . . .

Few of the largest public firms are controlled by a holder of a substantial block of shares. Few individuals have the wealth to take that large a position. Although banks, insurance companies, mutual funds, and pension funds have enough money, they do not take large positions. . . .

Certainly concentrated control by financial institutions is imaginable: Japanese and German corporate ownership is quite concentrated; their financial institutions are more actively involved in their companies than are financial institutions in the United States. . . .

I do not here argue that institutional control would have been better for the United States. Serious disadvantages would arise as well.[1] . . .

Four types of financial institutions dominate: commercial banks, mutual funds, insurance companies, and pension funds. They have assets, respectively, of $3.2 trillion, $548 billion, $1.8 trillion, and $1.9 trillion.

Clearly these financial institutions have enough assets to influence large corporations. But portfolio rules, antinetworking rules, and other fragmenting rules disable them from systematically taking control blocks. Demonstrating this requires regulatory detail, but the rules can be summarized quickly for those who do not want the detail: Banks and bank holding companies were repeatedly prohibited from owning control blocks of stock or from affiliation with investment banks that did. Insurance companies were for quite some time prohibited from owning any stock, and portfolio rules still restrict their ability to take control. Mutual funds cannot deploy more than a fraction of their portfolio in a concentrated position; buying more than 5 percent of a company triggers onerous rules. Pension funds are less restricted, but they are fragmented; rules make it difficult for them to operate jointly to assert control. Private pension funds are under management control; they are not constructed for a palace revolution in which they would assert control over their managerial bosses. . . .

Despite these legal roadblocks, I suspect some financial institutions determined to control or influence an industrial company could acquire a large

[1]Roe includes here conflicts of interest, banking instability, politically intolerable accumulations of power, and "capture" by management [EDITOR'S NOTE].

block. . . . And financial institutions could form joint ventures to manage the ownership.

However, four problems would afflict such joint ventures. First, not all organizational forms succeed. Second, legal restrictions on joint activity and control make many actions futile or costly. Third, institutions fear that their control would be publicized and criticized, and induce a reaction of increased regulation. Fourth, some financial institutions expect to make profits by trading; the securities laws inhibit trading by large blockholders who manage. . . .

[In addition to the effect on corporate structure of these direct prohibitions on financial institutions' equity ownership,] some influence was indirect: (1) fragmented investors talking to one another must act through the SEC's proxy machinery; (2) schedules have to be filed with the SEC; (3) groups that own 10 percent or more of an industrial company's stock risk imposition of 16(b) liability, forcing disgorgement of any short-swing profits; and (4) an institution wishing to obtain influence and control will be subject to enhanced duties and liabilities.[2] . . .

Populist antibank sentiment . . . is the backdrop to the interest group pressures for fragmentation. . . . Small financial institutions were the most important interest group militating for fragmenting finance. Small banks wanted to fragment large money-center institutions. Small businesses also wanted fragmentation, believing that small banks served them better than large ones. These groups have had great weight in Congress.

Modern eyes search for managers and labor behind the passage of fragmenting legislation, as they can be found behind modern antitakeover legislation. While we see them here and there, their role in fragmenting finance at the time of legislative passage is pale. . . . Perhaps historians one day will find plausible loose 1930s' support for fragmentation from managers, local controlling stockholders, who wanted equity capital without losing control to Wall Street, and labor. But whether or not such a coalition produced fragmentation, once fragmenting legislation passed, even if it passed without managers as necessary supporters, the *subsequent* stability of fragmentation probably has been due to managers as an interest group. Decades after passage, managers would oppose changes in fragmenting laws and would throw their weight in the way of change. When large financial institutions clash with managers, managers call upon politicians for aid, as they did when proxy contests heated up in the 1950s. Managers appealed to politicians to raise the costs of proxy contests, the Senate held hearings, and the SEC responded by promulgating rules that pulled informal joint discussions among institutions into the proxy ambit. . . .

This interest group effect need not reflect *conflict* between *operating* managers and *financial* managers. Financial managers themselves might have been happy to accept limitations. Monitoring is hard work; if all financial managers were precluded from the task, their job would be easier. Indeed the interest group effect can reflect conflict *among* financial institutions. Glass-Steagall's

---

[2]Roe includes here equitable subordination in bankruptcy, RICO liability or veil piercing for debts of the controlled entity [EDITOR'S NOTE].

separation of investment from commercial banking has been explained as resulting from the power of: (a) investment bankers, who wanted to thwart commercial bankers; (b) commercial bankers, who wanted to thwart investment bankers; (c) small-town bankers, who wanted to thwart money-center bankers; and (d) upstart, transaction-oriented money-center investment bankers, who wanted to thwart established money-center investment bankers. Indeed, once the legislation preventing financial institutions from taking influential blocks of stock is in place, the *regulated* could *themselves* resist easing of the regulation. Deregulation would allow competition on bases for which the incumbents lack an advantage.

Agents Watching Agents: The Promise of Institutional Investor Voice

BERNARD S. BLACK

The case for institutional oversight, broadly speaking, is that product, capital, labor, and corporate control market constraints on managerial discretion are imperfect, corporate managers need to be watched by someone, and the institutions are the only watchers available. The concerns about institutional oversight arise for two main reasons. First, controlling shareholders may divert funds to themselves at the expense of noncontrolling shareholders. Second, the institutions are themselves managed by money managers who need (and often don't get) watching and appropriate incentives. Mutual fund investors, for example, have little information about the corporate governance actions of mutual fund managers, little reason to care, and no power to do anything except sell their shares and invest in another fund. Public pension fund managers are watched as much by state politicians and the press as by fund beneficiaries.

Pure theory can't tell us whether we'd be better off if imperfectly watched money managers did more watching of corporate managers. Institutional detail matters. A complicating factor is that there are many different types of institutions: corporate pension plans; public pension plans; mutual funds; commercial banks; insurers; investment banks; foundations and endowments. Each has its own incentives, conflicts of interest, culture, history, and regulatory scheme. Some can take the lead in corporate governance initiatives. Conflicts of interest make others likely to be only followers.

Regulation is pervasive. It governs what the institutions, as currently constituted, can do. But it also determines, in substantial part, what the institu-

Originally published in 39 *UCLA Law Review* 811, Copyright 1992, The Regents of the University of California. All Rights Reserved. Reprinted by permission.

tions are. Banks, mutual funds, insurers, and pension funds are in significant part *defined* by a web of regulation. Each could be defined differently, if we so choose. Regulatory detail matters.

The benefits and costs of institutional oversight defy easy summary. Taking them as a whole, I believe that there is a strong case for measured reform that will facilitate joint shareholder action *not directed at control,* and reduce obstacles to particular institutions owning stakes *not large enough to confer working control.* Such reform will let six or ten institutions collectively have a significant say in corporate affairs, while limiting the power of any one institution to act on its own. I will call this limited role *institutional voice.* Institutional voice means a world in which particular institutions can easily own 5 to 10 percent stakes in particular companies, but can't easily own much more than 10 percent; in which institutions can readily talk to each other and select a minority of a company's board of directors, but can't easily exercise day-to-day control or select a majority of the board.

Institutional voice should be distinguished from *institutional control,* from a world where Citibank or Prudential could control General Motors in the way that Deutsche Bank controls Daimler-Benz. The far-reaching reform needed for concentrated control has large potential benefits, because with control can come strong oversight. But strong oversight is inevitably accompanied by strong potential for abuse of control. Moreover, the extensive legal reform needed for institutional control involves redefining our institutions, at least in part. It's hard to predict how the newly recreated institutions would behave. Thus, the relative costs and benefits of institutional control are unclear. . . .

Institutional voice means asking one set of agents (money managers) to watch another set of agents (corporate managers). Money managers have limited incentives to monitor because they keep only a fraction of the portfolio gains. But money managers also won't take the legal chances that an individual shareholder might, because they face personal risk if they breach their fiduciary duty or break other legal rules. The institution, however, realizes most of the gains from such misdeeds. That limits the downside risk from institutional voice.

Institutional voice requires a number of institutions, including different *types* of institutions, to join forces to exercise influence. That further limits the downside risk from institutional power, because money managers can monitor each others' actions to some extent. Reputation is a central element in this second form of watching. Diversified institutions interact over and over, at many different companies, over a span of years. Institutions that earn good reputations will elicit cooperation from other institutions; institutions that cheat will invite retaliation.

Corporate managers can watch their watchers. Corporate managers indirectly control the largest category of institutional investor, the corporate pension fund. Also, if other institutions abuse their power, corporate managers can complain—loudly and often—to state and federal lawmakers. If the costs to other shareholders, including smaller institutions, of abuse of power by the largest institutions exceed the other shareholders' gains from better monitoring, those shareholders will support corporate managers' efforts to clip the large institutions' wings. Political outcomes are hard to predict, but financial institutions have lost political battles before. Money managers know that, which limits their incentive to misbehave in the first place.

Much of the promise of shareholder monitoring lies in *informal* shareholder efforts to monitor corporate managers or to express a desire for change in a company's management or policies. That enhances corporate managers' ability to police money manager behavior. Corporate managers can cooperate only with those money managers who earn a reputation for promoting long-term company value.

. . . Diversification creates the opportunity for economies of scale in monitoring. Scale economies, in turn, affect the issues that the institutions will care about. Institutions are likely to devote more attention to process and structure issues, which promise scale economies, than to company-specific concerns. Process and structure issues include the value of confidential voting, the desirability of poison pills and other antitakeover devices, the composition and structure of the board of directors, the process by which directors are nominated, whether a company should have a nonexecutive chairman, and the form of management compensation.

Diversification also means that money managers interact repeatedly at different companies. That makes it easier for money managers to watch each other and makes reputation an important constraint on money manager behavior. Diversification also limits the risk that money managers will divert funds to themselves. Insider trading, for example, won't materially affect portfolio performance if done only occasionally. And a money manager who frequently trades on inside information runs a high risk of being caught.

Partly because institutional incentives push against direct, company-specific monitoring, a central reform goal should be to facilitate indirect monitoring through the board of directors. If the institutions can more easily select directors, at least for a minority of board seats, they can hire directors to watch companies on their behalf. Currently, directors are often more loyal to corporate officers than to the shareholders whom the directors nominally serve. Shareholder-nominated directors will owe more loyalty to shareholders, and may be more willing to ask tough questions when a company's performance lags, or when its CEO has dreams of grandeur.

Reform should focus on the *process* of voting, rather than substantive governance rules. For example, corporate boards may perform better if they include some directors who are nominated by large institutions. But we shouldn't legislate that result. Instead, we should empower the institutions to make their own decisions about optimal governance structures. They have incentives to make good choices—or at least better choices than lawmakers would make. Moreover, procedural reform won't force oversight or impose large regulatory costs on companies or shareholders. Oversight will take place only where the institutions conclude that the benefits of monitoring outweigh the costs.

Critics worry that the institutions will botch the job of monitoring corporate managers. There will undoubtedly be mistakes and false starts. Institutional shareholders won't develop monitoring skills overnight. But we should let them learn from their mistakes, like other actors in our market economy. In practice, legal change will occur gradually. That will make the inevitable mistakes less costly, and allow time to adjust the reforms if the initial efforts have unforeseen consequences. Given the level of congressional inertia, reform should, where possible, rely on federal agency action rather than legislative command.

Active Investors, LBOs, and the Privatization of Bankruptcy

MICHAEL C. JENSEN

The LBO Association: A New Organizational Form

It is instructive to think about LBO associations such as KKR[1] and Forstmann-Little as new organizational forms—in effect, a new model of general management. These organizations are similar in many respects to diversified conglomerates or to the Japanese groups of firms known as "keiretsu." It is noteworthy that the corporate sectors in Japan and Germany are significantly different from the American corporate model of diffuse ownership monitored by public directors. In both these economies, banks and associations of firms are more important than in the United States. Indeed, one way to see the current conflict between the Business Roundtable and Wall Street is that Wall Street is now a direct competitor to the corporate headquarters office of the typical conglomerate. Moreover, the evidence on the relative success of the *active investor* versus the *public director* organizational form seems to indicate that many CEOs of large diversified corporations have no future in their jobs; one way or the other many of those jobs are being eliminated in favor of operating-level jobs by competition in the organizational dimension. . . .

LBO associations such as KKR are one alternative to conglomerate organizations and, judging from their past performance, they apparently generate large increases in efficiency. . . . LBO associations . . . are run by partnerships instead of the headquarters office in the typical large, multibusiness diversified corporation. These partnerships perform the monitoring and peak coordination function with a staff numbering in the tens of people, and replace the typical corporate headquarters staff of thousands. The leaders of these partnerships have large equity ownership in the outcomes and direct

---

[1]KKR stands for Kohlberg Kravis Roberts, one of the major leveraged buyout firms [EDITOR'S NOTE].

Reprinted by permission from 27 *The Continental Bank Journal of Applied Corporate Finance* 35 (Spring 1989). © 1989 Stern Stewart Management Services, Inc.

fiduciary relationships as general partners to the limited partner investors in their buyout funds.

The LBO partnerships play a role that is similar in many ways to that of the main banks in the Japanese groups of companies. The banks (and LBO partnerships) hold substantial amounts of equity and debt in their client firms and are deeply involved in the monitoring and strategic direction of these firms. Moreover, the business unit heads in the typical LBO association, unlike those in Westinghouse or GE, also have substantial equity ownership that gives them a pay-to-performance sensitivity which, on average, is twenty times higher than the average corporate CEO. . . .

The proper comparison, however, of the pay–performance sensitivity of the compensation package of the conglomerate CEO is not with the CEOs of the LBOs but rather with the Managing Partner or Partners of the partnership headquarters (e.g., the KKRs of this world). Little is publicly known about the compensation plans of these partnerships, but the pay-to-performance sensitivity (including ownership interests, of course) appears to be very large, *even* relative to that of the managers of the LBOs. The effective ownership interest in the gains realized by the buyout pool generally runs about 20 percent or more for the general partners as a group. LBO business unit heads also have far less of a bureaucracy to deal with, and far more decision rights, in the running of their businesses. In effect, the LBO association substitutes incentives provided by compensation and ownership plans for the direct monitoring and often centralized decision making in the typical corporate bureaucracy. The compensation and ownership plans make the rewards to managers highly sensitive to the performance of their business unit, something that rarely occurs in major corporations.

In addition, the contractual relation between the partnership headquarters and the suppliers of capital to the buyout funds is very different from that between the corporate headquarters and stockholders in the diversified firm. The buyout funds are organized as limited partnerships, in which the managers of the partnership headquarters are the general partners. Unlike the diversified firm, the contract with the limited partners denies partnership headquarters the right to transfer cash or other resources from one LBO business unit to another. Generally all cash payouts from each LBO business unit must be paid out directly to the limited partners of the buyout funds. This reduces the waste of free cash flow that is so prevalent in diversified corporations.

Notes and Questions

1. What is the relation between Easterbrook and Fischel's explanation of why shareholders vote and Williamson's explanation of who the board represents?

2. Martin Lipton and Steven Rosenblum have proposed to eliminate shareholders' annual voting of directors, fixing board membership at five-year terms, and prohibiting all control changes between elections (except those agreed to by management). Lipton and Rosenblum, "A New System of Corporate Governance: The Quinquennial Election of Directors," 58 *University of Chicago Law Review* 187 (1991). What would be the probable impact on share value if this proposal were adopted? Why?

3. When would creating a dual class stock capital structure, which consolidates management's control, be value maximizing for the firm? Consider the finding of empirical studies that, in firms engaging in dual class stock recapitalizations, insiders already have control (averaging over 40 percent) prior to the transaction and that the stock price effect on the transaction's announcement is insignificant, or significantly negative but extremely small; for example, Gregg Jarrell and Annette Poulsen, "The Effects of Recapitalization with Dual Classes of Common Stock on the Wealth of Shareholders, 20 *Journal of Financial Economics* 129 (1988); Megan Partch, "The Creation of a Class of Limited Voting Common Stock and Shareholder Wealth," 18 *Journal of Financial Economics* 313 (1987). Do such capital structures protect the value of firm-specific investments by insiders? Note that because there is a control group, the public shareholders cannot expect a hostile takeover premium. Does this datum affect the plausibility of Gordon's thesis that management can engage in strategic behavior that coerces shareholders to approve dual class stock? Does it weaken his contention that shareholder acceptance of low voting stock in exchange for higher dividends is not a bona fide bargain? Does it explain the failure to find significant negative effects upon dual class stock recapitalizations?

4. Is an insider's threat to not adopt value-maximizing projects if a dual class stock plan is rejected credible? What could an insider do to make the threat credible? Would such behavior constitute a breach of fiduciary duty? Review Coffee's thesis, in Chapter III, concerning the important role of courts in filling in gaps in the corporate contract. Do you think a court would uphold a dissenter's challenge to a shareholder vote approving a dual class stock recapitalization in the context of an explicit threat? See *Lacos Land Co. v. Arden Group,* 517 A.2d 271 (Del. Ch. 1986).

5. What techniques are available to managers besides stock exchange listing requirements to bond the firm's capital structure? You might want to review the discussion in Chapter III of latecomer terms when considering this question.

6. Rule 19c-4 was adopted by the Securities and Exchange Commission in 1988, and as Gordon describes, it imposed a uniform listing rule for the major exchanges and the over-the-counter market (NASD) of one share–one vote with certain exceptions for shares issued pursuant to a registered public offering. The SEC rule was adopted in response to what was called a "race to the bottom" among exchanges, when the New York Stock Exchange decided

in 1986 to abandon its one share–one vote listing requirement and go the way of its competitors, under pressure from listed companies desiring to change their voting structure. This rule was, however, struck down in *Business Roundtable v. SEC,* 905 F.2d 406 (D.C. Cir. 1990) for exceeding the SEC's authority. The exchanges and NASD thereupon voluntarily adopted Rule 19c-4. But the accord was not long lasting. In February 1991, the American Stock Exchange adopted a rule permitting the listing of low or nonvoting shares created by dual class stock recapitalizations as well as those issued in public offerings, and in June 1992, the New York Stock Exchange sought members' comments on a proposed rule permitting the listing of super voting shares when approved by a majority of outside directors and shareholders and freely transferrable (under current practice, many dual class shares lose their super voting rights when transferred).

Consider again the arguments in Chapter III concerning state competition for corporate charters. Is a race to the bottom any more plausible in the context of securities exchanges? How would an exchange benefit from adopting rules that decreased the wealth of shareholders? For an insightful discussion of stock exchanges and their incentives, see Daniel R. Fischel, "Organized Exchanges and the Regulation of Dual Class Common Stock," 54 *University of Chicago Law Review* 119 (1987).

The SEC's approach in Rule 19c-4 drew upon Ronald Gilson's innovative characterization of dual class stock as a substitute for an LBO for concentrating control in management's hands. Gilson, "Evaluating Dual Class Common Stock: The Relevance of Substitutes," 73 *Virginia Law Review* 807 (1987). Gilson identifies when these disparate transactions will be efficient, noting that LBOs are appropriate for mature companies with steady cash flows whereas dual class stock recapitalizations are apt for companies with capital needs. This highlights the difference between the investment choices of LBO and venture capital firms, which are discussed, respectively, in the Jensen selection in this chapter and in the Sahlman selection in Chapter IV. Gilson notes, however, that it is also possible to distinguish the transactions in a less benign light, that LBOs are undertaken when management does not have enough stock to force the public shareholders to go along with a voting rights recapitalization. He maintains that because the two hypotheses are difficult to distinguish empirically, low voting shares should be permitted only if adopted in a new stock offering, rather than by a recapitalization of the rights of existing shares. Will the distinction always be difficult to make empirically given the characteristic differences of investment choices made by LBO and venture capital firms noted above?

7. Is Pound's analysis of biases in the proxy process consistent with Easterbrook and Fischel's? Do their assumptions about the relative merits of shareholder disagreements with management differ? How could corporation codes take account of multiple transactional roles and relations between investors and firms, and should they? Would confidential (i.e., anonymous) proxy voting mitigate the conflicts confronting institutional investors over which

Pound is concerned? Would this have undesirable side effects? Consider the selection by Black, and the question of agency problems for beneficiaries, in monitoring money managers' voting behavior.

8. What would you expect the abnormal stock returns of firms that experience proxy fights to look like? Would you expect different results depending on whether or not dissidents win the contest? For a recent study finding that proxy fight targets have significantly poor performance both before the contest and afterward when dissidents win, see David Ikenberry and Josef Lakonishok, "Corporate Governance Through the Proxy Contest: Evidence and Implications," *Journal of Business* (forthcoming). Would self selection explain the latter result (e.g., would managers be less likely to vigorously fight a proxy challenge if they believe firm prospects are poor)? Note that management typically turns over after a proxy contest regardless of the outcome, and that firms that are subject to a contest are often sold or liquidated shortly thereafter. See Harry DeAngelo and Linda DeAngelo, "Proxy Contests and the Governance of Publicly Held Corporations," 23 *Journal of Financial Economics* 29 (1989). For studies finding a positive stock price effect of proxy fights, regardless of outcome, see DeAngelo and DeAngelo, supra; Peter Dodd and Jerold B. Warner, "On Corporate Governance: A Study of Proxy Contests," 11 *Journal of Financial Economics* 401 (1983); J. Harold Mulherin and Annette B. Poulsen, "Does a Proxy Have Real Moxie? The Performance Effects of Proxy Contests in the 1980s," University of Georgia Working Paper (1992). These results are not necessarily inconsistent with Ikenberry and Lakonishok's finding. For example, the positive returns in the DeAngelo and DeAngelo study appear for a short interval around the contest, are largest when dissidents win representation but not control, and appear to be driven by subsequent sale or liquidation of the business as a result of the dissidents' efforts. In addition, Mulherin and Poulsen find that after the contest shareholder wealth declines and performance improves, but both results are statistically insignificant.

9. Consider the recommendations of a 1989 New York State gubernatorial task force report. The report stated that public pension fund managers should use broader guidelines for investing than achieving the best return, such as an investment's effect on other constituencies (employees, customers, and local communities). One explanation for the suggestion is fiscal exigency: a large budget deficit loomed in New York State, which limited the availability of state funds for economic development and other state programs. Does Roe suggest another explanation?

10. In recent years, institutional investors, especially public pension funds, have taken a more active role in corporate governance, and in particular, in making shareholder proposals. In 1991, for example, 169 shareholder proposals were submitted by institutional investors, a record high, and an increase of nearly 30 percent over the previous year's activity. See "Some Institutional Investors Turn to Negotiations, But High Level of Proposal Activ-

ity Predicted to Continue," *BNA's Corporate Counsel Weekly* 8 (Jan. 1, 1992). These investors formed associations that trade information on proxy issues and proposed reforms of SEC proxy regulations to facilitate their ability to influence corporate governance, which the SEC adopted in 1992. In addition, the number of private consulting firms advising these investors has mushroomed. Is this activity consistent with Roe's analysis?

Do managers of public pension funds (who are often elected officials) have the requisite expertise to be effective at monitoring? Is such expertise necessary if they are to perform the role that Roe ascribes to foreign banks? Should they delegate such responsibility to the professional money managers they employ? Would it be more efficacious for institutional investors to elect professional directors to represent them on their portfolio firms' boards rather than be active monitors themselves? See Ronald J. Gilson and Reinier Kraakman, "Reinventing the Outside Director: An Agenda for Institutional Investors," 43 *Stanford Law Review* 863 (1991).

11. John Coffee offers a skeptical view of calls for institutional activism and of Roe's thesis that political regulation has deterred it:

> Although those dissatisfied with the current state of shareholder power in American corporate governance may find foreign models of corporate governance more attractive, the danger in such comparisons is the "grass-is-always-greener" fallacy: one can easily idealize foreign systems and see only their benefits and not their costs. Monitoring by financial institutions does not necessarily mean net gains for other shareholders, but may improve only the position of the monitoring financial institution. Similarly, assigning primary causal significance to political constraints may lead to overlooking deeper and more fundamental reasons why financial monitoring has not developed in the United States. Put simply, the agents controlling institutional investors have considerable reason to remain "rationally apathetic" about corporate governance and little reason to become active participants. Why? . . . [One reason] stands out . . . : a trade-off exists and must be recognized between liquidity and control. Investors that want liquidity may hesitate to accept control.
>
> [T]he primary explanation for institutional passivity is not overregulation, but the insufficiency of existing incentives to motivate institutional money managers to monitor. Although proponents of institutional activism have analyzed at length the potential ability of institutional investors to hold corporate managers accountable, they have largely ignored the question of who holds institutional money managers accountable. The problem of who will guard the guardian is a timeless one, but it is particularly complicated when the proposed guardian is the institutional investor. Not only do the same problems of agency cost arise at the institutional investor level, but there are persuasive reasons for believing that some institutional investors are less accountable to their "owners" than are corporate managements to their shareholders. Put simply, the usual mechanisms of corporate accountability are either unavailable or largely compromised at the institutional level. This conclusion does not deny that there has been overregulation of institutional investors, but it suggests its impact may have been overstated by the new critics. More importantly, this perspective implies that deregulation alone is an inadequate policy response. If

the diagnosis that rational apathy will continue to prevail at the institutional level is accurate, then the law must intervene to correct the market's failure by creating adequate incentives for institutional managers to monitor.

Coffee, "Liquidity Versus Control: The Institutional Investor as Corporate Monitor." Copyright © 1991 by the Directors of the Columbia Law Review Association, Inc. All Rights Reserved. This article originally appeared at 91 *Columbia Law Review* 1277 (1991). Reprinted by permission. For another critique of the promise of institutional investor activism, see Edward B. Rock, "The Logic and (Uncertain) Significance of Institutional Shareholder Activism," 79 *Georgetown Law Journal* 445 (1991).

Coffee's policy proposals to improve institutions' incentives for activism include mandating the use of professional proxy advisers by investment managers, restricting portfolio diversification to a "level consistent with the institution's ability to monitor," which he views as between 20 to 100 stocks, and removing existing restrictions on incentive compensation for index fund managers (compensation formulas may follow percentage of assets under management because the Investment Advisors Act of 1940 prohibits compensation on the basis of asset appreciation unless the contract provides for compensation based on asset value in relation to the performance of an appropriate securities index). Coffee, supra, at 1353–55, 1363–65. What is your evaluation of such proposals? What would motivate professional proxy advisers to maximize shareholder welfare? Should they be compensated in stock of the corporations on whose votes they provide advice? If less widely diversified funds are more valuable to investors than existing funds because their managers can be better monitors of the firms held in the portfolio, why is regulation necessary to create such funds? Who would benefit from such a reform? Money managers, who are in increased demand as pension funds have to distribute smaller pieces of their portfolios to more firms? Why would compensation fixed to percentage of assets under control fail to provide fund managers with appropriate incentives to vote shares so as to maximize shareholder value? Would investors place much money in funds whose voting decisions favor corporate managers over shareholders?

12. Black's position on the desirability of institutional shareholder activism depends on removal of the constraints identified by Roe and Coffee in note 11. But he further maintains that only some institutional investors will be likely to be leaders in active monitoring. In particular, he contends that institutions' roles depend on the conflicts of interest to which the selection alludes and which he later summarizes as follows:

> Many institutional money managers face conflicts of interest. Money managers who vote against a company's proposals are likely to lose any business that they conduct with the company. Money managers who develop an antimanager reputation may lose corporate business, or find it harder to gain new business. These conflicts lead some institutions to vote promanager even when doing so is likely to decrease company value. The institutions face especially strong conflicts on the high-visibility decision to make a voting proposal or actively

oppose a manager proposal. . . . Legal rules haven't effectively controlled conflicts of interest. For example, only in the last couple of years has the Department of Labor begun to examine the voting records of ERISA [Employee Retirement Income Security Act] fiduciaries.

The single phrase "institutional investor" obscures important differences between institutions in the strength of their conflicts. Banks and insurers have strong conflicts because of their extensive business dealings with corporate managers. Corporate pension funds, even when they hire outside money managers, are still controlled by corporate managers. Public pension funds don't solicit corporate business, but respond to political pressure. No institution is completely beholden to corporate managers; no institution is conflict free.

Bernard S. Black, "Agents Watching Agents: The Promise of Institutional Investor Voice." Originally published in 39 *UCLA Law Review* 811, Copyright 1992, The Regents of The University of California. All Rights Reserved. Reprinted by permission. Note that this is also Pound's contention. Black goes on to maintain that public pension funds are the leaders in corporate governance because they have "weak promanager conflicts." Is this view justified? Reconsider the New York State task force report referred to in note 9. Do you think there would be a significant difference in shareholder voting on corporate governance issues when the fund manager is concerned with local employment levels rather than receipt of "corporate business"? What relationship would you expect to find between the political composition of public pension funds' boards of trustees and their performance? See Roberta Romano, "Public Pension Funds' Activism in Corporate Governance Reconsidered," 93 *Columbia Law Review* 795 (1993).

13. Can the LBO organization answer the agency problem, mooting the debate over the need for institutional investor activism? In answering this question, consider what kinds of businesses can be operated in the LBO form. In the 1990s, the number of new LBOs drastically declined and there were several spectacular LBO bankruptcies. Review the financial incentives of the LBO organization members for ensuring the success of the firms acquired, as detailed in Jensen. After 1985, the upfront fees for an LBO transaction paid to the investment banks operating LBO associations increased while their direct equity shares decreased. Jensen contends that this alteration in compensation created a contract failure (perverse incentives for the LBO entrepreneurs) that caused investors to overpay in these later deals and resulted in bankruptcies. Michael C. Jensen, "Corporate Control and the Politics of Finance," 4 *The Continental Bank Journal of Applied Corporate Finance* 13 (Summer 1991). Jensen further indicates that no LBO partnership participated in the biggest LBO failures (Revco, Interco, and Campeau), and hence there was no party with a long-term reputation to protect when these deals were priced.

VI

External Governance Structures: The Market for Corporate Control

The market for corporate control is often identified as the key backstop corporate governance device. By threatening managers with job loss if they do not maximize share value, the market for corporate control creates high-powered incentives for managers to further shareholder interests. However, because control changes are expensive, especially when opposed by managers who do not wish to be replaced, the agency problem is still only constrained, rather than eliminated, by an active market for control.

The readings in the first part of this chapter present a number of prominent explanations for acquisitive transactions, beginning with agency-based explanations in the selections by Henry Manne and Michael Jensen. Takeovers are a crucial disciplining device from the agency perspective because, unlike mergers, a takeover does not require management's approval. It is made directly to shareholders, who accept an offer by tendering their shares, thereby bypassing the board.

Takeovers that reduce agency costs are value-maximizing transactions that enhance efficiency. But it is possible for a takeover to provide substantial gains to shareholders without producing an efficiency gain by transferring wealth to them from other firm participants (just as an unanticipated increase in firm risk after debt has been incurred shifts wealth from bondholders to shareholders, as discussed in Chapter IV). The selection by Andrei Shleifer and Lawrence Summers develops one of the more widely held expropriation explanations of takeover gains, wealth transfer from labor.

The selections by Yakov Amihud and Baruch Lev and by Richard Roll offer a qualitatively different perspective from the others because they are non-value-maximizing explanations. The common theme is that acquisitive transactions are motivated by maximization of manager utility and not shareholder wealth. These selections thus suggest that takeovers are the paradigmatic agency problem, not its cure.

There is an immense cottage industry in empirical research on takeovers; event study results are summarized in the selections from two survey articles by Michael Jensen and Richard Ruback and by Gregg Jarrell, James Brickley, and Jeffry Netter. Sanjai Bhagat, Andrei Shleifer, and Robert Vishny provide additional insight from an analysis of firms' posttakeover changes. An undisputed fact is that takeovers generate substantial gains to target shareholders. All studies, whatever the time period or acquisitive form, find statistically significant positive abnormal returns to target shareholders on the announcement of a bid. The different explanations of takeovers are efforts at explaining the source of the gains.

The evidence on acquirer returns is less clear-cut: depending on time period and firm sample, acquirers experience positive, negative, or statistically insignificant abnormal returns. In general, returns to acquirers are of much smaller magnitude than target gains, and they have decreased over time. But when acquirers' losses are aggregated with target gains, the net is still a positive abnormal return. This suggests that non-value-maximizing transactions for bidders are socially beneficial (aggregate wealth is increased), assuming the source of target gains is increased efficiency and not a wealth transfer. The bulk of the empirical evidence supports efficiency rather than expropriation explanations of takeovers. But as there is no comprehensive theory that explains all acquisitions—each theory explains, at best, a subset of transactions—there are surely instances of takeovers that simply redistribute wealth to shareholders, or from shareholders to managers, rather than produce efficiency gains.

The substantial gains received by target shareholders pose a critical question concerning management's response to a bid, because there is an obvious conflict of interest if the reason for a bid is management's replacement. Takeover defenses have, however, two possible aims, which renders identification of management faithlessness at times extremely opaque. A manager who resists a bid may be seeking to entrench herself, but she may also be attempting to increase the target's share of the acquisition's gains. Opposition to a bid ups the ante, forcing a bidder to raise its price to obtain consent or face competition. This is because resistance delays a bid's completion, enabling other bidders to enter the fray and top the initial bidder's price.

The issue debated in the first three selections in part B involves the effect of higher premiums from bidder competition on the incidence of takeovers.

Frank Easterbrook and Daniel Fischel maintain that competition has an undesirable feedback on the market for corporate control: because identifying a
target is costly, if bidders must pay more or run the risk of losing in an auction,
they will not be able to recoup their investment in information, and they will
undertake less search. As a consequence, the number of takeovers will decrease, and correspondingly, the discipline of the market for control on management in general will be weakened. Ronald Gilson and Lucian Bebchuk
challenge this thesis, questioning whether search costs are high, investments
cannot be recouped, and the number of bids will decline. But the disagreement among these authors is actually over relatively narrow ground, the
efficacy of bidder competition. All would provide management with little
discretion otherwise to engage in maneuvering to defeat a bid, and thus they
all invite enhanced judicial scrutiny of takeover defenses.

The Jarrell, Brickley, and Netter selection in part B surveys the empirical
research on defensive tactics. For the most part, investors view such actions
unfavorably. Jonathan Karpoff and Paul Malatesta examine the impact of
state takeover laws, which present further barriers to bidders. The stock price
reaction is negative here as well. State regulation has led to calls for federal
preemption from commentators persuaded of the benefits of takeovers. Roberta Romano reviews the theoretical argument for national regulation of
takeovers, which revisits the question of state competition discussed in Chapter III. She then examines what Congress's response has been in practice and
concludes that the case for federal regulation is more apparent than real.

Theories and Evidence

Mergers and the Market for Corporate Control

HENRY G. MANNE

The Corporate-Control Market

The basic proposition advanced in this article is that the control of corporations may constitute a valuable asset; that this asset exists independent of any interest in either economies of scale or monopoly profits; that an active market for corporate control exists; and that a great many mergers are probably the result of the successful workings of this special market. . . . Perhaps the most important implications are those for the alleged separation of ownership and control in large corporations. So long as we are unable to discern any control relationship between small shareholders and corporate management, the thrust of Berle and Means's famous phrase remains strong. But, as will be explained below, the market for corporate control gives to these shareholders both power and protection commensurate with their interest in corporate affairs.

A fundamental premise underlying the market for corporate control is the existence of a high positive correlation between corporate managerial efficiency and the market price of shares of that company. As an existing company is poorly managed—in the sense of not making as great a return for the shareholders as could be accomplished under other feasible managements—the market price of the shares declines relative to the shares of other companies in the same industry or relative to the market as a whole. This phenomenon has a dual importance for the market for corporate control.

In the first place, a lower share price facilitates any effort to take over high-paying managerial positions. The compensation from these positions may take the usual forms of salary, bonuses, pensions, expense accounts, and stock options. Perhaps more important, it may take the form of information useful in trading in the company's shares; or, if that is illegal, information may be exchanged and the trading done in other companies' shares. But it is extremely doubtful that the full compensation recoverable by executives for managing their corporations explains more than a small fraction of outsider attempts to

Reprinted by permission from 73 *Journal of Political Economy* 110. © 1965 by The University of Chicago.

take over control. Takeovers of corporations are too expensive generally to make the "purchase" of management compensation an attractive proposition.

It is far more likely that a second kind of reward provides the primary motivation for most takeover attempts. The market price of shares does more than measure the price at which the normal compensation of executives can be "sold" to new individuals. Share price, or that part reflecting managerial efficiency, also measures the potential capital gain inherent in the corporate stock. The lower the stock price, relative to what it could be with more efficient management, the more attractive the takeover becomes to those who believe that they can manage the company more efficiently. And the potential return from the successful takeover and revitalization of a poorly run company can be enormous. . . .

But the greatest benefits of the takeover scheme probably inure to those least conscious of it. Apart from the stock market, we have no objective standard of managerial efficiency. Courts, as indicated by the so-called business judgment rule, are loath to second-guess business decisions or remove directors from office. Only the takeover scheme provides some assurance of competitive efficiency among corporate managers and thereby affords strong protection to the interests of vast numbers of small, noncontrolling shareholders. Compared to this mechanism, the efforts of the SEC and the courts to protect shareholders through the development of a fiduciary duty concept and the shareholder's derivative suit seem small indeed. It is true that sales by dissatisfied shareholders are necessary to trigger the mechanism and that these shareholders may suffer considerable losses. On the other hand, even greater capital losses are prevented by the existence of a competitive market for corporate control.

Takeovers: Their Causes and Consequences

MICHAEL C. JENSEN

Free Cash Flow Theory

More than a dozen separate forces drive takeover activity, including such factors as deregulation, synergies, economies of scale and scope, taxes, the level of managerial competence, and increasing globalization of U.S. markets. . . . One major cause of takeover activity, the agency costs associated with conflicts between managers and shareholders over the payout of free

Reprinted by permission from 2 *Journal of Economic Perspectives* 21. © 1988 American Economic Association.

cash flow, has received relatively little attention. Yet it has played an important role in acquisitions over the last decade.

Managers are the agents of shareholders, and because both parties are self-interested, there are serious conflicts between them over the choice of the best corporate strategy. Agency costs are the total costs that arise in such arrangements. . . . When these costs are large, the threat or actuality of takeovers can reduce them.

Free cash flow is cash flow in excess of that required to fund all of a firm's projects that have positive net present values when discounted at the relevant cost of capital. Such free cash flow must be paid out to shareholders if the firm is to be efficient and to maximize value for shareholders.

However, payment of cash to shareholders reduces the resources controlled by managers, thereby reducing the power of managers and potentially subjecting them to the monitoring by capital markets that occurs when a firm must obtain new capital. Further, managers have incentives to expand their firms beyond the size that maximizes shareholder wealth. Growth increases managers' power by increasing the resources under their control, and changes in management compensation are positively related to growth. . . .

Conflicts of interest between shareholders and managers over payout policies are especially severe when the organization generates substantial free cash flow. The problem is how to motivate managers to disgorge the cash rather than invest it at below the cost of capital or waste it through organizational inefficiencies.

The theory developed here offers a seeming paradox. Increases in financial flexibility that give managers control over free cash flow may actually cause the value of the firm to decline. This result occurs because it is difficult to assure that managers will use their discretion over resources to further the interests of shareholders. . . .

The Role of Debt in Motivating Organizational Efficiency

The agency costs of debt have been widely discussed but . . . the benefits of debt in motivating managers and their organizations to be efficient have largely been ignored.

Debt creation, *without retention of the proceeds of the issue,* enables managers effectively to bond their promise to pay out future cash flows. Thus, debt can be an effective substitute for dividends, something not generally recognized in the corporate finance literature: Debt reduces the agency cost of free cash flow by reducing the cash flow available for spending at the discretion of managers. By issuing debt in exchange for stock, managers bond their promise to pay out future cash flows in a way that simple dividend increases do not. In doing so, they give shareholder-recipients of the debt the right to take the firm into bankruptcy court if they do not keep their promise to make the interest and principal payments.

Of course, managers can also promise to pay out future cash flows by announcing a "permanent" increase in the dividend. But because there is no contractual obligation to make the promised dividend payments, such promises are weak. . . .

Issuing large amounts of debt to buy back stock sets up organizational incentives to motivate managers to pay out free cash flow. In addition, the exchange of debt for stock helps managers overcome the normal organizational resistance to retrenchment that the payout of free cash flow often requires. The threat of failure to make debt-service payments serves as a strong motivating force to make such organizations more efficient. . . .

The debt created in a hostile takeover (or takeover defense) of a firm suffering severe agency costs of free cash flow need not be permanent. Indeed, sometimes "overleveraging" such a firm is desirable. In these situations, levering the firm so highly that it cannot continue to exist in its old form creates the crisis to motivate cuts in expansion programs and the sale of those divisions that are more valuable outside the firm. The proceeds are used to reduce debt to a more normal or permanent level. This process results in a reexamination of an organization's strategy and structure. When it is successful, a much leaner, more efficient, and competitive organization results.

This control hypothesis does not imply that debt issues will always have positive control effects. For example, these control effects will not be as important for rapidly growing organizations with large and highly profitable investment projects but no free cash flow. Such organizations will have to go regularly to the financial markets to obtain capital. At these times the markets have an opportunity to evaluate the company, its management, and its proposed projects. Investment bankers and analysts play an important role in this monitoring, and the market's assessment is made evident by the price investors pay for the financial claims.

The control function of debt is more important in organizations that generate large cash flows but have low growth prospects, and it is even more important in organizations that must shrink. In these organizations the pressure to waste cash flows by investing them in uneconomic projects is most serious.

Leveraged Buyouts and Free Cash Flow Theory

Many of the benefits in going-private and leveraged buyout transactions seem to be due to the control function of debt. These transactions are creating a new organizational form that competes successfully with the open corporate form because of advantages in controlling the agency costs of free cash flow. . . .

Desirable leveraged buyout candidates are frequently firms or divisions of large firms that have stable business histories, low growth prospects, and high potential for generating cash flows; that is, situations where agency costs of free cash flows are likely to be high. . . .

Acquisitions are one way managers spend cash instead of paying it out to shareholders. Free cash flow theory implies that managers of firms with un-used borrowing power and large free cash flows are more likely to undertake low-benefit or even value-destroying mergers. Diversification programs gener-ally fit this category, and the theory predicts that they will generate lower total gains. Thus, some acquisitions are a solution to the agency problem of free cash flow while others, such as diversification programs, are symptoms of those problems.

The major benefit of diversification mergers may be that they involve less waste of resources than if the funds had been invested internally in unprofitable projects. Acquisitions made with cash or securities other than stock involve payout of resources to shareholders of the target company, and this can create net benefits even if the merger creates operating inefficiencies. To illustrate, consider an acquiring firm with substantial free cash flow that the market expects will be invested in low-return projects with a negative net present value of $100 million. If this firm uses up its free cash flow (and thereby prevents its waste) by acquiring another firm that generates zero synergies, the combined market value of the two firms will rise by $100 million. The market value increases because the acquisition eliminates the expenditures on internal invest-ments with negative market value of $100 million.

Breach of Trust in Hostile Takeovers

ANDREI SHLEIFER AND LAWRENCE H. SUMMERS

Consider three scenarios. In Scenario A, T. Boone Pickens takes over Plateau Petroleum and immediately lays off 10,000 workers, who immediately find work elsewhere at the same wage. Pickens also stops purchasing from numer-ous suppliers, who find that they can sell their output without any price reduction to other customers. The stock of Plateau Petroleum rises by 25 percent.

In Scenario B, Frank Lorenzo takes over Direction Airlines and immedi-ately stares down the union so that the wages of the existing workers are reduced by 30 percent and 10 percent of the work force is laid off and unable to find subsequent employment at more than 50 percent of their previous wage. Lorenzo does not change the airline's route structure or flight fre-quency. The stock of Direction Airlines rises by 25 percent.

In Scenario C, Carl Icahn takes over USZ. He closes down the corporate headquarters and lays off thousands of highly paid, senior employees, who had previously been promised lifetime employment by the now-displaced

Reprinted by permission from *Corporate Takeovers: Causes and Consequences*, A. Auerbach, ed., pp. 33–56. © 1988 by the National Bureau of Economic Research, published by University of Chicago.

managers. Icahn also shuts down factories that dominate the economies of several small towns. As a consequence numerous local stores, restaurants, and bars go bankrupt. The stock of USZ goes up by 25 percent.

All three takeovers yield equal private benefits to the shareholders of the target firms. Yet their social consequences are very different. In Scenario A society is better off because resources are diverted from less productive to more productive uses. The increased value of Plateau Petroleum approximately reflects the value of this gain. In Scenario B society is about equally well off. The gains to Direction shareholders are approximately offset by the losses to the human wealth of Direction employees. The redistribution is probably antiegalitarian. On the other hand, it may ultimately lead to advantages for customers of the airline. In Scenario C society is worse off. The gains to USZ shareholders are offset by the losses incurred by the laid-off employees and by the firms with immobile capital whose viability depended on the factories' remaining open. And other firms find that once their workers see what happened at USZ, they become less loyal and require higher wages to compensate them for a reduction in their perceived security. These firms also have a harder time inducing their suppliers to make fixed investments on their behalf.

These three examples make it clear that increases in share values in hostile takeovers in no way measure or demonstrate their social benefits. Scenario A is the only one in which share price increases capture the elimination of waste and the gains in social welfare. In contrast, shareholder gains in Scenarios B and C to a large extent come from losses of the value of employees' human capital. Even if some efficiency is realized from wages' coming more into line with marginal products, the efficiency is only a second order effect relative to the transfer from employees to shareholders. Scenario C has additional external effects of the acquisition which, while not resulting in gains to the acquirer, should enter the social calculation. The claim that the 25 percent takeover premium in Scenarios B and C measures social gains is simply incorrect. . . .

A corporation is a nexus of long-term contracts between shareholders and stakeholders. Because the future contingencies are hard to describe, complete contracting is costly. As a result, many of these contracts are implicit, and the corporation must be trusted to deliver on the implicit contracts even without enforcement by courts. To the extent that long-term contracts reduce costs, such trustworthiness is a valuable asset of the corporation. Shareholders own this asset and are therefore able to hire stakeholders using implicit long-term contracts.

The principal reason why long-term contracts between shareholders and stakeholders are needed is to promote relationship-specific capital investments by the stakeholders. An employee will spend time and effort to learn how to do his job well only if he knows that his increased productivity will be subsequently rewarded. A subcontractor exploring for oil will buy site-specific new equipment only if he believes that the contracting oil firm will not try to squeeze his profits once he sinks the cost. A sales representative will service past customers only if she is assured she will continue to benefit from their loyalty. In these and other cases it is important to the shareholders that the stakeholders do a good job, but shareholders may be unable to describe what specific actions this calls for, let alone to contract for them.

The necessary arrangement to ensure appropriate investment by stakeholders is a long-term contract, which allows them to collect some of the rewards of doing good work over time. The expense of writing a complete contingent contract ensures that these long-term contracts are implicit. Examples of such contracts are hiring an oil exploration company for the long haul, so that it acquires the equipment best suited for the long-term customer; lifetime employment for workers who then learn how to do the job efficiently; and surrender of customer lists to sales representatives who can then profit from repeated purchases. . . .

Although both shareholders and stakeholders benefit ex ante from implicit long-term contracts, ex post it might pay shareholders to renege. For example, it will pay shareholders to fire old workers whose wage exceeds their marginal product in a contract that, for incentive reasons, underpaid them when they were young. Or shareholders might profit from getting rid of workers whom they insured against uncertain ability and who turned out to be inept. Or shareholders might gain from refusing to compensate a supplier for investing in the buyer-specific plant, after this plant is built. Or an insurance company can repossess its sales representative's customer list. In all these cases implicit contracts specify actions that ex post reduce the firm's value, even though agreeing to these actions is ex ante value maximizing. Breach of contract can therefore raise shareholder wealth, and the more so the greater is the burden of fulfilling past implicit contracts. Conversely, the value of workers' human capital or of suppliers' relationship-specific capital stock suffers a loss.

To take advantage of implicit contracts, shareholders must be trusted by potential stakeholders. Otherwise, stakeholders would expect breach whenever it raises the firm's value and would never enter into implicit contracts. To convince stakeholders that implicit contracts are good, shareholders must be trusted not to breach contracts even when it is value maximizing to do so.

A standard solution to the problem of how implicit contracts are maintained is the theory of rational reputation formation. In this theory managers adhere to implicit contracts because their adherence enables them to develop a reputation for trustworthiness, and thus to benefit from future implicit contracts. If violating an implicit contract today would make the managers untrustworthy in the future, they will uphold the contract as long as the option of entering into future contracts is valuable enough. Conversely, if it is not important for the managers to be trusted in the future, that is, if a reputation is not valuable, they will violate the implicit contract. . . .

The position that the sole reason to trust a manager (or anyone else for that matter) is his reputation is not plausible. People commonly trust other people even when no long-term reputations are at stake. Most people do not steal not only because they fear punishment, but because they are simply honest. Those who leave their cars unlocked do so relying more on other people's integrity than on police powers. Waiters rely on their expectation that most people tip in restaurants even when they expect never to come back. In fact, evidence shows that travelers' tips are no smaller than those of patrons. . . .

To dispel the fear of breach on the part of stakeholders, shareholders find it value maximizing to *seek out* or *train* individuals who are capable of commitment to stakeholders, *elevate* them to management, and *entrench* them. To such managers, stakeholder claims, once agreed to, are prior to shareholder claims. Even when a rational reputation is not of high enough value to shareholders to uphold the implicit contracts with stakeholders, as would be the case if the company suffered a large permanent decline in demand, trustworthy managers will still respect stakeholder claims. From the ex ante viewpoint, such dedication to stakeholders might be a value-maximizing managerial *attribute* . . . . In a world without takeovers, potential stakeholders counting on such managers to respect their claims will enter into contracts with the firm. . . .

In some circumstances upholding the implicit contracts with stakeholders becomes a liability to shareholders. The incumbent managers are nonetheless committed to upholding stakeholder claims. In these cases ousting the managers is a prerequisite to realizing the gains from the breach. This is precisely what hostile takeovers can accomplish. As the incumbent managers are removed after the takeover, control reverts to the bidder, who is not committed to upholding the implicit contracts with stakeholders. Shareholders can then renege on the contracts and expropriate rents from the stakeholders. The resulting wealth gains show up as the takeover premia. Hostile takeovers thus enable shareholders to redistribute wealth from stakeholders to themselves.

Managers committed to upholding stakeholder claims will not concede to the redistribution. They will resist it, even though the shareholders at this point will withdraw their support from the managers to realize the ex post gain. Not surprisingly, then, takeovers that transfer wealth from stakeholders to shareholders must be hostile.

The importance of transfers in justifying the takeover premium does not imply that breach of implicit contracts is always the actual takeover motive. Breach can be the motive, as for example is the case in some takeovers explicitly aiming to cut wages. At other times the acquisition is motivated by the overinvestment or other free cash flows of the targeted firm. Even in these takeovers much of the gain must come from reducing the wealth of stakeholders, who did not count on changes in operations when agreeing to work for the firm. . . .

For breach to be an important source of gains, hostile takeovers must come as a surprise to stakeholders, who entered into implicit contracts expecting the firm to be run by trustworthy managers. For if the stakeholders anticipate a hostile takeover, they will realize the trustworthiness of the incumbent managers is worthless, since they will be duly removed when shareholder interest so demands. Implicit contracts based on trust are feasible only insofar as the managers upholding them are entrenched enough to retain their jobs in the face of a hostile threat. . . .

As we said at the start, the ability to enter into implicit contracts and to be trusted to abide by them may be one of the most valuable assets owned by shareholders. Takeovers may substantially reduce the value of these assets. In the popular literature this phenomenon has been called the decline of corporate loyalty, which is widely cited as a cost to firms. This cost can show up in explicit costly contracts with stakeholders (such as labor protection provisions), or in the need to pay them more now in return for their accepting uncertainty about future payments, or simply as forgone profitable trade. Whatever form this cost takes, it should *ultimately* show in the declining value of corporate equity.

Risk Reduction as a Managerial Motive for Conglomerate Mergers

YAKOV AMIHUD AND BARUCH LEV

Despite extensive research, the motives for conglomerate mergers are still largely unknown: What drives a substantial number of firms to engage in conglomerate mergers, given that a priori no real economic benefit (synergism) is expected from the combination of such functionally unrelated parties? The motive of risk reduction through diversification appeared at first to provide a natural explanation for the conglomerate merger phenomenon. However, it has been argued convincingly that in perfect capital markets such risk reduction cannot be beneficial to stockholders, since they can achieve on their own the desired level of risk through portfolio diversification. Even when market imperfections, such as transaction costs, are admitted, the risk-reduction benefits of conglomerate mergers from the stockholders' point of view seem highly questionable, given the relatively low cost of portfolio diversification in the capital market. Moreover, the call options pricing model suggests that adoption of projects which reduce the variance of the firm's income distribution (i.e., diversification) may adversely affect equity holders by inducing a wealth transfer from stockholders to bondholders. How, then, can the widespread and persisting phenomenon of conglomerate mergers be explained?

It is argued below that the fast-growing literature on "managerialism," and in particular the agency cost models, provide a possible explanation for the conglomerate merger phenomenon. In essence, such mergers may be viewed as a form of management perquisite intended to decrease the risk associated with managerial human capital. Accordingly, the consequences of such mergers may be regarded as an agency cost. . . .

Reprinted by permission from 12 *Bell Journal of Economics* 605 (1981).

Managers' income from employment constitutes, in general, a major portion of their total income. Employment income is closely related to the firm's performance through profit-sharing schemes, bonuses, and the value of stock options held by managers. Hence, the risk associated with managers' income is closely related to the firm's risk. Quite often, a firm's failure to achieve predetermined performance targets, or in the extreme case the occurrence of bankruptcy, will result in managers' losing their current employment and seriously hurting their future employment and earnings potential. Such "employment risk" cannot be effectively diversified by managers in their personal portfolios, since unlike many other sources of income such as stocks, human capital cannot be traded in competitive markets. Risk-averse managers can therefore be expected to diversify this employment risk by other means, such as engaging their firms in conglomerate mergers, which generally stabilize the firm's income stream and may even be used to avoid the disastrous effects bankruptcy has on managers. Thus, conglomerate mergers, while not of obvious benefit to investors, may benefit managers by reducing their employment risk, which is largely undiversifiable in capital or other markets. . . .

Managers' benefits from risk reduction through conglomerate mergers and other means may thus be viewed as a form of perquisite appropriated from the firm. The welfare loss to the principal, due to the real cost of mergers and the possible wealth transfer from stockholders to bondholders (suggested by the options model), thus constitutes an agency cost.

The Hubris Hypothesis of Corporate Takeovers

RICHARD ROLL

The mechanism by which takeover attempts are initiated and consummated suggests that at least part of the large price increases observed in target firm shares might represent a simple transfer from the bidding firm, that is, that the observed takeover premium (tender offer or merger price less pre-announcement market price of the target firm) overstates the increase in economic value of the corporate combination. . . .

Consider what might happen if there are no potential synergies or other sources of takeover gains but when, nevertheless, some bidding firms believe that such gains exist. The valuation itself can then be considered a random variable whose mean is the target firm's current market price. When the random variable exceeds its mean, an offer is made; otherwise there is no offer. Offers are observed only when the valuation is too high; outcomes in

Reprinted by permission from 59 *Journal of Business* 197. © 1986 by The University of Chicago.

the left tail of the distribution of valuations are never observed. The takeover premium in such a case is simply a random error, a mistake made by the bidding firm. Most important, the observed error is always in the same direction. Corresponding errors in the opposite direction are made in the valuation process, but they do not enter our empirical samples because they are not made public.

If there were no value at all in takeovers, why would firms make bids in the first place? They should realize that any bid above the market price represents an error. This latter logic is alluring because market prices do seem to reflect rational behavior. But we must keep in mind that prices are averages. There is no evidence to indicate that every individual behaves as if he were the rational economic human being whose behavior seems revealed by the behavior or market prices. We may argue that markets behave as if they were populated by rational beings. But a market actually populated by rational beings is observationally equivalent to a market characterized by grossly irrational individual behavior that cancels out in the aggregate, leaving the trace of the only systematic behavioral component, the small thread of rationality that all individuals have in common. Indeed, one possible definition of irrational or aberrant behavior is independence across individuals (and thus disappearance from view under aggregation).

Psychologists are constantly bombarding economists with empirical evidence that individuals do not always make rational decisions under uncertainty. Among psychologists, economists have a reputation for arrogance mainly because this evidence is ignored; but psychologists seem not to appreciate that economists disregard the evidence on individual decision making because it usually has little predictive content for market behavior. Corporate takeovers are, I believe, one area of research in which this usually valid reaction of economists should be abandoned; takeovers reflect individual decisions.

There is little reason to expect that a particular individual bidder will refrain from bidding because he has learned from his own past errors. Although some firms engage in many acquisitions, the average individual bidder/manager has the opportunity to make only a few takeover offers during his career. He may convince himself that the valuation is right and that the market does not reflect the full economic value of the combined firm. For this reason, the hypothesis being offered in this paper to explain the takeover phenomenon can be termed the "hubris hypothesis." If there actually are no aggregate gains in takeover, the phenomenon depends on the overbearing presumption of bidders that their valuations are correct. . . .

The hubris hypothesis is consistent with strong-form market efficiency. Financial markets are assumed to be efficient in that asset prices reflect all information about individual firms. . . .

Most other explanations of the takeover phenomenon rely on strong-form market inefficiency of at least a temporary duration. Either financial markets are ignorant of relevant information possessed by bidding firms, or product markets are inefficiently organized so that potential synergies, monopolies, or tax savings are being ineffectively exploited (at least temporarily), or labor markets are inefficient because gains could be obtained by replacement of inferior managers. Although perfect strong-form efficiency is unlikely, the concept should serve as a frictionless ideal, the benchmark of comparison by which other degrees of efficiency are measured. This is, I claim, the proper role for the hubris hypothesis of takeovers; it is the null against which other hypotheses of corporate takeovers should be compared. . . .

The hubris hypothesis might seem to imply that managers act consciously against shareholder interests. . . . But the hubris hypothesis does not rely on this result. It is sufficient that managers act, de facto, against shareholder interests by issuing bids founded on mistaken estimates of target firm value. Management intentions may be fully consistent with honorable stewardship of corporate assets, but actions need not always turn out to be right. . . .

An argument can be advanced that the hubris hypothesis implies an inefficiency in the market for corporate control. If all takeovers were prompted by hubris, shareholders could stop the practice by forbidding managers ever to make any bid. Since such prohibitions are not observed, hubris alone cannot explain the takeover phenomenon.

The validity of this argument depends on the size of deadweight takeover costs. If such costs are relatively small, stockholders would be indifferent to hubris-inspired bids because target firm shareholders would gain what bidding firm shareholders lose. A well-diversified shareholder would receive the aggregate gain, which is close to zero.

The Market for Corporate Control: The Scientific Evidence

MICHAEL C. JENSEN AND RICHARD S. RUBACK

Target Firm Stockholder Returns

Successful Target Returns

The thirteen studies summarized in table 3 [omitted] indicate that targets of successful takeover attempts realize substantial and statistically significant increases in their stock prices. The estimates of positive abnormal returns to targets of *successful tender offers* in the month or two surrounding the offer are uniformly positive ranging from 16.9 percent to 34.1 percent, and the weighted average abnormal return across the seven studies is 29.1 percent.

Reprinted by permission from 11 *Journal of Financial Economics* 5 (Amsterdam: Elsevier Science Publishers, 1983).

For targets of *successful mergers,* the estimated abnormal returns immediately around the merger announcement range from 6.2 percent to 13.4 percent, and the weighted average abnormal return is 7.7 percent. . . . The weighted average one-month return is 15.9 percent which is about twice the magnitude of the two-day abnormal returns. This comparison suggests that almost half of the abnormal returns associated with the merger announcements occur prior to their public announcement.

Abnormal returns from the first public announcement through the outcome day incorporate all effects of changing information regarding the offer that occur after the initial announcement. These returns are the most complete measures of the profitability of the mergers to target shareholders, but they underestimate the gains to target shareholders because they do not include the premium on shares purchased by the bidder prior to the completion of the merger. Dodd and Asquith report these total abnormal returns for successful targets as 34 percent and 15.5 percent respectively, and the weighted average of the two estimates is 20.2 percent.

Unsuccessful Target Returns

The weighted average abnormal returns to stockholders of target firms involved in *unsuccessful tender offers* is 35.2 percent. The comparable one-month abnormal return for targets of *unsuccessful mergers* is 17.2 percent. These weighted average abnormal returns for targets of unsuccessful takeover attempts are approximately equal to those for targets of successful takeovers. Hence, on average the market appears to reflect approximately equal expected gains for both successful and unsuccessful takeovers at the time of the first public announcement. However, one-month announcement abnormal returns are an insufficient measure of stock price changes associated with unsuccessful takeover attempts because they do not include the stock price response to the information that the offer failed. The correct measure of the wealth effects, therefore, is the cumulative return from the offer through the termination announcement. The weighted average return to *unsuccessful merger* targets from the initial announcement through the outcome date is -2.9 percent. Thus, all of the announcement gains are lost over the time that the merger failure becomes known.

In contrast to the behavior of stock prices of targets of unsuccessful mergers, stock prices of targets of *unsuccessful tender offers* remain substantially above their preoffer level even after the failure of the offer. Unfortunately, the tender offer studies do not present data on the cumulative abnormal return for unsuccessful tender offers from the initial announcement through the outcome date. Nevertheless, some information can be extracted from the abnormal returns following the initial announcement. Dodd and Ruback find an abnormal return of -2.65 percent for targets of unsuccessful tender offers in the month following the initial announcement, but the cumulative abnormal return over the entire year following the announcement is only -3.25 percent ($t = 0.90$).

Bradley, Desai, and Kim analyze the postfailure price behavior of a sample of 112 targets of *unsuccessful tender offers* that they segment into two categories: 86 targets that received subsequent takeover offers and 26 targets that did not receive such offers. Returns in the announcement month for the two subsamples are 29.1 percent and 23.9 percent respectively, and both are statistically significant. From the announcement month of the initial unsuccessful offer through the following two years, the average abnormal return for the targets that *received* subsequent offers is 57.19 percent ($t = 10.39$). In contrast, the average abnormal return over the same two-year period for targets that *did not receive* subsequent offers is an insignificant -3.53 percent ($t = -0.36$), and recall this return includes the announcement effects. Thus, the positive abnormal returns associated with unsuccessful tender offers appear to be due to the anticipation of subsequent offers; target shareholders realize additional positive abnormal returns when a subsequent offer is made, but lose the initial announcement gains if no subsequent offer occurs.

Summary: The Returns to Targets

In summary, the evidence indicates that targets of *successful tender offers* and *mergers* earn significantly positive abnormal returns on announcement of the offers and through completion of the offers. Targets of *unsuccessful tender offers* earn significantly positive abnormal returns on the offer announcement and through the realization of failure. However, those targets of unsuccessful tender offers that do not receive additional offers in the next two years lose all previous announcement gains, and those targets that do receive new offers earn even higher returns. Finally, targets of *unsuccessful mergers* appear to lose all positive returns earned in the offer announcement period by the time failure of the offer becomes known.

Bidding Firm Stockholder Returns

Successful Bidders

The abnormal returns for bidders in *successful tender offers* are all significantly positive and range from 2.4 percent to 6.7 percent with a weighted average return of 3.8 percent. Thus, bidders in successful tender offers realize significant percentage increases in equity value, although this increase is substantially lower than the 29.1 percent return to targets of successful tender offers.

The evidence on bidder returns in *mergers* is mixed and therefore more difficult to interpret than that for bidders in tender offers. On the whole it suggests that returns to bidders in mergers are approximately zero. The two-day abnormal returns associated with the announcement of a merger proposal differ considerably across studies. Dodd finds a significant abnormal return of -1.09 percent for 60 successful bidders on the day before and the day of the first public announcement of the merger—indicating that merger bids are, on average, negative net present value investments for bidders. However, over the same two-day period, Asquith and Eckbo report slightly positive, but statistically insignificant, abnormal returns—suggesting that merger bids are zero net present value investments. In contrast to the mixed findings for the immediate announcement effects, all five estimates of the one-month announcement effects are positive, but only the estimate of 3.48 percent by Asquith, Bruner, and Mullins is significantly different from zero. The weighted averages are 1.37 percent for the one-month announcement effects and -0.05 percent for the two-day announcement effects.

[Two studies] report the total abnormal return for successful bidding firms from the initial announcement day through the outcome announcement day. If the initial announcement is unanticipated, and there are no other information effects, this cumulative abnormal return includes the effects of all revisions in expectations and offer prices and therefore is a complete measure of the equity value changes for successful bidders. The weighted average of the two estimates is -1.77 percent, and the individual estimates are -7.22 percent for 60 successful bidders and -0.1 percent for 196 successful bidders.

The estimated abnormal returns to successful bidding firms in all six studies suggest that mergers are zero net present value investments for bidders—except for the Dodd estimates. . . .

Malatesta provides estimates of total abnormal *dollar* returns to the equity holders of successful bidding firms in the period 1969–74 that are consistent with Dodd's results. He reports an average loss of about \$28 million ($t = -1.85$) in the period four months before through the month of announcement of the merger outcome (indicated by announcement of board/management approval) of the merger.

Unsuccessful Bidders

Inferences about the profitability of takeover bids can also be made from the behavior of bidding firm stock prices around the time of termination announcements for unsuccessful acquisition attempts. Positive abnormal returns to a bidding firm in response to the announcement that a takeover attempt is unsuccessful (for reasons other than bidder cancellation) are inconsistent with the hypothesis that takeovers are positive net present value investments. Dodd reports insignificant average abnormal returns of 0.9 percent for 19 bidders on the day before and day of announcement of merger termination initiated by targets. If mergers are positive return projects, these target termination announcement returns should be negative. Dodd also reports positive termination announcement returns of 1.38 percent for 47 bidders in his bidder termination subsample. These positive returns are consistent with the hypothesis that bidders maximize shareholder wealth and cancel mergers after finding out they overvalued the target on the initial offer.

Ruback uses data on unsuccessful bidders to test directly for value-maximizing behavior of bidders. He argues that wealth-maximizing bidders will abandon takeover attempts when increments in the offer price would make the takeover a negative net present value investment. For 48 bidders in competitive tender offers (defined by the presence of multiple bidders), he finds the average potential gain to the unsuccessful bidder from matching the successful offer price is $-$91 million ($t = -4.34$). The potential gain is calculated as the abnormal bidder equity value change associated with the original announcement of the unsuccessful bid minus the additional cost if the higher successful bid were matched. Furthermore, 41 bidders did not match higher offer prices that would have resulted in a negative net present value acquisition. These results are consistent with value-maximizing behavior by bidding firms.

Problems in Measuring Bidder Returns

There is reason to believe the estimation of returns is more difficult for bidders than for targets. Since stock price changes reflect changes in expectations, a merger announcement will have no effect if its terms are fully anticipated in the market. Furthermore, targets are acquired once at most, whereas bidders can engage in prolonged acquisition programs. Malatesta and Schipper and Thompson point out that the present value of the expected benefits of a bidder's acquisition program is incorporated into the share price when the acquisition program is announced or becomes apparent to the market. Thus, the gain to bidding firms is correctly measured by the value change associated with the initial information about the acquisition program and the incremental effect of each acquisition. The abnormal returns to bidding firms associated with mergers measure only the incremental value change of each acquisition and are therefore potentially incomplete measures of merger value to successful bidders.

Bidding firms do not typically announce acquisition programs explicitly; this information is generally revealed as the bidders pursue takeover targets. However, Schipper and Thompson find that for some firms the start of a takeover program can be approximately determined. They examine the stock price behavior of thirty firms that announced acquisition programs during the period 1953 through 1968. The information that these firms intended to pursue an acquisition program was revealed either in annual reports or specific announcements to the financial press, or in association with other corporate policy changes. . . . Consistent with the hypothesis that mergers are positive net present value investments for bidding firms, they find abnormal returns of 13.5 percent ($t = 2.26$) for their sample of thirty firms in the twelve months prior to and including the "event month." However, the imprecise announcement month, the resultant necessity for measuring abnormal returns over a twelve-month interval, and contemporaneous changes in corporate policy make it difficult to determine with confidence the association between positive abnormal returns and initiation of the acquisition program. For example, suppose "good luck" provided bidder management with additional resources to try new projects such as mergers. As Schipper and Thompson discuss, in this case stock prices would show the pattern evidenced in their study even if the mergers have zero net present value. . . .

In addition to the problems caused by prior capitalization of the gains from takeover bids, measuring the gains to bidding firms is also difficult because bidders are generally much larger than target firms. Thus, even when the dollar gains from the takeover are split evenly between bidder and target firms, the dollar gains to the bidders translate into smaller percentage gains. Asquith, Bruner, and Mullins report that the abnormal returns of bidding firms depend on the relative size of the target. For 99 mergers in which the target's equity value is 10 percent or more of the bidder's equity value, the average abnormal return for *bidders* is 4.1 percent ($t = 4.42$) over the period twenty days before through the day of announcement. For the 115 remaining mergers in which the target's equity value is less than 10 percent of the bidder's equity value, the average abnormal return for bidders is 1.7 percent ($t = 2.00$). Furthermore, the precision of the estimated gains is lower for bidders than for targets because the normal variation in equity value for the (larger) bidder is greater, relative to a given dollar gain, than it is for the target. Thus, even if the gains are split equally, the relative sizes of bidding and target firms imply that both the average abnormal return and its t-statistic will be smaller for bidding firms. . . .

Summary: The Returns to Bidders

The reported positive returns to *successful bidders* in *tender offers* and the generally negative returns to *unsuccessful bidders* in both *mergers* and *tender offers* are consistent with the hypothesis that mergers are positive net present value projects. The measurement of returns to bidders in mergers is difficult, and perhaps because of this the results are mixed. The evidence suggests, however, that returns to *successful bidding* firms in *mergers* are zero. Additional work on this problem is clearly warranted.

The Total Gains from Takeovers

The evidence indicates that shareholders of target firms realize large positive abnormal returns in completed takeovers. The evidence on the rewards to bidding firms is mixed, but the weight of the evidence suggests zero returns are earned by successful bidding firms in mergers and that statistically significant but small positive abnormal returns are realized by bidders in successful tender offers. Since targets gain and bidders do not appear to lose, the evidence suggests that takeovers create value. However, because bidding firms tend to be larger than target firms, the sum of the returns to bidding and target returns do not measure the gains to the merging firms. The dollar value of small percentage losses for bidders could exceed the dollar value of large percentage gains to targets. . . . Bradley, Desai, and Kim report positive but statistically insignificant total dollar gains to bidders and targets in 162 tender offers of $17.2 million ($t = 1.26$). However, the average percentage change in total value of the combined target and bidder firms is a significant 10.5 percent ($t = 6.58$). This evidence indicates that changes in corporate control increase the combined market value of assets of the bidding and target firms. . . .

The Source of Takeover Gains [Market Power Tests]

Stillman and Eckbo use the equity price changes of firms that *compete* in product markets with the merged target to reject the hypothesis that takeovers create market power. The market power hypothesis implies that mergers increase product prices thereby benefiting the merging firms and other competing firms in the industry. Higher prices allow competing firms to increase their own product prices and/or output, and therefore the equity values of competing firms should also rise on the offer announcement.

Stillman examines the abnormal returns for rival firms in eleven horizontal mergers. The small sample size arises from his sample selection criteria. Of all mergers challenged under Section 7 of the Clayton Act, these eleven are the merger complaints in unregulated industries whose rivals were identified in the proceedings and for which constraints on data availability were met. While this screening process creates a small sample, it reduces ambiguity about the applicability of the test and the identity of rivals. He finds no statistically significant abnormal returns for rival firms in nine of the mergers examined. Of the remaining two mergers, one exhibits ambiguous results and the other is consistent with positive abnormal returns for rivals. Stillman's evidence, therefore, is inconsistent with the hypothesis that the gains from mergers are due to the acquisition of market power.

Eckbo uses the stock price reaction of rivals at the announcement of the antitrust challenge as well as at the announcement of the merger to test the market power hypothesis. Eckbo's final sample consists of 126 challenged horizontal mergers and, using product line classifications rather than records of court and agency proceedings, he identifies an average of 15 rivals for each merger. He also identifies rivals for 65 unchallenged horizontal mergers and 58 vertical mergers.

Eckbo's results indicate that rival firms have positive abnormal returns around the time of the first public announcement of the *merger*. Rivals of unchallenged mergers realized abnormal returns of 1.1 percent ($t = 1.20$) and rivals of challenged mergers realized abnormal returns of 2.45 percent ($t = 3.02$) in the period twenty days prior to and ten days following the first public announcement. These results are consistent with the market power hypothesis.

Eckbo uses the stock price reactions of *rivals* at announcement of the *antitrust challenge* to reject the market power hypothesis. The market power hypothesis predicts negative abnormal returns for rival firms at the time the complaint is filed because the complaint reduces the probability of completion of the merger (which, it is assumed, would have generated market power), and the concomitant increase in output prices is then less probable. In the period twenty days before through ten days after the antitrust challenge, the rivals to 55 challenged mergers realize statistically insignificant average abnormal returns of 1.78 percent ($t = 1.29$). This finding is inconsistent with the market power hypothesis, which implies the returns of rivals should be significantly negative at the complaint announcement. Furthermore, Eckbo reports that rivals with a positive market reaction to the initial merger announcement do not tend to have negative abnormal returns at the time of complaint. Thus, Eckbo's evidence is inconsistent with the market power hypothesis.

The Market for Corporate Control: The Empirical Evidence Since 1980

GREGG A. JARRELL, JAMES A. BRICKLEY, AND JEFFRY M. NETTER

Returns to Shareholders of Target Companies

Shareholders of target companies clearly benefit from takeovers. Jarrell and Poulsen estimate the premiums paid in 663 successful tender offers from 1962 to December 1985. They find that premiums averaged 19 percent in the 1960s, 35 percent in the 1970s, and from 1980 to 1985 the average premium was 30 percent. These figures are consistent with the thirteen studies of pre-1980 data contained in Jensen and Ruback which agree that targets of successful tender offers and mergers before 1980 earned positive returns ranging from 16 percent to 30 percent for tender offers.

Similar results are contained in studies of leveraged buyouts and going-private transactions. Lehn and Poulsen find premiums of 21 percent to shareholders in 93 leveraged buyouts taking place from 1980 to 1984. DeAngelo, DeAngelo, and Rice find an average 27 percent gain for leveraged buyouts between 1973 and 1980. . . .

Using a comprehensive sample of 225 successful tender offers from 1981 through 1984, including over-the-counter targets, OCE [SEC Office of Chief Economist] finds the average premium to shareholders to be 53.2 percent. OCE has updated these figures for 1985 and 1986 and finds a decrease over the last two years. OCE finds that the average premium is 37 percent in 1985 and 33.6 percent in 1986.

Reprinted by permission from 2 *Journal of Economic Perspectives* 49. © 1988 American Economic Association.

While the evidence reported thus far indicates substantial gains to target shareholders, it probably understates the total gains to these shareholders. In many cases events occur before a formal takeover offer, so studies that concentrate on the stock price reactions to formal offers will understate the total gains to shareholders.

Several recent empirical studies examine the stock market reaction to events that often precede formal steps in the battle for corporate control. Mikkelson and Ruback provide information on the stock price reaction to Schedule 13D filings. Schedule 13D must be filed with the SEC by all purchasers of 5 percent of a corporation's common stock, requiring disclosure of, among other things, the investor's identity and intent. Mikkelson and Ruback find significant price reactions around the initial announcement of the filing, and that the returns depend on the intent stated in the 13D. The highest returns, an increase of 7.74 percent, occurred when the filer in the statement of intent indicated some possibility of a control change. However, the abnormal returns were only 3.24 percent if the investor reported the purchase was for investment purposes. Holderness and Sheehan find a differential stock market effect to 13D filings depending on the identity of the filer. They show the filings of six "corporate raiders" increased target share prices by a significantly greater amount than a sample of other filers (5.9 percent to 3.4 percent). . . .

Returns to Shareholders of Acquiring Companies

The 1980s evidence on bidders comes from Jarrell and Poulsen with data on 663 successful tender offers covering 1962 to 1985. . . . For the entire sample period bidders on average realized small, but statistically significant, gains of about 1 to 2 percent in the immediate period around the public announcement. Most interesting is the apparent secular decline in the gains to successful bidders in tender offers. Consistent with the previous studies reviewed by Jensen and Ruback, [bidders experience] positive excess returns of 5 percent during the 1960s, and a lower, but still significantly significant, positive average of 2.2 percent over the 1970s. However, the 159 cases from the 1980s show statistically insignificant losses to bidders.

How the Distribution of Takeover Gains Is Determined

Companies that are targets of takeovers receive the bulk of the value created by corporate combinations and these gains are not offset by losses to acquirers. As one might predict, an important factor in determining how these takeover gains are split seems to be how many bidders are trying to acquire the target company. In fact, the secular decline in the stock returns to bidders probably reflects the increased competition among bidders and the rise of auction-style contests during the 1980s.

Conditions which foster an increase in multiple bidding tend to increase target premiums and reduce bidder returns. For example, Jarrell and Bradley demonstrate that federal (Williams Act) and state regulations of tender offers have this effect because they impose disclosure and delay rules that foster multiple-bidder auction contests and preemptive bidding. In addition to greater regulation, other factors contributing to this increased competition include court rulings protecting defensive tactics, the inventions of several defenses against takeovers, and the increase in sophisticated takeover advisers to implement them. . . .

Sources of Takeover Gains

Short-Term Myopia and Inefficient Takeovers

This theory is based on an allegation that market participants, and particularly institutional investors, are concerned almost exclusively with short-term earnings performance and tend to undervalue corporations engaged in long-term activity. From this viewpoint, any corporation planning for long-term development will become undervalued by the market as its resource commitments to the long-term depress its short-term earnings, and thus will become a prime takeover candidate.

Critics of this theory point out that it is blatantly inconsistent with an efficient capital market. Indeed, if the market systematically undervalues long-run planning and investment, it implies harmful economic consequences that go far beyond the costs of inefficient takeovers. Fortunately, no empirical evidence has been found to support this theory. In fact, a study of 324 high research and development firms and of all 177 takeover targets during 1981–84 by the SEC's Office of the Chief Economist shows evidence that (1) increased institutional stock holdings are not associated with increased takeovers of firms; (2) increased institutional holdings are not associated with decreases in research and development; (3) firms with high research and development expenditures are not more vulnerable to takeovers; and (4) stock prices respond positively to announcements of increases in research and development expenditures.

Further evidence opposing the myopia theory is provided by Hall in an NBER study and by McConnell and Muscarella. Hall studies data on acquisition activity among manufacturing firms from 1977 to 1986. She presents evidence that much acquisition activity has been directed toward firms and industries which are less intensive in R&D activity. She also finds that firms involved in mergers show little difference in their pre- and postmerger R&D performance compared with industry peers. McConnell and Muscarella, in a study of 658 capital expenditure announcements, show that stock prices respond positively to announcements of increased capital expenditures, on average, except for exploration and development announcements in the oil industry.

Undervalued Target Theory

Recalcitrant target management and other opponents of takeovers often contend that because targets are "undervalued" by the market, a savvy bidder can offer substantial premiums for target firms while still paying far below the intrinsic value of the corporation. By this theory, it becomes the duty of target managements to defend vigorously against even high premium offers since remaining independent, it is argued, can offer shareholders greater rewards over the long term than are offered by opportunistic bidders seeking short-term gains.

However, the evidence shows the promised long-term gains from remaining independent do not usually materialize. When a target defeats a hostile bid, its postdefeat value reverts to approximately the (market adjusted) level obtaining before the instigation of the hostile bid). . . . Bhagat, Brickley, and Lowenstein used option pricing theory to show that the announcement period returns around cash tender offers are too large to be explained by revaluations due to information about undervaluation.

This evidence indicates that the market does not, on average, learn much of anything that is new or different about target firms' intrinsic values through the tender offer process, despite the tremendous attention lavished on targets, and the huge amounts of information traded among market participants during takeover contests. If undervaluation had indeed been present, then the deluge of new information on the intrinsic value of targets should have caused fundamental price corrections even in the event of takeover defeats. But in the overwhelming majority of cases studied, prices dropped rather than increased for target firms that fought off takeovers.

Hostile Takeovers in the 1980s: The Return to Corporate Specialization

SANJAI BHAGAT, ANDREI SHLEIFER, AND ROBERT W. VISHNY

We examine the sample of all sixty-two hostile takeover contests between 1984 and 1986 that involved a purchase price of $50 million or more. In these contests, fifty targets were acquired and twelve remained independent. We use a sample of hostile takeovers exclusively to avoid using evidence from friendly acquisitions to judge hostile ones, as many studies have done. We examine such posttakeover operational changes as divestitures, layoffs, tax

Reprinted by permission from *Brookings Papers on Economic Activity: Microeconomics 1990 1* (Washington, D.C.: Brookings Institution, 1990), p. 1.

savings, and investment cuts to understand how the bidding firm could justify paying the takeover premium. We also examine the possibility of wealth losses by bidding firms' stockholders as the explanation for target shareholder gains.

The analysis of posttakeover changes is complicated because once the target and the bidding firms are merged, it becomes impossible to attribute to the target the changes recorded in joint accounting data. As a consequence, we do not use such data, but rather focus on discussion in annual reports, 10K forms, newspapers, magazines, Moody's and Value Line reports, and other such sources. . . . The advantage of this design is that we can attribute the changes we examine, such as layoffs and selloffs, to the target firm. The disadvantage is that most changes we examine are biased downward because some may not be reported.

Our calculations suggest that, on average, taxes and layoffs each explain a moderate fraction of the takeover premium. Layoffs, which disproportionately affect high-level white-collar employees, explain perhaps, 10 to 20 percent of the average premium, although in a few cases they are the whole story. Tax savings are usually somewhat smaller than savings from layoffs (although they are significant in a larger *number* of cases), since debt is typically repaid fairly fast. But tax reductions are very large in management buyouts, acquisitions by partnerships, and acquisitions by firms with tax losses. Large investment cuts occur infrequently in our sample, and do not appear to be an important takeover motive. Wealth declines of the bidding firms' shareholders, similarly, while important in a few cases, are usually small and cannot be a systematic source of target shareholders' gains.

Our most significant finding is that most hostile takeover activity results in allocation of assets to firms in the same industries as those assets. In most hostile takeovers, the bidding firm is in the same business as, or a business closely related to, that of the target firm. Similarly, the majority of selloffs, which amount to 30 percent of the acquired assets, are to buyers in the same business as the assets they acquire. Overall, of the assets that changed hands in our sample, 72 percent ended up owned by corporations with other similar assets. By and large, hostile takeovers represent the deconglomeration of American business and a return to corporate specialization.

These findings have significant implications for explaining the sources of gains in hostile takeovers. First, they suggest that the places to look for the gains are cost savings from joint operations, market power, or possibly overpayment by buyers of divisions and whole companies. Some of these gains might be from eventual layoffs that we document, but others we might not be able to capture. In any event, changes that result from consolidation of industries are essential for understanding takeover gains. Second, the findings suggest that incentive-intensive organizations, such as management buyout teams, investment companies, or raiders, are not very important in the long run. In our sample, only 20 percent of the assets ended up under control of such organizations after two to three years, and this fraction would surely dwindle if we looked at the assets over a longer period. Control by raiders or by MBO teams is often a transitory arrangement used to allocate assets to corporations managing other similar assets. . . .

The premiums in this sample vary significantly, and some of them are very small when adjusted for market movements. These are typically the cases of bids highly contested by the target's management where the bidder gained control in part by buying shares in the open market. In general, the premiums are somewhat smaller than those one would obtain without the market correction because the market rose during this period. . . .

The results are similar to the usual findings for bidders, except we find that bidder returns are negative in more than half the cases. On average, the bidders lose $15 million, a tiny fraction of the average acquisition price of $1.74 billion (including debt). Unavailability of data might bias these results toward finding poorer performance by bidders, since raiders typically bid through private firms and their returns are more likely to be positive. Note, however, that Irwin Jacobs, in two acquisitions by Minstar, earned a negative market-corrected return in both cases.

Bidder returns are very negative in some related acquisitions, such as Occidental's buying Midcon and Marriott's buying Saga. On the other hand, other strategic acquisitions, such as Citicorp's buying Quotron and Coastal's buying American Natural Resources, result in increases in the wealth of bidding shareholders. We cannot conclude from this sample that related acquisitions are systematically good or bad for the bidder, although earlier work by Morck, Shleifer, and Vishny suggests that in this period related acquisitions are better for the bidders than unrelated ones.

[The combined wealth change of the bidder and the target for the available observations] is positive in all but three cases, each of which is a strategic acquisition by a firm extending its product line. In general, most of the gains go to the target, and the bidder wealth change is relatively small, just as the other studies find. It is *not* the case in this sample that target gains can often be explained as bidder losses. . . .

[The authors reach the following tentative conclusions for policy analysis from the data—ED.] First, hostile takeovers do not result in massive employment cuts in acquired companies. State antitakeover laws that aim to stop takeovers to protect blue-collar workers are misguided. Since such laws probably stop some takeovers that foster specialization of corporations, they are more likely than not to reduce efficiency. Second, Reagan's lenient antitrust enforcement of the 1980s indirectly fostered deconglomeration of the U.S. economy. Since the experience with conglomerates seems almost uniformly disappointing, the move toward specialization probably on balance raises efficiency. In some cases, competition is probably reduced, but our case studies suggest that there are many business reasons for related acquisitions other than to raise prices. Unfortunately, we do not have the highly disaggregated market share data necessary to evaluate the precise scope for increased market power in our sample of acquisitions. On balance, however, the evidence suggests to us that the Reagan antitrust stance has had a positive influence on the economy.

Notes and Questions

1. A number of empirical studies support Manne's thesis. Management turnover is higher after a takeover than after a merger or in the ordinary course of events, and the targets whose managers are replaced earned negative abnormal returns, compared to their industries, before the takeover, unlike the targets whose managers are retained. See Eugene Furtado and Vijay Karan, "Causes, Consequences, and the Shareholder Wealth Effects of Management Turnover: A Review of Empirical Evidence," 19 *Financial Management* 60 (1990); Kenneth J. Martin and John J. McConnell, "Corporate Performance, Corporate Takeovers, and Managerial Turnover" 46 *Journal of Finance* 671 (1991). Although the high turnover rate does not prove that departing executives were poor managers, if we did not observe turnover at the top after a takeover, and if turnover were not related to target performance, then it would be difficult to accept Manne's thesis. In addition, targets of hostile takeovers (after which management is most likely to be replaced) are poor performers compared to targets of friendly acquisitions. Randall Morck, Andrei Shleifer, and Robert Vishny, "Characteristics of Targets of Hostile and Friendly Takeovers," in A. Auerbach, ed., *Corporate Takeovers: Causes and Consequences* 101 (Chicago: University of Chicago Press, 1988).

2. To what extent will competition in product or factor markets constrain managers from overinvesting free cash flow? Reconsider Easterbrook's thesis in Chapter IV concerning the monitoring function of dividends in light of Jensen's comparison with debt: can managers credibly commit themselves to maintaining a favorable distribution policy with dividends as well as they can with debt? In support of Jensen's thesis concerning the benefits of debt on management's use of free cash, Michael Maloney, Robert McCormick, and Mark Mitchell find that bidder returns increase with its leverage (presumably because there is less free cash or more creditor monitoring, ensuring that acquisitions are value maximizing). Maloney, McCormick, and Mitchell, "Managerial Decision Making and Capital Structure," 66 *Journal of Business* 189 (1993).

3. Would the free cash flow explanation lead you to expect takeovers financed with debt to produce higher returns than those financed by an exchange of stock? See James W. Wansley, William R. Lane, and Ho C. Yang, "Gains to Bidder Firms in Cash and Securities Transactions," 22 *Financial Review* 403 (1987). Jensen provides further evidence in his article consistent with the free cash flow theory: the high cash flow and low growth opportunities of the industries (exemplified by oil) that were the subject of significant takeover activity during the 1980s. There is also more direct evidence in support of the free cash flow explanation. In a study of LBOs, Kenneth Lehn and Annette Poulsen find that the likelihood of a firm going private is directly

related to the size of its free cash flows and the threat of a hostile takeover and inversely related to growth (favorable internal reinvestment opportunities). Lehn and Poulsen, "Free Cash Flow and Stockholder Gains in Going Private Transactions," 44 *Journal of Finance* 771 (1989). In addition, the size of the LBO premium is positively related to the size of the firms' undistributed cash flows.

4. A simpler explanation of LBO gains is the improved incentives from the transactions' ownership changes, in which management's equity dramatically increases. This is, of course, a straightforward application of the incentive compensation issues discussed by Smith and Watts in Chapter V. Management-led buyout firms experience significant increases in productivity postbuyout, and the value increases are directly related to management's ownership shares. See Frank R. Lichtenberg and Donald Siegel, "The Effects of Leveraged Buyouts on Productivity and Related Aspects of Firm Behavior," 27 *Journal of Financial Economics* 165 (1990); Abbie Smith, "The Effects of Leveraged Buyouts," 25 *Business Economics* 19 (April 1990); Mike Wright, Ken Robbie, and Steve Thompson, "Corporate Restructuring, Buy-outs, and Managerial Equity: The European Dimension," 3 *The Continental Bank Journal of Applied Corporate Finance* 46 (Winter 1991).

5. There is another efficiency explanation of takeovers besides agency cost reduction explanations: realization of synergy gains. According to this explanation, the combined firm is worth more than the value of the two firms taken separately. The increased value may come from real operating synergies (such as economies of scale or scope) or financial synergies (such as reduction in bankruptcy risk, improved usage of tax shields, or reduction in cost of capital from increased internal financing). An example of an economy of scale is a merger that spreads fixed costs over a larger volume of production; an example of an economy of scope is a merger that combines complementary resources, such as one firm's unique product and the other's sales organization to market it. For further discussion of synergy gain explanations, see Roberta Romano, "A Guide to Takeovers: Theory, Evidence and Regulation," 9 *Yale Journal on Regulation* 119 (1992); J. Fred Weston, Kwang Chung, and Susan Hoag, *Mergers, Restructuring and Corporate Control* (Englewood Cliffs, N.J.: Prentice-Hall, 1990).

6. Shleifer and Summers's breach of trust explanation should be considered in conjunction with Figure I.1 in Williamson, Chapter I. Why would stakeholders accept implicit contracts rather than explicit protective governance structures (locate at node *B* rather than *C*)? Would you expect parties at node *B* to charge shareholders more to participate in the firm (i.e., workers demand higher present wages) in exchange for the lack of protection that implicit contracts rather than explicit governance structures provide? Would such behavior undercut Shleifer and Summers's thesis?

7. Can you think of examples of implicit labor contracts that bidders could profitably breach? Consider overfunded pension plans. The assets of such plans

exceed the employees' promised benefits. Although most labor contracts do not provide for cost-of-living adjustments to pensioners, firms often use a plan's excess assets to provide such benefits voluntarily. Would you consider such a practice to be an implicit contract? The excess assets belong to the firm and not the employees, but to obtain the cash for its own use, the employer must terminate the plan under federal law (and since 1986 is subject to a punitive excise tax). Would a takeover that terminated a pension plan and reverted excess assets to the firm be a breach of trust? Does it matter to the analysis whether or not the employees covered by the plan have firm-specific skills?

One way to understand the difference between explicit and implicit contracts is their enforceability in court. An explicit contract's terms are observable to the contracting parties and courts (an outsider can verify whether it was performed). An implicit contract, however, is a contract whose terms are observable to the contracting parties but not verifiable (not observable to third parties like courts). That is, a contract is implicit because its terms cannot be enforced; otherwise, the parties would protect themselves through an explicit contract. Are pension benefits verifiable? Would it be difficult to draft an explicit contract providing contingency cost-of-living increases for pensioners? Is the firm's profitability an implicit term or condition in the use of overfunded pension plan assets to fund cost-of-living increases? Is it observable or verifiable?

Consider, as a second illustration of an implicit contract, Shleifer and Summers's paradigmatic example involving Carl Icahn's acquisition of Trans World Airlines (TWA). Icahn achieved substantial labor concessions from TWA's unions, and Shleifer and Summers estimate that the cost to labor equaled 38 percent of the premium paid to TWA shareholders. J. Fred Weston, Kwang Chung, and Susan Hoag suggest two additional sources of the reduction in labor cost besides extraction of the value of firm-specific capital: (1) union power (i.e., the union had been obtaining monopoly rents); and (2) incumbent management inefficiency caused by failure to bargain effectively with unions. Weston, Chung, and Hoag, *Mergers, Restructuring and Corporate Control* 214 (Englewood Cliffs, N.J.: Prentice-Hall, 1990). Which of these explanations do you think is most plausible?. What would be examples of human-specific capital in airlines? What would be the significance for Shleifer and Summers's argument if wages and benefits for the same jobs were lower at other airlines? Is this their Scenario B? Does a posttakeover decrease in employment or wage levels imply an efficiency loss? How is the analysis affected by the additional fact that TWA's labor contracts were negotiated under a regulatory regime and the takeover occurred after the airline industry was deregulated? In responding to this question, consider the finding of researchers that labor was a prime beneficiary of airline regulation, for example, Elizabeth Bailey, David Graham, and Daniel Kaplan, *Deregulating the Airlines* (Cambridge, Mass.: MIT Press, 1985). Should Shleifer and Summers consider airline consumers in their calculation of the social benefits from Icahn's bid? Is the experience of airline takeovers generalizable to other sectors of the economy?

8. Several studies have attempted to test the explanation of takeover gains as a wealth transfer from employees to shareholders. A summary of the findings follows:

> The labor expropriation explanation, in general or as refined by Shleifer and Summers, has scant empirical support. Pontiff, Shleifer, and Weisbach find, for instance, that pension fund asset reversions are too small to be a dominant motive for takeovers: although they are more frequent in hostile than friendly transactions, reversions occur in only 14 percent of takeovers, and they average only 10 to 13 percent of the premiums. Mitchell and Mulherin provide similar results: pension fund reversions occur in 12 percent of the takeovers in their sample, although they are not more frequent in hostile than in friendly bids, and the reversions account for a somewhat higher proportion of the premium (23 percent). More important, most fund reversions do not occur after a corporate takeover.
>
> Several studies examining the aftermath of takeovers more systematically do not find a similar dramatic impact on labor as in the TWA takeover. . . . [I]t is middle management (administrative staff), and not production plant employees, whose ranks are slimmed down after acquisitions. Consistent with these data, Blackwell, Marr, and Spivey find, in a sample of 286 plant closings, that very few (48) were announced by takeover targets, either before, after or during the bid, and only 22 of those were targets of hostile bids. Moreover, firms experiencing ownership changes have higher employment and wage levels, and increased productivity compared to firms that do not change control. Finally, Kaplan finds in a sample of leveraged buyout firms that employment increased after the transaction (although it is not as large an increase as that of their industries), while Lichtenberg and Siegel find leveraged buyout firms' employment declined, compared to their industries, but at a slower rate than before the buyout.
>
> Rosett tests Shleifer and Summers's breach of contract explanation more directly by examining union wage contracts before and after takeovers. He finds no support for their thesis: there is, in fact, a positive *gain* in union wealth levels after hostile acquisitions. Although there are losses after friendly acquisitions, even then the losses are insignificant relative to the premiums (when measured over eighteen years after the takeover, the union losses in friendly acquisitions equal approximately 5 percent of the shareholders' gain). Bhagat, Shleifer, and Vishny also find that layoffs occur infrequently, affect high-level white-collar workers, are higher when management successfully defeats a bid (either by remaining independent or by finding a white knight) than when a hostile bidder succeeds, and, most important, result in losses that are small compared to takeover premiums (10 to 20 percent). In sum, while we would need counterfactual data to test the labor expropriation hypothesis fully—we need to know how many workers would have been laid off or what the wage profile would have looked like if the firm had not been acquired—what we do know suggests that expropriation from labor does not motivate takeovers.

Roberta Romano, "A Guide to Takeovers: Theory, Evidence and Regulation," 9 *Yale Journal on Regulation* 119, 140–42 (1992). Reprinted by permission.

9. A second expropriation explanation of takeover gains is realization of tax savings, a wealth transfer from the fisc. The tax savings most frequently mentioned from acquisitions are the interest deduction on debt, accelerated depreciation deductions from a step-up in asset basis upon purchase, and use of favorable tax attributes, such as net operating losses. How compelling is this explanation? For instance, must a corporation engage in an acquisition to increase its debt level and obtain higher interest deductions? For development of the critique that mergers are not needed to realize the value of such tax benefits, see Ronald J. Gilson, Myron S. Scholes, and Mark A. Wolfson, "Taxation and the Dynamics of Corporate Control: The Uncertain Case for Tax Motivated Acquisitions," in, J. Coffee, L. Lowenstein, and S. Rose-Ackerman, eds., *Knights, Raiders, and Targets: The Impact of Hostile Takeovers* 271 (New York: Oxford University Press, 1988).

In a comprehensive empirical study of several hundred mergers from 1968 to 1983, Alan Auerbach and David Reishus find that reducing taxes is not a significant reason for the acquisitions. Auerbach and Reishus, "The Impact of Taxation on Mergers and Acquisitions," in A. Auerbach, ed., *Mergers and Acquisitions* 69 (Chicago: University of Chicago Press, 1988). Only 20 percent of the mergers could be classified as having tax benefits (transfer of losses or credits or a step-up in asset basis), and the estimated value of these benefits was substantially lower than the takeover premiums (average gain of 13.7 percent of acquired firm's market value). In addition, debt-equity ratios did not increase significantly after the mergers, so increased interest deductions are also an insignificant factor.

Auerbach and Reishus's sample period ends before highly leveraged transactions became a popular acquisition vehicle. Studies of the effects of LBOs obtain quite different results. As in Auerbach and Reishus's merger sample, neither net operating losses nor increased depreciation deductions from asset basis step-ups are significant in explaining LBO gains. But LBO premiums are significantly related to, and substantially explained by, the tax savings from interest deductions. Depending on the assumptions of marginal tax rates and debt refinancing, the tax savings explains from 13 to 130 percent of the premium. See Steven Kaplan, Management Buyouts: Evidence on Taxes as a Source of Value, 44 *Journal of Finance* 611 (1989); Katherine Schipper and Abbie Smith, "Effects of Management Buyouts on Corporate Interest and Depreciation Tax Deductions," 34 *Journal of Law and Economics* 295 (1991). Would this result be surprising to a believer in the efficient-market hypothesis?

Michael Jensen, Steven Kaplan, and Laura Stiglin suggest that estimates of the tax savings from increased interest deductions tell only part of the LBO tax story. They estimate that the net tax effect of LBOs is positive, and not negative, when the increased tax payments of the selling target shareholders and the new debtholders are included in the calculus, as well as the firm's future higher income from productivity increases. Jensen, Kaplan, and Stiglin, "The Effects of LBO's on Tax Revenues," 42 *Tax Notes* 727 (1989). Would their calculations be affected if tax-exempt institutions hold buyout debt? What is the bearing of the finding in the Bhagat, Shleifer, and Vishny selection that LBO firms rapidly reduce their debt levels? Consider also the evidence from reverse buyouts (firms that go public again) that postbuyout shareholders earn very high returns. Steven Kaplan, "Sources of Value in Management Buyouts," in Y. Amihud, ed., *Leveraged Management Buyouts: Causes and Consequences* 95 (Homewood, Ill.: Dow Jones-Irwin, 1989). Because prebuyout shareholders capture the entire tax savings, these data suggest that LBOs produce gains in value well beyond the tax benefits.

10. Another expropriation explanation for takeovers that are highly leveraged is a wealth transfer from bondholders to shareholders. There is not, however, much empirical support for this explanation. Outstanding bond ratings are typically lowered after an LBO, but studies of bond prices find either no significant effect or relatively small statistically significant negative returns, and, most important, the negative returns are nowhere near the magnitude of the gains to the shareholders in these transactions. See Paul Asquith and Thierry Wizman, "Event Risk, Covenants, and Bondholder Returns in Leveraged Buyouts," 27 *Journal of Financial Economics* 195 (1990) (−2.8 percent; but −5 percent for bonds with no covenant protection); Debra Denis and John McConnell, "Corporate Mergers and Security Returns," 16 *Journal of Financial Economics* 143 (1986) (no significant effect); Laurentius Marais, Katherine Schipper, and Abbie Smith, "Wealth Effects of Going Private for Senior Securities," 23 *Journal of Financial Economics* 155 (1989) (no significant effect). In addition, this is an explanation that is temporally bounded, for as the Lehn and Poulsen selection and note 2 in Chapter IV indicate, bondholders can (and do) protect themselves by event-risk indenture provisions.

11. There has not been substantial new research investigating a market power explanation of takeover gains beyond the studies discussed in the Jensen and Ruback selection. Randall Morck, Andrei Shleifer, and Robert Vishny find that the returns to unrelated conglomerate acquisitions in the 1980s are lower than the returns to related acquisitions (at a 10 percent significance level). Morck, Shleifer, and Vishny, "Do Managerial Objectives Drive Bad Acquisitions?" 45 *Journal of Finance* 31 (1990). Does this finding, in conjunction with Bhagat, Shleifer, and Vishny's finding that most hostile takeovers result in a reallocation of assets to related buyers, give more credence to a consumer expropriation explanation? Note, however, that research by Paul Healy, Krishna Palepu, and Richard Ruback suggests that improved cash flows of merged firms are due to increased asset productivity rather than monopoly rents: They find that sales margins do not increase postmerger. Healy, Palepu, and Ruback, "Does Corporate Performance Improve After Mergers?" 31 *Journal of Financial Economics* 135 (1992).

12. There is some evidence consistent with Amihud and Lev's non-value-maximizing diversification explanation of takeovers. They find, in an omitted section, as do William Lloyd, Naval Modani, and John Hand, "The Effect of the Degree of Ownership Control on Firm Diversification, Market Value, and Merger Activity," 15 *Journal of Business Research* 303 (1987), that manager-

controlled firms are more likely to engage in diversifying mergers than are owner-controlled firms. In addition, both of these studies find that the income streams and operations of manager-controlled firms are more diversified than those of owner-controlled firms.

13. Amihud and Lev's explanation of takeovers as a means to reduce managers' risk is a subset of what can more generally be described as managerialist explanations of takeovers. Risk may be a concern, but power, self-aggrandizement, or empire-building may be the goal as well. These are the more classical managerialist or non-value-maximizing explanations of takeovers, for example, Robin Marris, "A Model of the 'Managerial' Enterprise," 77 *Quarterly Journal of Economics* 185 (1963).

Mark Mitchell and Kenneth Lehn studied the relation between bidding firms and target firms and found that firms that engage in acquisitions with negative net present values are subsequently taken over by other firms. Mitchell and Lehn, "Do Bad Bidders Become Good Targets?" 98 *Journal of Political Economy* 372 (1990). This finding is a fascinating marriage of Marris's, Manne's, and Jensen's explanations: firms whose managers engage in non-value-maximizing acquisitions, wasting free cash flow, are themselves acquired, in keeping with the agency cost reduction's hypothesis that takeovers discipline inefficient managers. Takeovers are thus both the epitome of the agency problem and its solution.

Additional support of the managerialist version of Jensen's free cash flow explanation (that it is acquirers, not targets, that have the free cash) is provided in a recent study by Larry Lang, Rene Stulz, and Ralph Walkling. They find that an increase in free cash flows equal to 1 percent of a bidder's total assets is associated with a decrease in the bidder's gain from the takeover equal to 1 percent of its common stock and that free cash flow explains more of the variation in bidder returns than other variables, such as the number of bidders, target management's resistance, and the bidder's size in relation to the target. Lang, Stulz, and Walkling, "A Test of the Free Cash Flow Hypothesis: The Case of Bidder Returns," 29 *Journal of Financial Economics* 315 (1991). Their data do not provide support for characterizing Jensen's thesis as an expropriation explanation in which wealth is transferred from acquirers to targets, however, because target returns are not affected by the bidder's cash flow.

14. Roll's hubris hypothesis is related to the winner's curse phenomenon in auctions. When the value to the bidders of an auctioned object is uncertain, the winner will have overestimated the value. This is because, as Roll describes, a positive evaluation error produces a winning bid but a negative error does not. The intuition is that the winner pays too much—that is why she won. Winning is bad news (a "curse") because it signifies that all other bidders had lower estimates.

In support of Roll's thesis, Nikhil Varaiya finds evidence that bidders overpay: the difference between the bid premium (measured by the target's abnormal returns) and the combined market value of the bidder and target is positive and significant. Varaiya, "The 'Winner's Curse' Hypothesis and Corporate Takeovers," 9 *Managerial and Decision Economics* 209 (1988). Under what conditions would you expect overpayment—the winner's curse explanation—to be a long-run equilibrium in the takeover market?

The hubris hypothesis is also supported by the findings of negative returns to acquiring firms in the 1980s. While the Gregg Jarrell and Annette Poulsen study cited in Jarrell, Brickley, and Netter found insignificant negative returns to bidders in the 1980s, a study by Michael Bradley, Anand Desai, and E. Han Kim finds that acquirers experienced statistically significantly negative abnormal returns. Bradley, Desai, and Kim, "Synergistic Gains from Corporate Acquisitions and their Division between the Stockholders of Target and Acquiring Firms," 21 *Journal of Financial Economics* 3 (1988); Jarrell and Poulsen, "The Returns to Acquiring Firms in Tender Offers: Evidence from Three Decades," 18 *Financial Management Journal* 12 (1989). But like Jarrell and Poulsen, Bradley, Desai, and Kim find that the aggregate net return of bidders and their targets is still positive, a result reported in several other studies as well, for example, Steven Kaplan and Michael S. Weisbach, "The Success of Acquisitions: Evidence from Divestitures," 47 *Journal of Finance* 107 (1992); Larry H. P. Lang, Rene Stulz, and Ralph A. Walkling, "Managerial Performance, Tobin's *q,* and the Gains from Successful Tender Offers," 24 *Journal of Financial Economics* 137 (1989); Henri Servaes, "Tobin's *Q* and the Gains from Takeovers," 46 *Journal of Finance* 409 (1991). This indicates that takeovers involve more than simply a wealth transfer from bidder shareholders to managers and target shareholders: there is either an efficiency gain or a third-party wealth transfer accompanying the transactions.

15. Studies of ex post performance, like the event studies of acquirer returns, provide mixed evidence of acquirer value maximization. However, unlike event studies, the differences are methodological. In evaluating ex post performance, stock price data are less reliable because the interval over which returns are examined must be lengthy, and the event-study methodology will be inappropriate. Accounting data are therefore used instead. One difficulty in such a study is to determine the appropriate benchmark for evaluating performance, which entails constructing a counterfactual: what the firms' performance would have been had they not been combined. While early studies of postacquisition performance found no operating improvements in merged firms, more recent studies, which make more careful efforts in benchmark construction and adjust postmerger earnings for changes in accounting methods and acquisition financing, find that performance does improve postmerger. Compare Edward S. Herman and Louis Lowenstein, "The Efficiency Effects of Hostile Takeovers," in J. Coffee, L. Lowenstein, and S. Rose-Ackerman, eds., *Knights, Raiders, and Targets: The Impact of Hostile Takeovers* 211 (New York: Oxford University Press, 1988); and David J. Ravenscraft and F. M. Scherer, *Mergers, Selloffs and Economic Efficiency* (Washington, D.C.: Brookings Institution, 1987); with Paul M. Healy, Krishna G. Palepu, and Richard S. Ruback, "Does Corporate Performance Improve After Mergers?" 31 *Journal of Financial Economics* 135 (1992); and Sherry L.

Jarrell, "Do Takeovers Generate Value? Evidence on the Capital Market's Ability to Assess Takeovers," University of Chicago Graduate School of Business Ph.D. dissertation (1991). The difference in findings across studies is probably due to the earlier studies' failure to use an appropriate benchmark and earnings adjustment.

16. As Jensen and Ruback note, target shares experience significant abnormal returns at least a month before the bid announcement date. This pattern is commonly attributed to information leakage. Several researchers have tried to isolate the source of this price run-up: Is it due to illicit insider trading or legitimate trading informed by research and analysis of publicly available information, such as information on toehold block formation, 13D filings, and newspaper articles speculating about potential targets? While all researchers find that approximately 40 percent of the eventual takeover premium is anticipated in the run-up in stock price starting about twenty days before the announcement, their findings vary on the importance of insider trading. Gregg Jarrell and Annette Poulsen conclude that the run-up data are consistent with a legitimate market for information because the most significant explanatory variable of price run-up is media speculation, and an insider trading variable has the wrong sign and is not always significant. Jarrell and Poulsen, "Stock Trading Before the Announcement of Tender Offers: Insider Trading or Market Anticipation?" 5 *Journal of Law, Economics, and Organization* 225 (1989). Lisa Muelbroek, however, concludes that insider trading is a far more important phenomenon, as she finds that 43 percent of the price run-up occurs on insider trading days (as opposed to public news announcement days), and the run-up on such days is greater than on all other sample days. Muelbroek, "An Empirical Analysis of Illegal Insider Trading," 47 *Journal of Finance* 1661 (1992).

Who gains from prebid leakage? Hostile bidders, as the shares move into the hands of arbitrageurs who are likely to support a bid against a strategy of remaining independent? Target shareholders who decide to hold on to their shares because of the unexpected price movement? Incumbent managers who obtain advance warning of a bid from the price movement? Can the SEC do much to prevent illegal trading in this setting? Note that the SEC has adopted Rule 14e-3, which prohibits trading on nonpublic information regarding a tender offer, in order to avoid the fiduciary requirement imposed by the Supreme Court for insider trading when it failed to hold liable a financial printer who traded in target stocks, whose identities he was able to decipher from the information he was typesetting. *Chiarella v. U.S.*, 445 U.S. 222 (1980). The rationales for the rules against insider trading are discussed in the Scott selection in Chapter VII. Reconsider the desirability of prohibiting insider trading in takeovers in conjunction with the readings in the next part of this chapter by Easterbrook and Fischel, Gilson, and Bebchuk, debating the effect on bidder incentives of takeover auctions.

17. There have been very few hostile takeovers in Japan. One explanation, which is suggested by the Roe reading in Chapter V, is that there are

alternative governance structures that substitute as a monitor of managers. This explanation has been advanced by J. Mark Ramseyer, who identifies the following alternative monitoring devices: an extensive pattern of cross-share-holding among corporations and active monitoring by banks, which own stock in the corporations to which they lend. Ramseyer, "Takeovers in Japan: Opportunism, Ideology and Corporate Control," 35 *UCLA Law Review* 1 (1987). A more recent work that comprehensively details a similar thesis is W. Carl Kester, *Japanese Takeovers* (Boston: Harvard Business School Press, 1991). Is cross-holding a superior governance mechanism from the outside shareholders' perspective? Who monitors the Japanese banks?

18. A common perception is that mergers occur in waves, but some economists question this characterization. Compare William F. Shughart II and Robert D. Tollison, "The Random Character of Merger Activity," 15 *RAND Journal of Economics* 500 (1984); with Devra L. Golbe and Lawrence J. White, "Mergers and Acquisitions in the U.S. Economy: An Aggregate and Historical Overview," in A. Auerbach, ed., *Mergers and Acquisitions* 25 (Chicago: University of Chicago Press, 1988). Whatever the answer to this fairly technical debate over the properties of time-series data on mergers, the 1980s was the era of the hostile takeover, which was basically unheard of in prior merger "wave" episodes, and the burst of intensive acquisition activity dramatically slowed in the 1990s. The collapse of the junk bond market and corresponding credit crunch, caused by weakness in the financial services sector and new government policies restricting financial institutions' holding of high-yield debt, surely contributed to the decline in transactions. For evidence that these market conditions, and not the increased use of defensive tactics discussed in part B of this chapter, caused the decline, see Robert Comment and G. William Schwert, "Poison or Placebo? Evidence on the Deterrent and Wealth Effects of Modern Antitakeover Measures," University of Rochester Graduate School of Business (manuscript 1993). However, the international acquisitions market continued to be active, particularly in Europe, where firms sought to position themselves to compete successfully in a fully integrated market.

Would you expect the explanations for foreign acquisitions to be the same as those in the United States? Consider the following data: Frank Lichtenberg finds that, unlike U.S. acquisition targets, foreign acquisition targets do not experience pretransaction profitability declines and are not poorer performers relative to their industry. Lichtenberg, *Corporate Takeovers and Productivity* 100 (Cambridge, Mass.: MIT Press, 1992). In addition, foreign firms' ownership is much more concentrated than that of U.S. corporations. Mike Wright, Ken Robbie, and Steve Thompson, "Corporate Restructuring, Buy-Outs, and Managerial Equity: The European Dimension," 3 *The Continental Bank Journal of Applied Corporate Finance* 47 (Winter 1991).

Management's Fiduciary Duty and Takeover Defenses

The Proper Role of a Target's Management in Responding to a Tender Offer

FRANK H. EASTERBROOK AND DANIEL R. FISCHEL

A cash tender offer typically presents shareholders of the "target" corporation with the opportunity to sell many if not all of their shares quickly and at a premium over the market price. Notwithstanding the apparent benefit both to shareholders of the target and to the acquirer when such offers succeed, the target's management may oppose the offer, arguing that the premium is insufficient or that the corporation would be harmed by its new owners. To defeat the offer, management may file suits against the offeror, sell new shares to dilute the offeror's holdings, manufacture an antitrust problem by acquiring one of the offeror's competitors, or engage in a wide variety of other defensive tactics. Sometimes the resistance leads to the target's being acquired at a price above the initial bid, either by the original bidder or by a "white knight," and sometimes the resistance defeats the takeover attempt altogether.

The ability of management to engage in defensive tactics in response to a cash tender offer is a relatively recent development in contests for corporate control. Prior to the enactment of the Williams Act in 1968, offerors were free to structure offers in a manner designed to force shareholders to decide quickly whether to sell all or part of their shares at a premium. The target's management consequently had little time to mobilize a defensive strategy to impede the offer. The Williams Act, however, has deprived the offeror of this advantage of speed by regulating the conditions under which the offer can be made. More than half the states have enacted tender offer statutes that go beyond the Williams Act in placing restrictions on the ability of an offeror to wage a tender offer. The effect of state and federal regulation of tender offers has been to give the target's management the time to undertake a defensive strategy.

The reaction of shareholders to managerial resistance depends on the outcome. Few protest when resistance leads to a takeover at a higher price.

When resistance thwarts the takeover attempt altogether, however, litigation usually follows. Although defeat of the takeover attempt may deprive the target's shareholders of a substantial premium, shareholders' suits against management to recover this loss are almost always unsuccessful. Relying on the business judgment rule, courts typically have held that the target's management has the right, and even the duty, to oppose a tender offer it determines to be contrary to the firm's best interests. Commentators generally have applauded the results of these cases.

We argue that current legal rules allowing the target's management to engage in defensive tactics in response to a tender offer decrease shareholders' welfare. The detriment to shareholders is fairly clear where defensive tactics result in a defeat of a takeover, causing shareholders to lose the tender premium. Even where resistance leads to a higher price paid for the firm's shares, however, shareholders as a whole do not necessarily benefit. The value of any stock can be understood as the sum of two components: the price that will prevail in the market if there is no successful offer (multiplied by the likelihood that there will be none) and the price that will be paid in a future tender offer (multiplied by the likelihood that some offer will succeed). A shareholder's welfare is maximized by a legal rule that enables the sum of these two components to reach its highest value. Any approach that looks only at the way in which managers can augment the tender offeror's bid, given that a tender offer has already been made, but disregards the effect of a defensive strategy on the number of offers that will be made in the future and the way in which the number of offers affects the efficiency with which corporations are managed, ignores much that is relevant to shareholders' welfare. . . .

Tender offers are a method of monitoring the work of management teams. Prospective bidders monitor the performance of managerial teams by comparing a corporation's potential value with its value (as reflected by share prices) under current management. When the difference between the market price of a firm's shares and the price those shares might have under different circumstances becomes too great, an outsider can profit by buying the firm and improving its management. The outsider reduces the free-riding problem because it owns a majority of the shares. The source of the premium is the reduction in agency costs, which makes the firm's assets worth more in the hands of the acquirer than they were worth in the hands of the firm's managers.

All parties benefit in this process. The target's shareholders gain because they receive a premium over the market price. The bidder obtains the difference between the new value of the firm and the payment to the old shareholders. Nontendering shareholders receive part of the appreciation in the price of the shares.

More significantly for our purposes, shareholders benefit even if their corporation never is the subject of a tender offer. The process of monitoring by outsiders poses a continuous threat of takeover if performance lags. Managers will attempt to reduce agency costs in order to reduce the chance of takeover, and the process of reducing agency costs leads to higher prices for shares. . . . If the company adopts a policy of intransigent resistance and succeeds in maintaining its independence, the shareholders lose whatever premium over market value the bidder offered or would have offered but for the resistance or the prospect of resistance. This lost premium reflects a foregone social gain from the superior employment of the firm's assets.

The target's managers, however, have a substantial interest in preserving their company's independence and thus preserving their salaries and status; the less effective they have been as managers, the greater their interest in preventing a takeover. They may disguise a policy of resistance to all offers as a policy of searching for a better offer than any made so far. Extensive manuals describe both the stratagems of resistance and the methods of disguise. There is no signal that separates intransigent resistance from honest efforts to conduct an auction for the shareholders' benefit. The fact that the first tender offer or any subsequent offer is defeated supplies little information, because any auctioneer understands that determined efforts to collect the highest possible price may lead to no sale at all in the short run.

Even resistance that ultimately elicits a higher bid is socially wasteful. Although the target's shareholders may receive a higher price, these gains are exactly offset by the bidder's payment and thus by a loss to the bidder's shareholders. Shareholders as a group gain nothing; the increase in the price is simply a transfer payment from the bidder's shareholders to the target's shareholders. Indeed, because the process of resistance consumes real resources, shareholders as a whole lose by the amount targets spend in resistance plus the amount bidders and any rivals spend in overcoming resistance. These additional costs can be substantial.

This argument may appear to be inconsistent with fiduciary principles. If resistance touches off an auction, it may drive up the price paid for the target's shares. Ordinarily managers are charged with the duty of maximizing the returns to the firm's shareholders without regard to adverse consequences to other firms' shareholders or to society at large. . . .

But this is not invariably true. . . . The resulting increase in the prices paid for target firms will generally discourage prospective bidders for other targets: when the price of anything goes up, the quantity demanded falls. Changes in the incentives of bidders affect the utility of monitoring by outsiders, and that affects the size of agency costs and in turn the preoffer price of potential targets' stock. In order to explore the nature of these effects, it would be useful to ask what rational shareholders would do if, *before* a tender offer was in prospect, they could bind the management to resist or to acquiesce in any offer.

Consider the effects of two polar rules. Under the first rule, management is passive in the face of tender offers. If there are no competing bidders, the first tender offeror will prevail at the lowest premium that will induce the shareholders to surrender their shares. Under the second rule, management uses all available means to resist the offer. This resistance creates an auction, so that no bidder can acquire the target without paying a price almost as high as the shares would be worth under the best practicable management. . . .

Which of these rules maximizes the welfare of . . . shareholders? If the question is asked ex post, after a tender offer has been made, then plainly the shareholders would prefer the second rule and the bidding war. But if the shareholders were asked which strategy the managers should pursue ex ante, prior to the offer, they would have substantial reasons to choose acquiescence. It is easy to see why. If the target's shareholders obtain *all* the gains from the transaction, no one has an incentive to make a tender offer, and thus no one will offer a premium for the shares. . . .

Prospective offerors must do substantial research to identify underpriced corporations and to determine how their management can be improved. They may engage investment banking houses and investigate the affairs of many corporations before finding one whose management could be improved. The position of a tender offeror is particularly precarious because, at the time it makes a bid, its investment in information about the target is sunk. . . .

Once the offeror announces its bid, however, other potential acquirers learn the target's identity. The bid itself, and the accompanying disclosures under federal and state law, may reveal much of what the offeror has learned. If the offeror does not supply other bidders with valuable information, the target's management may do so as part of a strategy to set up an auction. But any other bidder need not bear costs as high as those already incurred by the first bidder. The subsequent bidders take a free ride. . . . As a result, no firm wants to be the first bidder unless it has some advantage, such as speed, over subsequent bidders to compensate for the fact that only it had to incur monitoring costs. And, of course, if there is no first bidder there will be no later bidders and no tender premium.

Perhaps most important of all, requiring bidders to pay a high premium will lead to a decrease in the price of the target's shares. A bidder facing the prospect of paying a high premium is less likely to monitor other firms, and the decrease in searching for targets leads to a decrease in the number of bids. Then the price of . . . stock is likely to decrease because, with the reduction in monitoring, agency costs rise. . . .

The Meaning of Managerial Passivity

Although we have concluded that shareholders would want management to be passive in the face of a tender offer, we have not attempted to define precisely what we mean by passivity. Doubtless, managers must carry out the corporation's ordinary business. Perhaps, too, management should be able to issue a press release urging shareholders to accept or reject the offer. The offeror also will convey its views to the shareholders, who can act on these messages in light of the self-interest of both the management and the offeror. But almost any other defensive actions expend the target's resources and produce no gain to investors. Thus, management should not propose anti-takeover charter or bylaw amendments, file suits against the offeror, acquire a competitor of the offeror in order to create an antitrust obstacle to the tender offer, buy or sell shares in order to make the offer more costly, give away to some potential "white knight" valuable corporate information that might call forth a competing bid, or initiate any other defensive tactic to defeat a tender offer.

Seeking Competitive Bids Versus Pure Passivity in Tender Offer Defense

RONALD J. GILSON

Unlike my more limited rule barring defensive tactics designed to prevent the offer but not barring the facilitation of competitive bids, Easterbrook and Fischel would prohibit both. . . .

As originally put, Easterbrook and Fischel argued that auctioneering was undesirable because of the sunk costs in information incurred by the original bidder. If a competitive bid was successful, the unsuccessful first bidder would be unable to recover these costs; the risk of this occurrence would reduce the incentive to invest in information in the first place. Therefore, monitoring would decrease and agency costs would increase.

I responded by noting that a first bidder could hedge its risk of lost information costs by buying a block of the target's stock that could be sold at a profit to a subsequent higher bidder. I also argued that any loss from a reduction in monitoring would be offset by the increased efficiency that would result from allocating "target assets to their most efficient user." Professors Easterbrook and Fischel now concede the existence of a hedge, but argue that the hedge is imperfect. In their view, even if first bidders both recover their sunk costs and earn a return on the investment, the increase in takeover prices associated with competitive bidding will nonetheless reduce the return on investment in information below what would have been earned in the absence of competitive bidding, with the same undesirable results: a reduction in monitoring and an increase in agency costs. . . . I argue that competitive bidding may *increase* rather than decrease the return on investment in information. If this is correct, then the choice between the two rules turns on their efficiency at resource allocation and on their susceptibility to abuse. . . .

To understand the potential that competitive bidding has for increasing the return on investment in search, it is necessary to decompose a first bidder's investment in a takeover and examine the return associated with each portion of that investment. One portion of a bidder's investment is search costs, incurred to identify a target whose value can be increased by displacing inefficient management or through some form of synergy. The second portion is the amount paid to secure control and implement the strategy necessary to take advantage of the identified opportunity. . . . A successful acquisition requires two different sets of attributes: one involving information production skills and not very much capital, the other involving the operating skills required for implementing the takeover and substantially more capital. I see no reason to expect that both sets will always be present in a single entity. And if, as this analysis suggests, specialization does occur, information producers would prefer a rule allowing target management to facilitate competitive bidding, since competitive bidding would increase the return on their investment in information. Implementers, on the other hand, would prefer a rule of pure passivity since that rule, by making competitive bidding less likely, would reduce takeover prices and enable implementers to secure higher returns by capturing some of the value associated with information production. In short, implementers, not information producers, are made worse off by competitive bidding. But that should not result in less than the appropriate number of takeovers, since the implementers will always capture the entire value of their investment in implementation.

Easterbrook and Fischel take issue with my argument at this point by correctly pointing out that another alternative is available to the specialized information producer: Rather than initiating an auction by giving the information away, the producer could, if no competitive bidding were allowed, sell the information to an implementer. . . .

Evaluation of the sale alternative puts the issue in a somewhat different context. Assuming that a specialized information producer earns a higher return on its investment in information than an integrated acquirer, the question becomes how the information producer can best exploit its advantage—by selling the information to an implementer or by fomenting competitive bidding. In both alternatives the problem confronting the information producer is verification; whichever is pursued, in order for the information producer to secure any return on its investment in search, it must convince potential acquirers that its information is of a quality that warrants investment. . . . An information producer's preference between the two alternatives, then, turns on a transaction cost analysis of available solutions to the verification problem.

Consider first the sale alternative. Information producers can respond to the verification problem by strategies—characterizable either as bonding or signaling—designed to ensure the authenticity of the information to the buyer. For example, one would expect information producers who hope to sell information to make substantial investments in reputation, thereby both signaling that their product is of a quality to warrant repeat purchases and putting their investment in reputation at risk should the information prove to be inaccurate. One might also expect the information producer to allow payment for the information to be conditioned on the success of the transaction and, in order to avoid creating a conflict of interest that could dilute the signal sent by investment in reputation, to voluntarily limit speculation in the identified target.

The pattern described, of course, is that of the major investment banking houses. . . .

But the sale alternative is not the only strategy open to information producers. Indeed, for producers without either a preexisting investment in reputation or the capital and time to adopt the verification techniques used by investment bankers, this strategy may not even be available. For them, disclosure of their information in order to promote competitive bidding may be the only verification mechanism available and, therefore, the only way to appropriate the return on their investment in information. First, the bidding process itself acts as a verification technique. Second, it can also be used to cause the producer's information to be verified by the best possible source: the target. An information producer might adopt the strategy of announcing, together with its information and its stock position, that it intended to cause the target to be acquired by someone. This would create an incentive for target management to select the ultimate acquirer—to seek a white knight. In this case, verification of the producer's information is provided by the target through its attempts to prove to those potential acquirers that it approaches the existence of the very opportunity the information producer has disclosed.

Institutional arrangements reflect this pattern as well. For example, Carl Icahn does not sell his information; he profits by reselling target stock to implementers. . . .

Allocational Efficiency

There is agreement that tender offers serve an allocational role, and that competitive pricing generally facilitates the shifting of assets to their most productive users. What separates Easterbrook and Fischel's position from mine is their claim that a series of independent sales can cause assets to be shifted to their most productive users as efficiently as competitive bidding in connection with a single sale. I disagree. While resolution of this issue turns on a comparative analysis of transaction costs, which would require a good deal more effort than has been undertaken by either Easterbrook and Fischel or myself, it is critical to note that a rule allowing target management to solicit competitive bids cannot be less efficient than a rule requiring managerial passivity. If would-be second bidders have a choice between entering a competitive bid or acquiring the target later from the first bidder, they will presumably choose the cheaper method. Therefore, the fact that many competing bids are made is powerful empirical evidence that a rule of passivity is inefficient. Indeed, unless we are certain that a series of sales is cheaper in *every* transaction, choice is preferable. My rule gives acquirers that choice. . . .

Rule Efficiency

Easterbrook and Fischel finally argue that the rule I recommend—prohibiting defensive tactics but permitting management to solicit competitive bids—will allow defensive tactics to be undertaken under a guise of neutrality and, to that extent, will result in a loss not incurred under a pure passivity rule. I expect that such a loss, if it occurred at all, would be small. Moreover, the potential for loss must be compared to the similar problems that would arise under a pure passivity rule. . . . Under a pure passivity rule, one must draw *some* line specifying when an offer has gone "too" far to allow the solicitation of competitive bids. This exercise, however, has the same potential for creating uncertainty and unintended costs as Easterbrook and Fischel claim would be created by my more limited prohibition.

The Case for Facilitating Competing Tender Offers: A Reply and Extension

LUCIAN A. BEBCHUK

I support a legal rule that: (1) regulates offerors in order to provide time for competing bids and (2) allows incumbent management to solicit such bids by supplying information to potential buyers. In my previous article, and in this exchange, I refer to this rule as "the auctioneering rule." . . . The question of whether management should be allowed to provide information to potential buyers is indeed of limited importance: As long as a regulatory delay is provided, an active competition among acquirers will take place even if management must remain passive. . . .

Easterbrook and Fischel correctly observe that targets were not destined to be targets and could have become acquirers. They incorrectly assume, however, that every company is *exactly* as likely to turn into an acquirer as to turn into a target. . . . A company is unlikely to face an equal probability of becoming a target and an acquirer. Some companies are more likely to become an acquirer than a target, others are more likely to become a target than an acquirer. . . .

The Effect of Facilitating Competing Bids on Search

Easterbrook and Fischel rest their case on a claim that, because the auctioneering rule decreases the amount of search by prospective acquirers, it reduces the number of beneficial acquisitions. . . . I suggested, however, that the rule's overall effect on the number of offers might be desirable; and that in any event whatever undesirable effect the rule might have is unlikely to be substantial. I made three claims: (1) the decrease in prospective acquirers' search is unlikely to produce a substantial decrease in buyer-initiated beneficial acquisitions; (2) the decrease in prospective acquirers' search may be desirable; and (3) the auctioneering rule produces a desirable increase in prospective sellers' search and consequently in the number of seller-initiated beneficial acquisitions. . . .

A General Framework

I should like to place the auctioneering rule that I advocate in a broader perspective. The rule should be viewed as an element of a legal framework that is intended to enable the dispersed shareholders of a potential seller to function as a sole owner would. Many public companies are characterized by a separation between management and ownership and by dispersed ownership. As a result, the market for corporate assets will not function without legal intervention as does the market for "ordinary" assets that have a sole owner-manager. This should be addressed by a legal framework consisting of three elements, each corresponding to a capability possessed by a sole owner facing an offer to buy.

First, a sole owner is free to accept any offer made to him to buy his assets. Where management and ownership are separate, however, management's actions might threaten the shareholders' freedom to accept offers. Management might use the powers that it has for running the company's business to preclude shareholders' acceptance of an offer. This problem should be addressed by the first element of the proposed framework . . . that bars management from obstructing offers made to the shareholders.

Second, a sole owner who receives an offer to buy his assets is capable of seeking better offers. He can delay accepting the original offer, and can provide information about his assets to other potential buyers. But a target's dispersed shareholders, under pressure to tender, cannot act in concert to secure a delay, and they have no access to the company's internal information. The auctioneering rule—the second element of the framework—addresses this problem. The rule secures a delay by regulating offerors and allows management, which has access to internal documents, to provide information to potential buyers.

Third, a sole owner is free not only to accept any offer but also to reject, at least temporarily, all offers made to him. The pressure to tender, however, impairs the ability of a target's dispersed shareholders to take this course of action. This problem should be addressed by the framework's third element—a set of legal rules that would enable a target's shareholders to decide freely whether or not a sale is in their interest. Putting forward such a set of rules is a goal toward which future research should be directed.

The Market for Corporate Control:
The Empirical Evidence Since 1980

GREGG A. JARRELL, JAMES A. BRICKLEY,
AND JEFFRY M. NETTER

The Effects of Defending Against Hostile Takeovers

Defensive strategies against hostile takeovers have always been controversial since they pose a conflict of interest for target management. After all, takeovers can impose significant welfare losses on managers, who may be displaced and lose their organization-specific human capital. These conflicts may tempt some managers to erect barriers to hostile takeovers, thus insulating themselves from the discipline of the outside market for control at the expense of their shareholders and the efficiency of the economy.

However, providing target management with the power to defend against hostile takeover bids might also help target shareholders during a control contest. Target management can in certain cases defeat bids that are "inadequate." Although this rationale is popular, the evidence . . . shows that in very few cases do these alleged long-term gains of independence actually materialize. The other benefit of resistance comes when resistance by target management helps promote a takeover auction. Litigation and other blocking actions can provide the necessary time for the management of the target firm to "shop" the target and generate competing bids. This auction rationale for resistance is harder to reject statistically. Evidence on occasional shareholder losses after the defeat of a takeover attempt does not in itself disprove the auction theory. This negotiating leverage can be expected to fail in some cases, with the sole bidder becoming discouraged and withdrawing. It is a gamble. The hypothesis is rejected only if the harmful outcome of defeating all bids is sufficiently frequent and costly to offset the benefits of inducing higher takeover prices. One must also consider the social cost of tender offers that never occur because of the presence of defensive devices. Unfortunately, this deterrence effect is very difficult to measure and we present no direct evidence of the extent of these costs.

Evidence on the effects of defensive measures by target management is obtained mainly from two approaches, the event-type study and the outcomes-type study. The event-type study recognizes that an efficient market must judge this cost–benefit trade-off when it adjusts the market value of a firm in response to the adoption of a charter amendment or some other kind of resistance. Alternatively, the outcomes-type study examines the actual outcomes of con-

Reprinted by permission from 2 *Journal of Economic Perspectives* 49. © 1988 American Economic Association.

trol contests over a significant time horizon among firms using a common kind of resistance—say all firms adopting poison pills. That is, an event study measures the stock price reaction to the introduction of defensive devices while outcomes studies follow the use of defensive devices in control contests to determine their effects on the outcomes of the contests. . . .

Defensive Measures Approved by Shareholders

Proposed antitakeover amendments are very rarely rejected by voting shareholders; Brickley, Lease, and Smith find for a sample of 288 management-sponsored antitakeover proposals in 1984 that about 96 percent passed.

Supermajority Amendments

Most state corporation laws set the minimum approval required for mergers and other important control transactions at either one-half or two-thirds of the voting shares. Supermajority amendments require the approval by holders of at least two-thirds and sometimes as much as nine-tenths of the voting power of the outstanding common stock. These provisions can apply either to mergers and other business combinations or to changing the firm's board of directors or to both. Pure supermajority provisions are very rare today, having been replaced by similar provisions that are triggered at the discretion of the board of directors. This allows the board to waive the supermajority provisions allowing friendly mergers to proceed unimpeded.

Five years ago, Jensen and Ruback found mixed evidence on the effect of supermajority amendments passed before 1980. However, a more recent study by Jarrell and Poulsen covers 104 supermajority amendments passed since 1980 and reports significant negative stock price effects of over 3 percent around the introduction of the proposals. They also show that firms passing supermajority amendments have relatively low institutional stock holdings (averaging 19 percent) and high insider holdings (averaging 18 percent), which they interpret as helping to explain how these amendments received voting approval despite their harmful wealth effect. That is, firms proposing these amendments have fewer blockholders with incentives to invest in the voting process. Jarrell and Poulsen further conjecture that the increased shareholder resistance to harmful supermajority amendments helps explain their declining popularity in contrast to the success of the fair price amendment which appears less likely to harm shareholders (as discussed below).

Fair Price Amendments

The fair price amendment is a supermajority provision that applies only to nonuniform, two-tier takeover bids that are opposed by the target's board of directors. Uniform offers that are considered "fair" circumvent the supermajority requirement, even if target management opposes them. Fairness of the offer is determined in several ways. The most common fair price is defined as the highest price paid by the bidder for any of the shares it has acquired in the target firm during a specified period of time. Jarrell and Poulsen report that 487 firms adopted fair price charter provisions between 1979 and May 1985, with over 90 percent of these coming [from] 1983 to May 1985.

The stock price effects reflect the low deterrence value of the fair price amendment. Jarrell and Poulsen report an average loss of 0.73 percent around the introduction of these amendments, which is not statistically significant. They also show that firms adopting fair price amendments have roughly normal levels of insider holdings (12 percent) and of institutional holdings (30 percent). They interpret this evidence as supporting the view that shareholder voting retards adoption of harmful amendments, especially when insider holdings are low and institutional holdings are high. Further support for this view is provided by Brickley, Lease, and Smith who document that "no" votes on antitakeover amendments (especially ones that harm shareholders) increase with institutional and other outside blockholdings, while "no" votes decrease with increases in managerial holdings. . . .

Reduction in Cumulative Voting Rights

Cumulative voting makes it possible for a group of minority shareholders to elect directors even if the majority of shareholders oppose their election. Dissidents in hostile takeovers and proxy contests will often attempt to elect some board members through the use of cumulative voting. Bhagat and Brickley examine the stock price reaction to 84 management-sponsored charter amendments that either eliminate or reduce the effect of cumulative voting. Since these amendments decrease the power of dissident shareholders to elect directors, they increase management's ability to resist a tender offer. Bhagat and Brickley find statistically significant negative abnormal returns of about 1 percent at the introduction of these charter amendments.

Defensive Measures That Do Not Require Shareholder Approval

Four general kinds of defensive measures do not require voting approval by shareholders: general litigation, greenmail, poison pills, and the use of state antitakeover laws. With the exception of general litigation, these defensive actions are associated on average with negative stock price reactions indicating that in most cases they are economically harmful to stockholders of companies whose management enacted them.

Litigation by Target Management

As described earlier, litigation can be expected to hurt shareholders of some target companies by eliminating takeovers and to help shareholders of other companies by giving their management time and weapons to cut a better deal. Jarrell examines 89 cases involving litigation against a hostile suitor based on charges of securities fraud, antitrust violations, and violations of state or federal tender offer regulations. His results show that litigation usually delays the control contest significantly and that litigating targets are frequently the beneficiaries of auctions. The 59 auction-style takeovers produced an additional 17 percent excess return to shareholders over the original bid, while the 21 targets that remained independent lost nearly all of the original average premium of 30 percent. Overall, Jarrell concludes that this evidence cannot reject the theory that on average target litigation is consistent with shareholder wealth maximization. . . .

Targeted Block Stock Repurchases (Greenmail)

Greenmail occurs when target management ends a hostile takeover threat by repurchasing at a premium the hostile suitor's block of target stock. This controversial practice has been challenged in federal courts, in congressional testimony, and in SEC hearings, and it has brought negative publicity both to payers and to receivers of greenmail. In reviewing earlier studies, Jensen and Ruback conclude that greenmail repurchases are associated with significantly negative abnormal stock returns for the shareholders of the repurchasing firms (probably because they eliminate potential takeover bids) and significantly positive abnormal stock returns for shareholders of the selling firms. These negative effects of greenmail repurchases contrast sharply with the normally positive stock price effects associated with nontargeted offers to repurchase a company's own stock.

Since then, three new empirical studies have contributed to a more complex and less conclusive discussion of greenmail transactions. These studies indicate that it is not necessarily in the interests of shareholders to ban greenmail payments. Such a ban has the potential to discourage outside investment in the potential target's stock by investors anticipating greenmail payments and hence reduces the incentives of outsiders to monitor managers.

Mikkelson and Ruback examine 39 cases of greenmail (based on 13Ds filed during 1978 to 1980). They find a significant stock price loss of 2.3 percent upon the announcement of the repurchases. However, they also report an average gain of 1.7 percent over the entire period including the original stock purchase by the hostile suitors. Holderness and Sheehan's outcome-type study includes 12 cases of greenmail, and they report a pattern of returns consistent with the evidence of Mikkelson and Ruback. Although the greenmail transaction itself harms target shareholders, the net returns to stockholders resulting from the initial purchase and related events is positive. A more comprehensive sample of targeted block stock repurchases is covered by OCE [SEC Office of Chief Economist]. This study includes 89 cases of large repurchases (blocks greater than 3 percent of the outstanding common stock) from 1979 to 1983. The initial announcement of investor interest induces a positive return averaging 9.7 percent, while the greenmail transaction is associated with a stock price loss of 5.2 percent.

Poison Pills

Since its introduction in late 1982, the "poison pill" has become the most popular and controversial device used to defend against hostile takeover attempts. Poison pill describes a family of shareholder rights agreements that, when triggered by an event such as a tender offer for control or the accumulation of a specified percentage of target shares by an acquirer, provide target shareholders with rights to purchase additional shares or to sell shares to the target at very attractive prices. These rights, when triggered, impose significant economic penalties on a hostile acquirer.

Poison pills are considered very effective deterrents against hostile takeover attempts because of two striking features. First, pills can be cheaply and quickly altered by target management if a hostile acquirer has not pulled the trigger. This feature pressures potential acquirers to negotiate directly with the target's board. Second, if not redeemed, the pill makes hostile acquisitions exorbitantly expensive in most cases. As an obstacle to hostile takeover attempts, the poison pill is unmatched except by dual-voting recapitalizations or direct majority share ownership by incumbent management. . . .

The most comprehensive study of poison pills is Ryngaert. . . . The Ryngaert study features an exhaustive collection of 380 poison pills adopted from 1982 to December 25, 1986. . . . Ryngaert divides his sample into discriminatory pills (the most restrictive) and flip-over pills (the least restrictive). He also accounts for whether firms are subject to takeover speculation and whether confounding events occur close to the announcement of the pill that contaminate the data. The stock price effect over the 283 cases with no confounding events is a statistically significant $-.34$ percent. Focusing on 57 cases subject to takeover speculation, the average loss is 1.51 percent, also statistically significant. These results are supported by the findings of Malatesta and Walkling. . . . Ryngaert reports that pill-adopting managements own a surprisingly low average of around 3.0 percent of their firms' outstanding stock. This fact, together with high institutional holdings, suggest that many of these firms would have difficulty obtaining shareholder voting approval if it were required.

Ryngaert also examines the stock price effects of important court decisions emanating from legal battles involving pill defenses during 1983 to 1986. He shows that 15 of 18 protarget, propoison pill decisions have negative effects on the target's stock price, and 6 of 11 proacquirer decisions have positive effects on the target stock price. This evidence is inconsistent with the theory that pill defenses improve shareholder wealth by strengthening management's bargaining position in control contests.

Although these losses are not large in percentage terms, these empirical tests suggest that poison pills are harmful to target shareholders. . . .

Summing Up Defensive Tactics

Four years ago Jensen and Ruback reviewed empirical studies of antitakeover charter amendments, shark repellents, changes of incorporation, and greenmail. They conclude: "It is difficult to find managerial actions related to corporate control that harm stockholders; the exception are those actions that eliminate an actual or potential bidder, for example, through the use of targeted large block repurchases or standstill agreements."

Since their review, the defensive arsenal available to target management has been strengthened. These defensive tactics have been developed through a fascinating process of sequential innovations, as specific defenses arise to counter improved bidder finances and other tactics. In 1983, the now common fair price amendment was a novel idea and the poison pill was not yet invented. Financial economists in academia and government have kept close pace with these developments, providing timely analyses of new charter amendments, poison pill defenses, greenmail transactions, and so on. While Jensen and Ruback were correct in predicting this area would be a "growth industry," we cannot reiterate their then-accurate conclusion that harmful defensive tactics are rare.

The Wealth Effects of Second-Generation State Takeover Legislation

JONATHAN M. KARPOFF AND PAUL H. MALATESTA

From 1982 through 1988, thirty-four states passed more than sixty-five major laws regulating corporate takeovers. These so-called second-generation state takeover laws were enacted after a 1982 U.S. Supreme Court decision (*Edgar v. MITE Corp.*) effectively invalidated thirty-seven preexisting laws. Many additional takeover laws were passed after another Supreme Court ruling in April 1987 (*CTS Corp. v. Dynamics Corp. of America*) upheld an Indiana takeover law and created the presumption that other takeover laws are also valid. Some of the more recent laws are called third-generation laws by some researchers. In this article, however, we refer to all state takeover laws enacted since 1982 as second-generation laws.

We present evidence about the effects of second-generation state takeover laws on stockholder wealth. Because these laws can increase the cost of hostile corporate acquisitions, they provide takeover defenses to firms subject to

Reprinted by permission from 25 *Journal of Financial Economics* 291 (Amsterdam: Elsevier Science Publishers, 1989).

their jurisdiction. One hypothesis holds that, by making hostile takeovers more costly, these laws entrench incumbent managers at the expense of stockholders and the efficient deployment of resources. If this hypothesis is true, second-generation state takeover laws should decrease stockholder wealth. Under a competing hypothesis, takeover defenses increase expected takeover premiums even while they decrease the probability of takeover, because they enable managers to obtain a higher control premium from bidders in the event of a takeover. This hypothesis implies that second-generation state takeover laws should increase stockholder wealth.

A related issue concerns takeover defenses that are available to individual firms, such as antitakeover charter amendments and poison pills. One hypothesis contends that state takeover laws and firm-level defenses are substitutes, and that the effects of state takeover laws will be attenuated by the presence of firm-level defenses. An alternative hypothesis, however, holds that the effects of state takeover laws will be more pronounced in firms with takeover defenses. Firms that choose to adopt poison pills and antitakeover charter amendments may do so because they are particularly vulnerable to takeover bids. If so, these firms may be more likely to be affected by takeover laws.

To test these hypotheses, we analyze stock returns on portfolios of affected firms in states where second-generation takeover laws were introduced in the state legislatures. First, we examine the average stock price reaction to all initial press announcements of state takeover legislation from 1982 through 1987. Second, we test the hypothesis that firm-level takeover defenses can substitute for state takeover laws by examining separately the announcement-period stock returns of firms with and those without firm-level takeover defenses. We also present evidence on several related issues, including the stock price effects of different types of second-generation takeover laws and the stock price reactions to press announcements of several court decisions that affect some of the laws.

Previous examinations of the stock price effects of second-generation state takeover laws have produced conflicting results. Schumann and Sidak and Woodward report significant declines in the stock prices of firms incorporated in New York and Indiana on important days in the legislative histories of laws adopted in those states. Romano on the other hand, finds no significant effects of laws adopted in Connecticut, Missouri, and Pennsylvania on the stock prices of firms incorporated in those states. Ryngaert and Netter conclude that the 1986 Ohio law decreased stockholder wealth for firms incorporated in Ohio, but this result is challenged by Margotta, McWilliams, and McWilliams, who conclude that the stock price effect is insignificant. Similarly, Margotta and Badrinath report no significant effect of the 1986 New Jersey takeover law on the stock prices of firms incorporated in New Jersey, whereas Broner interprets his data as indicating that the New Jersey law decreased the stock prices of affected companies. Pugh and Jahera conclude that the Ohio, Indiana, New York, and New Jersey laws had no significant stock price effects on samples of firms incorporated in these states.

A major source of the discrepancies among previous findings is that estimates of any single law's effect are heavily influenced by idiosyncratic characteristics of the researcher's event window and sample of affected firms. For example, one study [Conner] is unlikely to be informative because stock returns are examined on the days two laws were signed by state governors, not on days the legislation become public news. Quirin argues that Sidak and Woodward's results are driven by confounding events that influence the stock price of a single firm in the sample. Similarly, Margotta, McWilliams, and McWilliams argue that Ryngaert and Netter's results are due to inappropriate selections of the event window and sample firms.

To circumvent these problems, we examine the average stock price effect of *all* second-generation takeover laws covered in the press through 1987, controlling for preexisting antitakeover charter amendments and poison pills. . . .

The evidence indicates that, on average, the announcement of a state takeover law is associated with a small but statistically significant decrease in the stock prices of the affected firms. The significance of this result, however, is largely attributable to firms lacking antitakeover charter amendments and poison pills. The stock prices of firms with preexisting firm-level takeover defenses are not significantly affected by the enactment of a state takeover law. . . .

Second-Generation State Takeover Laws

Control share acquisition laws require a target company's shareholders to preapprove acquisitions of voting rights above a stipulated level. . . . Fair price laws are similar to fair price charter amendments adopted by many firms, and regulate the back-end price in a two-tiered takeover bid or other significant business combination involving a large shareholder. . . . Freeze-out laws prohibit a bidder from engaging in any business combination with the target firm for a specified number of years unless approval is obtained from the target firm's directors before the bidder acquires more than a specified fraction of target shares. Even after the mandatory waiting period, most freeze-out laws allow the business combination to proceed only if the transaction satisfies fair price provisions. Thus, the typical freeze-out law is like a fair price law with a forced delay. . . .

Empirical Results

Our empirical analysis involves estimating abnormal stock returns for the portfolios of firms associated with states where takeover bills were introduced. . . . In each case, a state portfolio's abnormal returns are calculated for each legislative event. The estimated abnormal returns are then aggregated and hypothesis tests performed. We also assess the impact of individual court decisions bearing on takeover laws, though here we follow a slightly different procedure. We dispense with the formation of multiple portfolios, one for each state affected by a court's decision, and form instead an equally weighted portfolio of potentially affected firms. In some of our analyses of court decisions, the portfolio contains firms associated with a single state. In other instances, however, the portfolio includes firms associated with all states having laws similar to the law before the court. Hypothesis tests are then based on abnormal returns estimated for the given portfolio of potentially affected firms. . . .

Table 3 [omitted] reports the average stock return forecast errors for all forty state takeover legislative events for which a unique news publication date was found. Event time is measured from this date. . . . [T]he average forecast error for the two-day announcement interval $[-1,0]$ is -0.294 percent with a z-statistic[1] of -2.430, which has a p-value of 0.0075. . . . Average forecast errors for the two-day intervals immediately before and after the announcement interval are insignificantly different from zero. These results indicate that, on average, a press announcement of a state takeover law is associated with a decrease in stockholder wealth, and that the decrease occurs during the announcement interval.

The average stock price effect of state takeover legislation is small in relation to the market value of the affected firms. The laws in our sample, however, cover approximately 88 percent of all firms listed on the New York and American Stock Exchanges, which in turn had an aggregate market value of $2.315 billion at the end of 1987. Thus, the 0.294 percent decline translates into a roughly $6 billion loss in the market value of equity of the subset of affected firms that are listed on the NYSE or Amex. . . .

To examine how firm-level takeover defenses affect the stock price reactions to state takeover laws, we identified firms that had poison pill defenses or antitakeover charter amendments in place before the relevant state takeover legislation was introduced. . . . Of the 1,505 firms, 1,107 firms had no antitakeover charter amendments or poison pill defenses before the introduction date of the *last* takeover legislation considered in their states through 1987. . . . Of the 1,505 original firms, 368 had at least one antitakeover charter amendment or poison pill plan before the introduction date of the *first* takeover legislation considered in their states. . . . Thirty firms adopted a firm-level takeover defense between the introduction dates of the first and last takeover legislation considered in their states. . . .

The results indicate that the average decrease in stockholder wealth is more pronounced among firms with no prior takeover defenses. Among such firms, the average forecast error for the two-day announcement interval is -0.388 percent with a z-statistic of -2.541, which is significant at the 1 percent level. The average forecast error for the two-day announcement interval is -0.126 percent for the sample of firms with prior takeover defenses,

---

[1]The significance test for a z-statistic, which comes from a standardized normal distribution, functions identically to the test for the t-statistic discussed in Chapter I, part B, note 4 [EDITOR'S NOTE].

which, with a z-statistic of -0.865, is not statistically different from zero at traditional significance levels. . . .

We interpret these results as indicating that state takeover legislation protects incumbent managers of firms that are potential takeover targets *and* that have no significant firm-level takeover defenses. The state laws appear to substitute for firm-level defenses. The average announcement interval forecast error differs insignificantly from zero for firms with takeover defenses, which is consistent with the hypothesis that the legislation is not expected to provide additional protection for these firms. The evidence does not support the hypothesis that takeover laws have the largest effects on firms with preexisting takeover defenses.

Announcement Effects of Different Types of Takeover Laws

Different types of legislation could correspond to different mean abnormal stock returns if the laws are expected to differ in their ability to deter takeovers. . . .

To investigate the differences, if any, in the wealth effects of different types of takeover laws, we calculated average announcement interval forecast errors with the sample broken down by type of law. . . . The average forecast error is negative for control share acquisition, fair price, and freeze-out laws. Among these three types of laws, the average forecast error is significantly different from zero at the 5 percent level only for freeze-out laws. The two-day average forecast errors for all three types of laws do not differ significantly from each other.

Among the less common types of laws, the average forecast errors are negative and significant at the 5 percent level for poison pill laws. . . . The average forecast errors for the other types of laws are not statistically different from zero at the 5 percent level. . . .

Announcement Effects for Three Important States

It is also possible that the stock price reactions are larger for laws that represent important innovations or that are passed by states in which many firms are incorporated, and that therefore influence legal developments in other states. The data do not support this conjecture, either. Although any such selections are, in part, arbitrary, we examined the two-day forecast errors for legislation announced in New York and Delaware (because many firms are incorporated in these states) and Indiana (because the control share acquisition law is a prototype for laws enacted in many other states). . . . [T]he average two-day forecast error for two announcements in New York is -0.217 percent, with a z-statistic of -0.598. For the Delaware law, the two-day forecast error is -0.437 percent with a z-statistic of -1.103. Of the three states, only Indiana's two-day forecast error is large (-2.138 percent) and statistically significant ($z = -3.456$). . . .

Announcement Effects for Introduction, Passage, Signing, and Veto Dates

We also examined the return forecast errors for intervals around a takeover bill's introduction in the state legislature, its final passage, and its signing by the governor. The data indicate that the average stock price reactions to these events are approximately zero. . . . These results indicate that the valuation effects of the legislation occur primarily in response to the first public announcement that takeover legislation is being considered.

There is weak evidence, however, that when bills in Arizona and New York were vetoed by state governors, the prices of firms incorporated in those states *increased.* For these two events, the average forecast error for days −1 and 0 in relation to the veto date is 1.080 percent, with a z-statistic of 1.581. . . .

Court Decisions and Control Share Acquisition Laws

Court decisions provide an additional opportunity to assess the wealth effects of state takeover laws, and in particular, control share acquisition laws. In 1985 and 1986, five separate U.S. District Courts ruled that five different control share acquisition laws were unconstitutional. Two of these rulings were upheld at the appellate court level. No court upheld a second-generation takeover law until the landmark *CTS Corp. v. Dynamics Corp. of America* decision on April 21, 1987, in which the U.S. Supreme Court overturned two lower court rulings and upheld the Indiana control share acquisition law. Six days later, on April 27, the U.S. Supreme Court ruled that Ohio's control share acquisition law, which had previously been ruled unconstitutional in lower courts, should be reconsidered by an appellate court in light of the *CTS* decision. . . .

Press dates for two of the lower court rulings and both of the Supreme Court rulings were obtained from a search of newspapers. . . . Press dates were not found for the other decisions. . . . For none of the portfolios [for the lower court rulings] is the average forecast error significantly different from zero. . . .

The average forecast errors for the portfolio of firms incorporated in Indiana are positive but insignificant for the April 22, 1987, report that the Indiana law was upheld. . . . Likewise, the average two-day forecast errors are insignificant for the portfolio of firms incorporated in states with existing control share acquisition laws. . . . The average forecast error [for two days centered on the April 28 announcement that the Ohio law had been upheld] is negative and insignificant for the sixty-seven firms incorporated in Ohio, and is negative and significant at the 5 percent level for the 185 firms incorporated in states with existing control share acquisition laws. This result is consistent with our earlier findings, and with the managerial-entrenchment hypothesis. . . .

The Future of Hostile Takeovers: Legislation and Public Opinion

ROBERTA ROMANO

The Path of State Regulation of Takeovers

The political history of second-generation takeover statutes is similar across the states. The statutes are typically enacted rapidly, with virtually unanimous support and little public notice, let alone discussion. They are frequently pushed through the legislature at the behest of a major local corporation that is the target of a hostile bid or apprehensive that it will become a target. . . .

The statutes are not, however, as some might intuit, promoted by a broad coalition of business, labor, and community leaders who fear that a firm's takeover will have a detrimental effect on the local economy. While some legislators may be concerned about such an effect, labor and community groups are not at the forefront in the attack on takeovers. In fact, the organization most actively involved in promoting the legislation besides corporate management and business groups, in nearly all states, is the local bar association. Although the bar has been involved in drafting legislation in some states, in others, such as Connecticut, it was deliberately bypassed by the statute's corporate sponsors, in order to ensure passage of the legislation without revision. There is no doubt that the corporate bar's interest can differ from that of managers and shareholders. For example, corporate lawyers profit from takeover litigation, and a statute that prevented all hostile takeovers would, presumably, also eliminate the lawsuits. A factor mitigating the incentive for maintaining some modicum of takeover activity is that a merged firm typically retains the acquirer's legal counsel. Because the acquirer and, correspondingly, its counsel are quite often out-of-state entities, the local bar's interest will be similar to that of incumbent management in seeking to block takeovers. As we can conjecture plausible, diverse incentives for corporate lawyers independent of their clients' interests, we cannot identify a priori what motivates their behavior.

Perhaps the most important reaction to *CTS*,[1] and the key to the resurgence of the federalism debate, is that Delaware, the leading incorporation state, enacted a second-generation statute. To its credit, Delaware's legislative process was more open and deliberative than that of many other states. . . . Several commissioners of the SEC, shareholder organizations, and institutional investors, some of which had voiced objections to the bar committee,

[1] In CTS Corp. v. Dynamics Corp. of America, 481 U.S. 69 (1987), the Supreme Court upheld Indiana's control share acquisition statute, ending a period of uncertainty over the constitutionality of state regulation of takeovers [EDITOR'S NOTE].

Reprinted by permission from 57 *University of Cincinnati Law Review* 457 (1988).

actively opposed the bill in the legislature. At the same time, corporations also increased their lobbying efforts. . . .

Delaware's second-generation statute, like its first-generation statute, is considerably less hostile to acquiring firms than the laws of other states. It provides bidders with greater flexibility to complete an unfriendly transaction. This is explained, in large part, by the Delaware corporate bar's representation of a more diverse constituency, consisting of both targets and bidding firms, than is found in other states. For instance, large Delaware firms average more acquisitions over their lifetime than large firms incorporated in other states, and more firms that have undertaken a hostile acquisition are incorporated in Delaware than in any other state. But because so many corporations are under Delaware's jurisdiction—approximately half of the Fortune 500 manufacturing firms and over 40 percent of New York Stock Exchange listed firms—even a comparatively mild takeover statute is a matter of serious concern. The enactment of a Delaware takeover statute therefore provides momentum to the movement calling for federal preemption of state takeover laws. . . .

When Is Federal Regulation Preferable to State Regulation?

Economic theories of federalism provide the foundation for advocating that takeover regulation should be shifted to the jurisdiction of the national government. Analytically, the starting point is the theoretical justification of the state, that a principal function of government is to correct market failures. Markets fail to allocate resources efficiently when an activity produces externalities or when a commodity or service is a public good. An activity produces external economies or diseconomies—externalities—when it yields benefits or costs to individuals or firms other than the actor, and those third parties cannot be excluded from enjoying the benefits or bearing the costs. When this occurs, because actors only consider the benefits or costs that directly affect themselves, the socially optimal level of activity will not be undertaken. . . .

The economic theory of federalism focuses on the extent of the externality, . . .to identify which level of government, if any, is the appropriate one to intervene in a market. A government should have control over an activity whose externalities fall completely within its borders, for then the costs and benefits will accrue solely to the citizens to which that government is accountable and the allocation of resources will be efficient. When the costs and benefits spill over jurisdictional boundaries, allocational inefficiencies can arise. For example, a government's regulation may not induce enough of an activity by failing to count benefits accruing to citizens of other states in addition to its own citizenry. It might also induce too much of an externality-producing activity when the activity benefits its citizens and citizens in other states bear the costs. In a spillover situation, a higher-level government will contain all benefited and harmed citizens within its borders and will therefore be the more appropriate authority. . . .

[I]n the takeover context, the target's management has its livelihood at stake. The heightened differential calculus for managers considering hostile takeovers in contrast to other issues of corporation law highlights [a problem] of the . . . politics of state takeover laws: the benefits of takeover statutes are concentrated on local citizens—managers and, arguably, locally employed workers of targets, and local businesses and charities with relations to targets—but the costs are dispersed among shareholders and bidding firms who typically do not reside in the legislating state.

At a higher-level government, this failure to count all costs and benefits should not occur because all parties are included within the jurisdiction. But Delaware is also able to internalize many of the costs and benefits: not only are a larger number of hostile acquirers incorporated in Delaware than any other state, but also, few Delaware corporations are physically present in the state. The very large number of potential targets in Delaware makes it more worthwhile for nonresident bidders and shareholders to lobby in Delaware than other states, as would be equally true of a national forum. Although they do not vote in the jurisdiction, their attorneys do. Unlike the national government, however, Delaware is subject to competition for revenues from other states whose calculations favor enacting takeover statutes. There is thus an externality, albeit of a different sort, affecting Delaware politics that would not be present at the national level, where income from corporate franchise taxes would be an insignificant factor. In this context of stark shareholder-manager conflict, state competition may not be for the better.

Yet it would be a mistake to infer automatically that national regulation would be an improvement. One source of the political market failure regarding takeover legislation, the organizational difference between managers and shareholders, is relatively constant across government level. Managers are easier to coordinate across firms than shareholders whether the forum is Congress or a state legislature. Managers already interact through trade associations and positions on boards, and they clearly stand to realize substantial benefits—job protection—from takeover legislation. And . . . when a major firm is the target of a hostile bid, its managers, who are far fewer in number than its shareholders, are benefited by a takeover statute by more than the lobbying cost, and thus have individual incentives to bear the cost of furnishing the collective good—the statute—to all group members—all managers.

Just as the benefits of takeover regulation at the national level are still concentrated, the costs are still diffuse, for shareholders are dispersed across the states and most investors' holdings are so small that the free-rider problem is severe. Large shareholders, especially institutional investors, have more of an incentive to lobby but they lack some of the organizational advantages available to managers. Business organizations, such as trade associations, typically provide valuable information to their member firms, which induces individual participation, and have served as the mechanism for management's lobbying for takeover statutes. Collective action by shareholders is harder to sustain because they have less need for a centralized organization to share information. It is well established that the markets in which the stocks of concern (stock of companies subject to hostile takeovers) are traded—the national stock exchanges—are efficient, such that stock prices include all publicly available information. There is, accordingly, no benefit for one investor—especially an institution competing for clients—to share its private information with all other investors in an umbrella shareholder organization. Without political entrepreneurs or the group's production of private goods for its individual members, such as shared information, collective action is unlikely to succeed.

In addition, there is another important asymmetry affecting the taking of collective action in the takeover context: individuals are more likely to co-ordinate their actions to avoid public "bads" than to obtain public goods. That is because most people are risk averse—they care more about preventing losses than achieving gains of equal dollar magnitude. In the takeover context, this asymmetry favors the lobbying effort of managers over share-holders, as managers may be harmed by takeovers while target shareholders profit. Moreover, management typically stands to lose more than even large shareholders stand to gain, which exacerbates the collective action problem. Accordingly, an analytical case can be made for national regulation of corporations in the takeover context, but it is considerably more problematic than at first glance.

The Probable Output of Federal Action

Efforts at Federal Legislation, 1963–1987

Content of congressional proposals. An externality analysis of takeover laws suggests that congressional proposals could conceivably differ significantly from the states' protectionism. The data indicate otherwise. From 1963 to 1987, over 200 bills regulating corporate takeovers, excluding bills directed exclusively at acquisitions of banks, were introduced in Congress. . . . The most striking characteristic of the takeover-related bills is that, like state statutes, the vast majority aim at making acquisitions more difficult. . . .

In addition to the substantive tilt against bidders, unlike virtually all of the state second-generation statutes, all of the congressional bills restricting bids establish mandatory regimes. Firms can neither opt-out of nor opt-in to these regulatory schemes. . . . In short, shareholders would fare no better, and in all likelihood far worse given the absence of an opt-out provision, under the vast majority of congressional proposals. . . .

In a comprehensive study of policy initiation, John Kingdon found that successful law reform in the national arena involves the recombination of old elements rather than the invention of completely new proposals. When Congress has acted to regulate takeovers in any major way, the recombinant aspect of legal change stressed by Kingdon has been a prominent feature. Senator Williams introduced the first proposal regulating cash tender offers and requiring disclosure of ownership of stock blocks in 1965, although it was not until 1968 when a modified version, originating in his 1967 bill, was enacted. The Hart-Scott-Rodino Antitrust Improvements Act of 1976, which requires premerger notification and waiting periods for certain acquisitions, including cash tender offers, had first been proposed twenty years earlier. It had also been periodically introduced in succeeding Congresses. The new elements found in bills of the 1980s have now been repeatedly introduced over several Congresses, and they continue the trend of restricting bidders. An educated prediction, accordingly, is that any future federal action on takeovers will regulate bidders still further.

The federal political process. The similarity in prospective legislative output between Congress and the states suggests that, notwithstanding the externality analysis, the inputs of the politics at the federal level may not be that different from the states. . . .

The most important evidence of the common political process for congressional bill introductions and state enactments is that the principal factor motivating political action at the state level—the presence of a large-sized target firm in the state—also appears to be the primary impetus for congressional interest in regulating takeovers. . . . Besides spurring the introduction of legislation, takeovers of local firms prompt members of Congress to testify in support of other members' bills and to continue to introduce, cosponsor, and testify in support of takeover regulation in subsequent sessions after the outcome of their local problem has been resolved. . . .

[A]s in the politics of state statutes, labor unions have not been extremely active in the congressional politics. . . . [I]n seventy-seven congressional hearings on takeover bills or particular hostile bids held during 1963 to 1987, while forty-six had at least one target manager as a witness, only nineteen had a union representative or member as a witness. . . . [S]hareholders are the one group affected by takeovers that is nearly always absent from the witness roster, having participated in only two hearings. It is, of course, possible that other witnesses, such as raiders (present at 15 hearings), academics (37), investment banks (15), or government agencies (62), are advocates, at least in part, of shareholders' interests.

Behavioral explanations. Despite all this congressional activity, hearings and takeover-related bills rarely result in legislation. This is consistent with research findings that the subjects of congressional hearings and reports do not rank very high as priority agenda items. One might therefore wonder why members of Congress engage so frequently in activity which appears altogether futile, and why target managers even bother with the federal route? From the congressional members' perspective, holding hearings and introducing bills on takeovers are equivalent to participating in roll call votes, performing casework, and making home appearances and speeches. These activities are forms of congressional "advertising, credit claiming, and position taking," which are important for reelection because they identify the incumbent with particularized benefits to constituents and popular messages associated with little issue content or controversy. It is thus not surprising that many hearings have senators and representatives among the witnesses. Also consistent with this credit-claiming explanation of congressional activity is the bipartisan politics. As in state legislatures, in Congress state congressional delegations virtually unanimously support an individual member's takeover-related proposal.

The congressional politics on takeovers shares features with a particular form of constituent service that has been termed the "fire alarm" approach to executive branch oversight. In this approach, Congress does not sniff out fires itself but instead responds to constituents' pulling of the fire alarm box. By calling or participating in a hearing on a takeover, members of Congress can receive credit for responding to a constituent's problem, while the constituent has borne the cost of informing the member. But despite the structural similarity, the member's efforts typically do not eliminate the cause of the constituent's complaint in this context. The low rate of actual legislation makes the benefit to managers from lobbying, and, by implication, the actual credit a member of Congress can receive, difficult to identify. However, the expenditures involved in seeking federal action are negligible. Managers may simply believe it worthwhile to cover all bases, given the large potential loss they face from a successful hostile takeover, and when the constitutionality of state laws became more uncertain after *MITE*[2] it was all the more important to engage in such activity. It also appears that a congressional hearing has benefits besides the production of legislation: it may delay the outcome of a bid, which is a tactic that sometimes enables management to thwart a hostile offer. In addition, even if the credit received is scant, the cost to the member of Congress of service (introducing a bill or holding a hearing) is small, while the cost of inaction is potentially large (loss of support of not only the constituent firm but also of the rest of the local business community).

This analysis is not meant to suggest that congressional efforts at takeover legislation are uniformly perceived as futile or insincere gestures. Members of Congress do not necessarily intend all of their takeover-related action to be symbolic, and, of course, federal takeover legislation has periodically been enacted. . . .

The data do raise a potentially puzzling question: why, particularly given . . . the increasing pressure for federal action in the 1980s after *MITE*, was no general takeover legislation enacted? To frame the question more broadly, the persistence of policy alternatives on the decision agenda raises the question when does an idea become one "whose time has come?" To state the obvious, there is no theory powerful enough to predict political change. Cognizant of this caveat, researchers of congressional decision making have nonetheless identified as important variables in the timing of reform: (1) changes in presidential administration; (2) changes in the composition of Congress; and (3) shifts in national mood.

The principal reason why the substantial constituent pressure in the 1980s

---

[2]In Edgar v. MITE, 457 U.S. 624 (1982), the Supreme Court invalidated an Illinois takeover statute for unconstitutionally burdening interstate commerce [EDITOR'S NOTE].

for federal action on takeovers did not produce major legislation is, in my estimation, that one of the most important players, the Reagan administration, was opposed to it. The SEC Chairman throughout the period, John Shad, was a vigorous proponent of the administration's position in favor of a deregulated market. Agency heads are quite important in successful policy initiation, and the takeover field has been no exception. The Williams Act was vigorously supported by then-SEC Chairman Manuel Cohen, just as the Carter administration's Justice Department enthusiastically supported the Hart-Scott-Rodino Act. This fact no doubt goes some way in explaining the popularity of federal government personnel as witnesses in the takeover-related hearings, for without agency support, congressional policy objectives can be undermined by their inadequate or unenthusiastic implementation.

With the relevant regulatory agency and the president opposed, production of a takeover statute would clearly have been viewed by most members of Congress as a losing proposition, making efforts at logrolling unlikely to be pursued. . . . [T]here are important institutional features of the presidency [apart from ideology, that enter into the calculus]: unlike members of Congress, the president is elected by and represents all citizens and is thereby considered to be more removed from pork barrel politics. The Reagan administration's opposition to takeover regulation could be an instance of the operation of this institutional factor. . . .

An important background factor that can strengthen the executive branch's resistance to congressional pork barrel on takeover legislation is the third factor appearing in the policy initiation literature, national mood. A conventional measure of national mood is public opinion polls. . . .

Attitudes toward takeovers are stable not only across polls, but also over time and across different classes of respondents. Given the constancy in responses over the polls, public opinion on takeovers can be summarized . . . as follows: (1) a majority of the public is indifferent to, and at best casually informed about, takeovers; (2) a decisive plurality, and sometimes a majority, of the public has a negative opinion of takeovers; and (3) the public identifies shareholders and executives as winners, and workers as losers, in takeovers. . . .

The low saliency of takeovers as an issue for most voters, evidenced by the public's widespread lack of interest and knowledge concerning corporate acquisitions, most likely . . . contributes to Congress's one-sided yet dilatory progress toward takeover regulation. . . . Low visibility issues provide lobbyists with a large voice in policymaking because legislators can satisfy the most concerned parties with little personal cost, as evidenced by the speedy, near-unanimous passage of state takeover statutes. The public opinion poll data reinforce the prediction that were Congress to act, it would not be able to internalize fully the costs of takeover regulation, and it would behave in a similar fashion to the states. Given public attitudes toward and ignorance about takeovers, the policy recommendation of this article is that advocates of preemption might best serve their cause by seeking to educate the public concerning the theoretical and empirical findings on the beneficial effects of takeovers and a competitive market for corporate control. Otherwise they will, in all likelihood, be sorely disappointed in the legislation that is produced.

Notes and Questions

1. Easterbrook and Fischel predict that takeover auctions will decrease the number of bids, thereby increasing managerial slack and, hence, depressing share prices. Could lower share prices have an additional offsetting effect of increasing the gains available from a control change? How desirable is a passivity rule, which limits target returns in takeovers, if targets search for bidders? See David Haddock, Jonathan Macey, and Fred McChesney, "Property Rights in Assets and Resistance to Tender Offers," 73 *Virginia Law Review* 701 (1987). Is Easterbrook and Fischel's claim that ex ante, investors would choose to maximize the number of bids bolstered if investors' stock portfolios follow modern portfolio theory (see Chapter I) and are broadly diversified across firms? What if acquirers are privately held firms?

2. As Gilson notes, line drawing will be required under Easterbrook and Fischel's passivity rule as well as his own proposal to permit managers to engage in auctions. What criteria could a court use to distinguish between bona fide business strategies and illicit defensive tactics? For instance, could management continue with plans for a merger after a takeover bid is made, although the bid is contingent on the merger's termination?

3. In the late 1980s the Delaware courts seemed to have sided with Bebchuk and Gilson in the auction debate, as they upheld poison pill defenses that facilitated management's ability to hold an auction and required management to run a fair auction when it put the company up for sale. *Moran v. Household International,* 500 A.2d 1346 (Del. 1985); *Revlon v. MacAndrews & Forbes Holdings,* 506 A.2d 173 (Del. 1986). But by the 1990s, the trend to an "auctions always" rule had stalled. Auctions could be circumvented by a "passive" auction strategy, in which management publicly announced and held open a bid without actively soliciting competing offers, *Barkan v. Amsted Industries,* Civ. No. 8224 (Del. 1989), and by mergers that the shareholders approved, *Wheelabrator Technologies Shareholder Litigation,* Civ. No. 11495 (Del.Ch. 1990). In addition, a preplanned merger was held not to come under the *Revlon* rule in *Paramount Communications v. Time,* 571 A.2d 1140 (Del. 1989), despite the case's egregious circumstances: Confronted with a hostile takeover, management changed acquisition form in order to avoid a shareholder vote. For a view of the court's shifting position on auctions as tracking changing public sentiment toward takeovers, see Jeffrey N. Gordon, "Corporations, Markets, and Courts," 91 *Columbia Law Review* 1931 (1991).

4. The American Law Institute has endorsed the promotion of auctions in the context of management-led leveraged buyouts (MBOs). For the reporter's analysis of the proposal, see Ronald J. Gilson, "Market Review of Interested Transactions: The American Law Institute Proposal on Management Buyouts," in Y. Amihud, ed., *Leveraged Management Buyouts: Courses and Consequences* 217 (Homewood, Ill.: Dow Jones-Irwin, 1989). How do the policy considerations toward auctions differ in the MBO context compared to a third-party offer? Reconsider this question when reading the selections on insider trading regulation in Chapter VII.

5. The pressure to tender over which Bebchuk expresses concern refers both to front-end-loaded two-tier offers, in which a bidder tenders for 50.1 percent and, upon obtaining control, merges out the remaining shares, with the price in the second-step merger lower than that paid in the first step, and to any-or-all offers, in which the shareholder believes the offer price is lower than the target's true value but feels coerced into tendering because she believes that a majority will tender, leaving her worse off holding minority shares. Bebchuk has proposed, as the solution to pressure to tender, a variant of the approach of control share acquisition statutes: Legislation requiring a shareholder vote for a takeover to succeed while permitting shareholders who vote no to simultaneously tender their shares in case they are in the minority and the bid succeeds. Lucian A. Bebchuk, "The Pressure to Tender: An Analysis and a Proposed Remedy," in J. Coffee, L. Lowenstein, and S. Rose-Ackerman, eds., *Knights, Raiders, and Targets: The Impact of Hostile Takeovers* 371 (New York: Oxford University Press, 1988). How plausible is the hypothesized pressure to tender in an any-or-all bid? Would you expect the sophisticated arbitrageurs who acquire target shares when a bid is announced to experience such pressure? Can a public shareholder have a reasonable basis for believing the target's value is higher than the bid premium? Why wouldn't she make a competing bid and obtain shares at a discount?

There is scant empirical evidence of pressure to tender in front-end-loaded two-tier bids. Robert Comment and Gregg Jarrell find that the blended premium in two-tier bids is indistinguishable from the premium in any-or-all offers, and there is no evidence of front-end-loaded bids with lower blended premiums defeating any-or-all offers. Comment and Jarrell, "Two-tier and Negotiated Offers: The Imprisonment of the Free-Riding Shareholder," 9 *Journal of Financial Economics* 283 (1987). Note that two-tier bids disappeared by the late 1980s with the widespread availability of junk bond financing.

6. The following selection reviews the empirical evidence on auctions:

The economic literature cannot readily arbitrate this debate. To do so in a compelling fashion we would need to know the unknowable—how many takeovers there would have been had auctions not been permitted. But there is some suggestive research on the effect of regulations encouraging auctions, and it tends to support auction opponents. Bidder returns tend to be negative after state takeover laws were enacted and poison pill defensive tactics jelled. In addition, firms with active acquisition programs experienced negative stock price effects upon the enactment of the Williams Act. Furthermore, auctions increase premiums, and the number of auctions and the size of premiums increased after both the Williams Act and state takeover statutes were enacted. This suggests that bidders pay more in regimes facilitating auctions, which lowers their returns. Hence, these regimes will deter acquirers from making bids. However, these premium increases may not be due to the Williams Act. Nathan and O'Keefe suggest that any impact of the act was, at best, delayed, as they find that takeover premiums significantly increased after 1974 and not after 1968. In addition, Franks and Harris find that takeover premiums increased in the United Kingdom after 1968, which suggests that the increases Jarrell and Bradley found in the United States after 1968 may not be attributable to the Williams Act. Some studies provide evidence that more directly addresses the effect of auctions on the incidence of bids. Jarrell and Bradley find that the number of tender offers declined more sharply after the Williams Act than did the number of acquisitions. . . .

Hackl and Testani further find that states adopting second-generation takeover statutes had a smaller increase in the number of takeovers than states with no statutes (controlling for the number of firms incorporated in a state). This suggests that regulation promoting auctions chills takeovers. But they also find that bid premiums do not vary significantly across regulating and nonregulating states, and that regulating states do not experience more auctions than states without statutes. While the similar level of auction activity may explain the insignificant difference in premium levels across the two sets of states, these data imply that the predicted trade-off of a higher premium for a reduced probability of takeover is not realized through regulation. . . . In any event, a finding of no trade-off of higher premiums for fewer bids undercuts the position of auction proponents, because it indicates that a policy encouraging auctions may not maximize target revenues either ex ante or ex post.

Roberta Romano, "A Guide to Takeovers: Theory, Evidence and Regulation," 9 *Yale Journal on Regulation* 119, 158–60 (1992). Reprinted by permission.

7. Peter Cramton and Alan Schwartz apply the results of an extensive economic literature analyzing auctions to takeover bids and emphasize that optimal auction policy is a function of three factors: the auction environment, the policy goal of the auction, and anticipated behavior of target management. Cramton and Schwartz, "Using Auction Theory to Inform Takeover Regulation," 7 *Journal of Law, Economics, and Organization* 27 (1991). Auction environments vary according to whether the auctioned object's value is the same to all bidders ("common value" auction) or whether each bidder values it differently ("independent value" auction). Cramton and Schwartz show that under an efficiency goal, auctions are good policy in independent value settings (for they identify the highest-valuing user), whereas in common value settings, restricting the number of bidders and, in particular, negotiating with one buyer only is preferable. Under a target revenue-maximizing goal, however, auctions may be appropriate even in the common value setting because they increase revenues, compared to a regime of unregulated offers to buy stock from dispersed shareholders.

Cramton and Schwartz maintain that different sources of takeover gains indicate different auction environments. They contend that a takeover undertaken to achieve synergy gains is an independent value setting (i.e., synergistic gains are considered to be unique to the particular bidder), whereas a takeover motivated by agency costs, such as eliminating free cash flow or replacing management, is a common value setting (i.e., all bidders adopt the same strategy and receive the same value from owning the target). If efficiency is the goal, this suggests that the optimal approach to auctions will vary in conjunction with takeover type. However, Cramton and Schwartz further note that if target managers are disloyal agents, they could use the discretion granted under a context-dependent rule to defeat bids by holding auctions when they should not or by not holding auctions when they should. Because Cramton and Schwartz are persuaded that the agency problem is widespread in the takeover context, they recommend adopting a uniform policy, based on the most prevalent auction type, which they characterize as common value settings, and they therefore propose a ban on takeover auctions.

Do you agree with Cramton and Schwartz's categorization of takeover types into auction environments? How would the efficacy of their "auctions never" proposal be affected if the auction environment has varied dramatically over time? Consider, that is, the changing patterns in takeover activity: in the 1960s, conglomerate mergers were predominant, whereas bust-up takeovers dominated the 1980s. Would the difficulty with a rule giving management discretion to hold an auction be overcome by adopting a shareholder choice rule (i.e., let shareholders vote on all defensive tactics, including solicitation of competing bids)?

8. How significant is the finding mentioned in Jarrell, Brickley, and Netter that most shark repellent amendments are approved? Is it evidence of the problems in shareholder voting discussed in Chapters III and V? Consider the well-established practice of management's consulting a proxy specialist regarding the likelihood of a proposal's success before putting it on the agenda. If success is unlikely, the proposal is never proffered, and consequently, any study of amendments will suffer from selection bias. Jarrell, Brickley, and Netter's key point is, however, a different one: the defensive tactics subject to shareholder approval are less likely to result in negative abnormal returns than those for which no shareholder vote is required. For articles discussing why fair price and related defensive charter amendments may, at least under certain circumstances, benefit shareholders, see William J. Carney, "Shareholder Coordination Costs, Shark Repellents, and Takeout Mergers: The Case Against Fiduciary Duties," 1983 *American Bar Foundation Research Journal* 341; Barry Baysinger and Henry Butler, "Antitakeover Amendments, Managerial Entrenchment, and the Contractual Theory of the Corporation," 71 *Virginia Law Review* 1257 (1985); Roberta Romano, "The Political Economy of Takeover Statutes," 73 *Virginia Law Review* 111 (1987). For the case against all defensive charter amendments, see Ronald J. Gilson, "The Case Against Shark Repellent Amendments: Structural Limitations on

the Enabling Concept," 34 *Stanford Law Review* 775 (1982). How might stock price returns on the announcements of defensive charter amendments vary with firms' ownership composition? See Victoria McWilliams, "Managerial Share Ownership and the Stock Price Effects of Antitakeover Amendment Proposals," 45 *Journal of Financial Economics* 1627 (1990).

9. Can you reconcile Jarrell, Brickley, and Netter's characterization of the elimination of cumulative voting as a defensive tactic with Easterbrook and Fischel's view (see Chapter V) that cumulative voting hinders control changes?

10. Two defensive tactics that some commentators maintain benefit share-holders are golden parachutes and greenmail. Golden parachutes are executive compensation contracts that provide extraordinarily high severance pay upon a control change. Is this what Williamson (see Chapter V) has in mind when referring to severance pay as a method of protecting managerial investments in firm-specific human capital? Golden parachutes have been attacked as a defensive tactic for increasing the bidder's cost (by the amount of the compensation). But by providing managers with financial incentives to accept a takeover, golden parachutes align their interests with those of shareholders and thus can be seen as value-maximizing strategies. See, for example, David Baron, "Tender Offers and Management Resistance," 38 *Journal of Finance* 331 (1983). There is a competing concern: managers may be willing to accept an inadequate bid in order to exercise their parachutes. However, investors may not regard the latter as a significant problem: stock prices significantly increase on the announcement of golden parachute plans. Richard Lambert and Donald Larckner, "Golden Parachutes, Executive Decision-making and Shareholder Wealth," 7 *Journal of Accounting and Economics* 179 (1985). An alternative explanation of this finding is that shareholders positively reevaluate the stock price because they consider the plan a signal that management expects a takeover bid to be in the offing.

Greenmail can be a value-maximizing tactic if management believes there is another bidder who will pay more than the first but who will not enter into a takeover auction with the first bidder. See, for example, Jonathan Macey and Fred McChesney, "A Theoretical Analysis of Corporate Greenmail," 95 *Yale Law Journal* 13 (1985); Andrei Shleifer and Robert Vishny, "Greenmail, White Knights, and Shareholders' Interest," 17 *RAND Journal of Economics* 293 (1986). Because sometimes no second bidder will be forthcoming, the stock price may decline on the payment of greenmail (as reported in the Jarrell, Brickley, and Netter selection). Thus, an ex ante efficient (value-maximizing) decision turns out to be a negative one ex post. Can advocacy of greenmail be reconciled with Easterbrook and Fischel's argument against defensive tactics that encourage auctions?

The analysis in defense of greenmail is more controversial than that for golden parachutes because of the difficulty in distinguishing when greenmail is used by faithful managers as in the hypothesized scenario and when it is used by self-serving managers who do not expect a second bid to be forthcoming.

See Jeffrey Gordon and Lewis Kornhauser, "Takeover Defense Tactics: A Comment on Two Models," 96 *Yale Law Journal* 295 (1986). Can you think of any empirical test that could resolve this question? Should there be a higher frequency of control changes after greenmail repurchases than after adoption of other defensive tactics? Note that Wayne Mikkelson and Richard Ruback find in a sample of targeted stock repurchases that 29 percent of the firms paying greenmail experienced control changes within three years of the repurchase. Mikkelson and Ruback, "Targeted Repurchases and Common Stock Returns," 22 *RAND Journal of Economics* 544 (1991).

11. Jarrell, Brickley, and Netter repeat the widely shared wisdom of corporate law practitioners that poison pills are the most effective deterrent strategy. Are poison pills as effective as they suggest? The Ryngaert study they discuss found that poison pills raised the probability of remaining independent to 31 percent from 16 percent (the probability for firms without pills). But consider the following data. From 1986 to 1989, there were eighteen Delaware cases in which a court denied an unsolicited hostile tender offeror's motion to compel redemption of a poison pill. See "Poison Pills as Value Enhancer?: The Outcome of Delaware Redemption Cases," 4 *Bowne Digest for Corporate and Securities Lawyers* 8 (March 1990) (reproducing chart prepared by Henry Lesser and Ann E. Lederer). In fifteen of these cases, the targets were subsequently acquired. If managements sought to retain independence through use of the poison pill, they were unsuccessful. In addition, the original hostile bidder was the successful acquirer in eleven of the cases. The final price was, on average, more than 20 percent above the initial bid. These data indicate that the poison pill strategy may be more likely to raise premiums than to defeat bids. Of course, these data are conditional on a bid; they do not account for how many acquirers were deterred from making bids for firms with poison pills.

What if bidders anticipate that target management will use a pill to demand a higher price, and courts are hesitant to mandate redemption until there has been upward revision? Could bidders simply discount their initial bids, thereby paying no more after negotiation than they would have paid in the absence of the pill? What risk would such a strategy pose for a bidder? Would data showing an upward trend in premiums after the introduction of poison pills refute this alternative interpretation of the Delaware redemption cases?

As a last consideration on poison pills, is there any means by which managers can signal their faithfulness as agents in adopting a poison pill apart from requesting shareholder approval? Should shareholders' assessment of the tactic differ according to whether managers have traded in the corporation's stock prior to adopting the pill? See Charmen Loh, "Poison Pill Securities: Shareholder Wealth and Insider Trading," 27 *Financial Review* 241 (1992).

12. Is Karpoff and Malatesta's failure to find a significant negative price effect from Delaware's second-generation takeover statute surprising under Romano's analysis? Given the Supreme Court's takeover statute jurispru-

dence, should Karpoff and Malatesta have expected to find significant abnormal returns for second-generation statutes enacted after *MITE* but prior to *CTS?* Do studies' findings, referred to in the Karpoff and Malatesta selection, of significant negative returns upon enactments of statutes prior to *CTS* evidence that investors are oblivious to the constitutionality question?

13. What does the prevalence of state takeover laws suggest about the efficacy of state competition? Does the externality analysis and politics of takeover laws provided by Romano suggest that such laws are different from other corporation laws? Is it significant that Delaware is a laggard in legislating in this area and that its laws are less hostile to bidders? Is this evidence of the market for corporate charters working? Before reaching a conclusion on state competition, consider the following events in Pennsylvania:

In 1990, Pennsylvania enacted one of the most Draconian takeover statutes, which has, among other provisions, a disgorgement clause that prevents an unsuccessful bidder from selling target shares (in the market or otherwise) at a profit. The enactment of the statute had a significant negative effect on stockholder wealth. Samuel H. Szewczyk and George P. Tsetsekos, "State Intervention in the Market for Corporate Control: The Case of Pennsylvania Senate Bill 1310," 31 *Journal of Financial Economics* 3 (1992). The statute permitted boards to opt out, in whole or in part, within ninety days. Under pressure from institutional investors, many corporations opted out of at least part of the statute, including companies constituting over 60 percent of the equity value of Pennsylvania firms (approximately 25 percent of the public companies subject to the statute). Szewczyk and Tsetsekos find that firms experience significant positive abnormal returns upon opting out of the statute.

Does the political analysis in Romano support the criticism of other constituency statutes presented in Chapter V, part A, note 1? Could such statutes provide shareholders with a way to precommit to not breach implicit labor contracts in a takeover? For a skeptical appraisal of the statutes, see William J. Carney, "Does Defining Constituencies Matter?" 59 *University of Cincinnati Law Review* 385 (1990). For the view that most such statutes are consistent with the common law, see Charles Hansen, "Other Constituency Statutes: A Search for Perspective," 46 *Business Lawyer* 1355 (1991).

If state takeover laws are a concern because they adversely affect shareholders, are there other, less restrictive solutions than national regulation? Would it be useful to require corporation code provisions to have an opt-in rather than opt-out structure when a provision is one in which we can intuit a clear conflict of interest between managers and shareholders (such as takeover regulation)? What other governance rules would need revision to make such a scheme effective? Consider, for example, Delaware General Corporation Law §242(b), which indicates whose approval is necessary for charter amendments.

VII

Securities Regulation

This chapter focuses on the nationally mandated system of corporate disclosure and insider trading regulation under the Securities Act of 1933 and the Securities Exchange Act of 1934. Regulation of corporate financial and proxy statements aims to enhance shareholders' ability to monitor management as well as the quality of their investment decisions. It also imposes additional fiduciary duties on management, as does the national insider trading regime. In this regard, securities regulation supplements, in important ways, the principal mechanisms of shareholder monitoring, litigation, and voting examined in Chapter V.

The important policy question concerning disclosure regulation is whether a mandatory system is necessary or worth the expense: Do firms, operating in competitive capital markets, have sufficient incentives to disclose voluntarily the information investors desire? Although they are sympathetic to the claim that voluntary disclosure should suffice, Frank Easterbrook and Daniel Fischel offer a justification for mandatory disclosure regulation at the federal rather than state level based on third-party effects. The information disclosed by any one firm benefits shareholders in many other firms. Because a firm cannot charge other firms' investors for the cost of producing information, it will be underproduced. This is a variant on the public goods explanation for government (see Chapter III, part B, note 1). John Coffee questions the value added by the third-party effects explanation compared to a more straightforward externality analysis concerning suboptimal production of information by firms for their own investors.

One way to resolve the debate over the efficacy of mandated disclosure is to examine whether the value of listed securities increased upon the adoption of the federal requirements. George Stigler and Gregg Jarrell found that the returns of newly issued securities did not significantly increase after the 1933 Act and concluded that the disclosure regulation did not improve the quality of new issues. Stigler, "Public Regulation of the Securities Markets," 37 *Journal of Business* 117 (1964); Jarrell, "The Economic Effects of Federal Regulation of the Market for New Security Issues," 24 *Journal of Law and Economics* 613 (1981). It is, however, possible that the information about investment quality provided by the 1933 Act reduced uncertainty about security values; this could produce an effect on the variance of stock returns, rather than the returns themselves. The selection by Carol Simon, accordingly, investigates the impact of the 1933 Act on securities' risk as well as returns and reassesses the earlier studies' conclusions.

Insider trading regulation raises a similar issue. Is a national prohibition necessary, or should insider trading restrictions be one of the many provisions in the corporate contract subject to hypothetical bargaining between managers and shareholders? The federal prohibition, as Kenneth Scott explains, can best be justified as an extension of traditional fiduciary concepts (the prohibition of self-dealing) in that it prevents managers from using corporate property (inside information) to favor themselves over shareholders. This rationale does not, of course, address the underlying issue, whether the ban should be mandatory. Scott also reviews what he considers to be less satisfying rationales for the prohibition of insider trading, fairness across investors requiring equal access to information and improving market efficiency. The case against a mandatory insider trading ban is presented in the selection by Dennis Carlton and Daniel Fischel, which refines the iconoclastic contribution of Henry Manne, *Insider Trading and the Stock Market* (New York: Free Press, 1966), who was the first to question the wisdom of this regulation. The primary objections are that it hinders market efficiency (insider trades carry information that move stock prices in the right direction) and eliminates an effective form of incentive compensation (the substantial magnitude of potential trading profits encourages managerial risk taking).

Disclosure Regulation

Mandatory Disclosure and the Protection of Investors

FRANK H. EASTERBROOK AND DANIEL R. FISCHEL

[The securities laws] have two basic components: a prohibition against fraud, and requirements of disclosure when securities are issued and periodically thereafter. The notorious complexities of securities practice arise from defining the details of disclosure and ascertaining which transactions are covered by the disclosure requirements. There is very little substantive regulation of investments. . . .

In securities markets only a limited amount of information can be verified at all. Investors cannot "inspect" a business venture in a way that enables them to deduce future profits and risks. Investors do not even want to inspect; they seek to be passive recipients of an income stream, not to be private investigators. When investors spend time and resources inspecting, each one's effort will duplicate another's. A system of inspection by buyers would forfeit much of the benefit of the division of labor. . . .

High-quality firms must take additional steps to convince investors of their quality. One traditional step is to allow outsiders to review the books and records and to have these outsiders certify the accuracy of the firms' representations. The accountant who certifies the books of many firms has a reputational interest—and thus a possible loss—much larger than the gains to be made from slipshod or false certification of a particular firm. Similarly, firms may sell their securities through investment bankers who inspect the firm's prospects, put their money on the line in buying the stock for resale, and put their reputations on the line in making representations to customers. . . .

The firms themselves can take actions that render their disclosures more believable. One such action is to ensure that their managers hold substantial quantities of their stock. This can be accomplished by stock options or by "cheap stock" when the firm goes public, as well as by inducing managers to buy stock in the market. Then if the firm does poorly, the managers lose with the other investors. . . . Another action open to the firm is to issue debt, which could lead to bankruptcy. This strategy might seem implausible, but consider that bankruptcy imposes very high costs on managers' portfolios and

Reprinted by permission from 70 *Virginia Law Review* 669 (1984).

careers. By using leverage to increase the risk of bankruptcy, the firms with the best prospects and thus the lowest bankruptcy costs can certify themselves to investors. . . . Finally, managers could warrant their statements in the traditional way: They could make legally enforceable promises (perhaps backed up by insurers) to pay the investors if the firm does worse than promised (perhaps, say, in a comparison against a market index). The person vouching for the payment obligation would look very carefully at the firm's claims, so that only high-quality firms could find solvent guarantors, and the investors would be protected.

Even in a market without a rule against fraud, these methods of verification would offer investors substantial protection and make it possible for high-quality firms to raise money. Investors, after all, need not donate cash to new firms. They can put their money in government securities or bank accounts with no risk; they can invest in regulated public utilities that have very little risk; they can purchase land or other productive assets. New or less-known firms can obtain money only if they offer packages more attractive than those already existing. . . .

For the reasons we have spelled out, a rule against fraud is not an essential ingredient of securities markets. Each of these certification methods is costly, however. Auditing, investment banking, and underwriting firms are very expensive to establish and operate; debt and dividend strategies entail transaction costs; managers must be paid extra to induce them to hold undiversified portfolios, and their risky position may lead them to make inferior investment decisions later on; direct verification of claims by thousands of buyers may be the most expensive of all.

A rule against fraud can reduce these costs, especially for new firms. The penalty for fraud makes it more costly for low-quality firms to mimic high-quality ones by making false disclosures. An antifraud rule imposes low or no costs on honest, high-quality firms. Thus it makes it possible for high-quality firms to offer warranties at lower cost. . . .

Antifraud rules also impose costs of their own. Enforcement costs (investigative, prosecutorial, and judicial staffs) are obvious. The costs of overenforcement or inaccurate enforcement are harder to see but no less real. . . .

Why Is the Rule Against Fraud National?

In 1933 every state had a rule against fraud. What, then, was the point of the many new rules contained in the 1933 and 1934 Acts? The Acts' supporters usually say that the national rules were necessary because the state rules were "ineffective" (witness the discovery of frauds), but this is not a good explanation. The existence or even increase of reported frauds no more proves that the state laws against fraud were "ineffective" than the existence of murder shows that state criminal law is "ineffective" and should be replaced with a national murder statute enforced by a Federal Homicide Commission.

The justification of federal legislation lies, rather, in the efficiency of enforcing in one case all claims that arise out of a single transaction. Many new issues of securities are sold to purchasers in several states; even issues initially sold within a single state ultimately find their way into the hands of out-of-state owners, if only because the owners move. Thus almost all substantial firms' securities sell in interstate markets. If claims arising out of these securities were litigated where plaintiffs resided, there would be multiple cases for every security, with the possibility of inconsistent decisions and inconsistent legal standards. . . . The securities laws create nationwide service of process and have a liberal venue rule that permits litigation to consolidate all defendants and all claims in a single forum. The class action device created by Rule 23 of the Rules of Civil Procedure makes it easy to bring all plaintiffs together. . . .

Disclosure and the Public Goods Aspect of Information

In a world with an antifraud rule but no mandatory disclosure system, firms could remain silent with impunity. . . . A mandatory disclosure system substantially limits firms' ability to remain silent. . . .

What does a mandatory disclosure system add to the prohibition of fraud? The implicit public-interest justification for disclosure rules is that markets produce "too little" information about securities when the only rule is one against fraud. One often hears the assertion that information is a "public good," meaning that it can be used without being used up and that the producer of information cannot exclude others from receiving the benefits. If the producer of information cannot obtain all of its value, too little will be produced. It seems to follow that there are virtues in a rule requiring production of all information that would be forthcoming were gains fully appropriable.

This rationale gets us only so far. For one thing, it proves too much. No one can fully appropriate the value of information about toothpaste, but there is no federal rule about disclosing the efficacy of toothpaste in preventing cavities. Why are securities different? We leave the other products to competitive markets because of a conclusion that people who make or use a product (or test it as a Consumers' Union does) will obtain enough of the gains from information to make the markets reasonably efficient. . . .

The more sophisticated version of the public goods explanation is that although investors produce information, they produce both too much and too little. They produce too little because the benefits are imperfectly appropriable. . . . Investors produce too much information, though, if several create the *same* ten dollars bit of information (redundant production). Mandatory disclosure will prevent redundant production of information, the argument concludes.

[A] source of excessive production is the gain available from forecasting the future. Some information, such as the quarterly earnings of a firm, offers opportunities for trading gains; the person who learns the news first can make great profits. In one important sense, though, the information is worthless.

Trading on news that is bound to come out anyway does not change the future or lead to better investment in new securities. The price will ultimately change to reflect the true earnings. That it changes a day or so quicker is not of much moment for allocative efficiency. The lure of trading profits may induce people to spend a lot of effort and other resources "beating the market." Much of this is waste because the profit opportunity is larger than the efficiency gains from expediting the transition of prices. The argument concludes by observing that the prompt disclosure of information by the affected firm will extinguish the trading opportunity. When everyone knows the truth, no one can speculate on it. Investors as a group would pay to have these trading gains (and the costly search for information) eliminated. What better way to do this than mandatory disclosure by the firm that knows the truth?

These arguments have a common problem: they do not link the benefit of disclosure and the benefit of *mandatory* disclosure. If disclosure is worthwhile to investors, the firm can profit by providing it. The firm is in privity with its investors, and the Coase Theorem suggests that firm and investors can strike a mutually beneficial bargain. A decision by the firm effectively "coordinates" the acts of many investors who could not bargain directly. . . .

The principle of self-induced disclosure as a solution to the lack of property rights in information applies to trading in the secondary market as well as to the initial issuance of stock. The firm's investors always want to be able to sell their stock in the aftermarket for the highest price. Their ability to do so depends on a flow of believable information (otherwise potential buyers reduce the bid prices, assuming the worst). For most information about a firm, the firm itself can create and distribute the knowledge at less cost than the shareholders, and the firm's decision, because it reflects the value to all shareholders, will be correct at the margin. A firm that wants the highest possible price when it issues stock must take all cost-justified steps to make the stock valuable in the aftermarket, so it must make a believable pledge to continue disclosing.

The evidence bears this out. Firms have been disclosing the most important facts about themselves—and certifying these facts through third parties—as long as there have been firms. It is possible to trace the use of auditors back to the beginning of the corporation, and at the time the 1934 Act, which created the requirement of annual disclosure by listed companies, became law, every firm traded on the national markets made voluminous public disclosures certified by independent auditors. Between 1934 and 1964, annual disclosure was required only of those firms traded on national exchanges. (In 1964 the statute was amended to cover all firms with more than a specified number of investors.) Firms could avoid disclosure by delisting or not listing initially. Nonetheless, firms eagerly listed themselves on an exchange and disclosed; firms that were not listed also disclosed substantial amounts of data, following the pattern set by those covered by the statute. Even today, the securities of state and local governments are exempt from the mandatory disclosure rules, yet these issuers routinely supply voluminous information to purchasers.

Disclosure for the purpose of stilling investors' doubts also reduces (to the appropriate degree) investors' incentives to search too much for trading information. The problem is that knowing the future creates profit opportunities without making investors as a group better off. Because searching out such information is costly, investors as a group gain if firms disclose so as to minimize the opportunities and thus the incentives to search. The net return on a security is its gross return (dividends plus any liquidating distribution) less the cost of information and transactions in holding the security. A firm can increase this net return as easily by reducing the cost of holding the stock as well as by increasing its business profits. Firms that promise to make disclosures for this purpose will prosper relative to others, because their investors incur relatively lower costs and can be more passive with safety. The more convincing the promise, the more investors will pay for the stock. . . .

Controlling Third-Party Effects

There is one other reason why firms' disclosures may not be optimal: third-party effects. The information produced by one firm for its investors may be valuable to investors in other firms. Firm A's statements may reveal something about the industry in which Firm A operates . . . that other participants in the industry can use in planning their own operations. There may be other collateral benefits to investors in rival firms. Yet Firm A cannot charge the investors in these other firms for the benefits, although they would be willing to pay for them. Because they cannot be charged, the information will be underproduced. . . .

Or suppose there is an optimal format for communicating information to investors. Some disclosures are easier to understand, verify, and so forth, than others, while some disclosures tend more to hide than to reveal information. If contracts among all investors in society could be written costlessly, the investors would require all firms to identify and use the optimal format of disclosure. The costs may be too high, though, for one firm acting on its own. The optimal form of disclosure may entail use of some specialized language (one can think of accounting principles, with their detailed definitions, as a specialized disclosure language), yet no one firm can obtain a large share of the benefits of inventing and employing this language; others will be able to use the format without charge. Sometimes, too, the ease of using a given method of disclosure will depend on other firms adopting the same format, so as to facilitate comparisons across investments. Other firms may not be anxious to cooperate.

Mandatory disclosure rules promulgated by the government are one means to achieve standardization, but it does not follow that mandatory disclosure is necessary. Markets frequently devise ingenious solutions to problems of information. Indeed, the problems faced by sellers of securities are not much different from those involved in bringing new products to market. . . . Color television was not feasible until manufacturers and broadcasters agreed on a standard method of transmission. . . . Sometimes trade associations may devise such standards, as the electronics industry and, in part, the accounting industry have done. Whether standardization may be achieved more cheaply by private or governmental responses is an empirical question. . . . Private organizations cannot compel adherence, so there will be holdout problems. Competition among the states cannot obtain all benefits because of the interstate nature of some of these effects; if being a holdout is in the interest of some firms, it could pay states to be havens to the holdouts. . . .

[The] difficulty with coordinating disclosure through competition among the states reflects the fact that such jurisdictional competition is most effective when the consequences of a decision will be experienced in one jurisdiction. Because only one state's law governs the "internal affairs" of a corporation, competition can be effective. Disclosure rules for a firm chartered in State D, in contrast, affect many firms incorporated elsewhere. Indeed, the multistate nature of securities markets creates opportunities for states to attempt to exploit investors who live elsewhere.

Consider the problem of a firm incorporated in State D, which prescribes disclosure of facts X, Y, and Z. Suppose that this amount of disclosure is optimal; additional disclosures would cost more than the benefits. . . . A group of investors living in state N might bring litigation there, contending that under the law of N, the firm should have disclosed Q. It may well be in state N's interest to sustain this claim and order the firm to pay damages to N's residents, even if N's officials know that the disclosure of Q is counterproductive. This is because most of the investors live outside state N and will not receive damages. The money to pay to state N's residents will come from residents of other states. State N may seize the occasion to "exploit" the residents of other states, once the firm is underway. State N's residents gain more from the award of damages than they lose from future "inefficiently large" disclosures, since the costs of these disclosures will be borne in the other forty-nine states. State N's residents end up with 100 percent of the benefits of this transfer payment. . . .

If state N attempts such exploitation, other actors in the markets will adjust. The firms may sell less stock in state N, but they cannot prevent it from migrating there. Firms may start to disclose Q, but by hypothesis this is not optimal, and the disclosure of Q will not prevent state N from insisting on some new disclosure tomorrow. Finally, other states may retaliate. . . . Everyone is better off if the states desist from such attempts to exploit one another's residents, but it may be costly to control this exploitation. . . . Only federal regulation may be able to prevent states from engaging in exploitation in securities transactions.

Market Failure and the Economic Case for a Mandatory Disclosure System

JOHN C. COFFEE, JR.

A simpler theory can justify a mandatory disclosure system [than Easterbrook and Fischel's third-party effects hypothesis]. . . . Essentially, this response will make four claims.

First, because information has many characteristics of a public good, securities research tends to be underprovided. This underprovision means both that information provided by corporate issuers will not be optimally verified and that insufficient efforts will be made to search for material information from nonissuer sources. A mandatory disclosure system can thus be seen as a desirable cost reduction strategy through which society, in effect, subsidizes search costs to secure both a greater quantity of information and a better testing of its accuracy. Although the end result of such increased efforts may not significantly affect the balance of advantage between buyers and sellers, or even the more general goal of distributive fairness, it does improve the allocative efficiency of the capital market—and this improvement in turn implies a more productive economy.

Second, a substantial basis exists for believing that greater inefficiency would exist without a mandatory disclosure system because excess social costs would be incurred by investors pursuing trading gains. Collectivization minimizes the social waste that would otherwise result from the misallocation of economic resources to this pursuit.

Third, the theory of self-induced disclosure, now popular among theorists of the firm and relied upon by Professors Easterbrook and Fischel, has only a limited validity. A particular flaw in this theory is that it overlooks the significance of corporate control transactions and assumes much too facilely that manager and shareholder interests can be perfectly aligned. In fact, the very preconditions specified by these theorists as being necessary for an effective voluntary disclosure system do not seem to be satisfied. Although management can be induced through incentive contracting devices to identify its self-interest with the maximization of share value, it will still have an interest in acquiring the shareholders' ownership at a discounted price, at least so long as it can engage in insider trading or leveraged buyouts. Because the incentives for both seem likely to remain strong, instances will arise in which management can profit by giving a false signal to the market.

Fourth, even in an efficient capital market, there remains information that the rational investor needs to optimize his securities portfolio. Such information seems best provided through a mandatory disclosure system. . . .

Reprinted by permission from 70 *Virginia Law Review* 717 (1984).

Plausibility is not the ultimate test of a theory. As a result, the seeming flaws in the theory of market-induced disclosure still do not eliminate the more difficult question of whether any empirical evidence is available by which to gauge the comparative efficacy of voluntary and mandatory disclosure. . . . When social scientists cannot do controlled experiments, they frequently look for "natural experiments" that can be interpreted. . . . Correspondingly, we can look today at the differences between the disclosure level within the public securities market subject to SEC regulation and the level that prevails within the one major securities market that is exempt from registration—the municipal bond market. . . . [I]f the recent experiences with the New York City bond offerings in the 1970s and the Washington Public Power System's failure in the 1980s are indicative, critical information is not being disclosed to investors. Most observers would agree with this statement, but the neoclassical theorist will respond that little information need reach investors because they are protected instead by the bond rating agencies—Moody's or Standard and Poor's; these agencies digest the relevant information, which in the case of a debt security consists only of its risk level, and assign a rating to each security.

If one examines the securities markets only at a distance and through the telescope of neoclassical economic theory, this rebuttal may sound persuasive. If one examines the institutional structure more closely, however, disturbing problems begin to appear. First, in the New York City fiscal crisis, Moody's did not reduce New York's rating until the crisis was universally acknowledged. Second, because the issuer pays the bond rating agency to be rated, there is a conflict of interest problem. Third, the bond rating agencies are not themselves investigating agencies. Instead, they depend on the data that the issuer gives them. Yet a recent survey by Arthur Young & Co., the auditing firm, suggests that this data's accuracy is in serious doubt. . . . In part, these problems may stem from the still-rudimentary state of the accounting principles applicable to governmental and nonprofit bodies. . . .

Where then are we left? Notwithstanding these criticisms, the theory of voluntary disclosure does seem to have some validity as applied to initial public offerings and, to a lesser extent, to all primary distributions. This theory has far less persuasive force, however, when applied to secondary market trading, which the 1934 Act chiefly governs. Here, high agency costs currently exist (as the persistence of high takeover premiums averaging between 50 percent and 70 percent in recent years arguably seems to show), thus sheltering opportunistic managerial behavior.

The Effect of the 1933 Securities Act on Investor Information and the Performance of New Issues

CAROL J. SIMON

The economic effects of the 1933 Securities Act have previously been studied by George Stigler and Gregg Jarrell. Motivated by the assertion that misrepresentation and fraud were consequences of unregulated markets, both studies focused on whether the mandated disclosure of financial information required by the Act increased the average return earned by new-issues investors. Neither study finds evidence of a significant increase in average returns following disclosure regulation, leading both authors to conclude that federal regulation of new-issues markets was ineffective, or at least superfluous given existing private market sources of financial information.

The existence of substantial uncertainty about the true value of a security need not imply that the issue will be, on average, overvalued or undervalued. Rather, the expectations of rational investors should be unbiased. The availability of quality information will, however, affect the riskiness of the purchase. As such, the effects of legislation aimed at increasing investor information should be reflected in changes in the *dispersion* of market-adjusted returns. Accordingly, this study examines regulation-induced changes in both the *means and variances* of the distributions of returns earned by new-issues investors.

This article also evaluates the extent to which private sources of investment-quality information were available in the absence of regulated disclosure. In general, consumers may obtain quality information directly from sellers, through experience with the good, or from third-party appraisers. Prior to SEC regulation, investors formed expectations of future returns by relying on information obtained directly from brokers and underwriters, by observing a security's historic performance (if any) and/or through the reports and actions of independent appraisers—most notably the Listing Committee of the NYSE. The economic effects of minimum disclosure would be expected to be the greatest where the private costs of obtaining and verifying information were highest. Specifically, this article examines the effectiveness of the Act conditional upon the prior market seasoning of a security (experience) and whether the issue had been approved for listing by the NYSE (third-party appraisal). . . .

Two samples of issues are constructed. The "preregulation" sample contains new issues from the period 1926 to 1933. The "post" sample is composed

Reprinted by permission from 79 *American Economic Review* 295. © 1989 American Economic Association.

of common stock issues floated between 1934 and 1939. Monthly returns for the five-year period following the date of issue have been collected for all issues in the sample. Both samples contain seasoned and unseasoned issues as well as stocks traded on the NYSE and stocks listed exclusively on regional exchanges. Recall, an issue is "seasoned" if the stock traded on an exchange prior to the offering. . . .

The data used in this study include virtually all new issues of common stock exceeding $1.95 million sold by manufacturing firms, railroads, retail, and service establishments between 1926 and 1940. . . .

Briefly, there is no evidence that, *on average,* either seasoned or unseasoned issues traded on the NYSE were significantly over- or under-priced. . . . For both the samples of NYSE issues and the sample of seasoned issues traded on regional exchanges no significant excess returns are measured prior to the SEC. There is no evidence that investors were systematically misinformed in these markets.

The evidence is quite different for unseasoned issues traded on the smaller regional exchanges. Prior to 1933, unseasoned, non-NYSE issues suffer statistically significant risk-adjusted losses. . . . Over 85 percent of the firms in the sample (thirty of thirty-five) suffer significant losses. Cumulative losses are statistically significant.

In contrast, there is no evidence of abnormal gains or losses among seasoned or unseasoned, NYSE or regional issues following 1933. . . . The results of this study suggest that only unseasoned issues floated on exchanges other than the NYSE earn significantly greater risk-adjusted returns following the 1933 SEC Act. . . . The results suggest that there was not a universal rise in the return earned by new-issues investors following the 1933 Act. Investors—exclusive of those in non-NYSE, unseasoned issues—held unbiased expectations of future returns both before and after the SEC Act. The analysis of average returns, however, is inadequate for assessing potential changes in risk borne by investors.

. . . Owing to differences in the costs of obtaining prior information we would expect that the variance of excess returns is higher for unseasoned than for seasoned issues, and that investors' forecasts of issue performance are less informed for non-NYSE issues than they are for NYSE securities. . . . To provide a benchmark for comparing variance changes between periods, the cross-sectional variance of excess returns for a randomly selected sample of NYSE firms (not issuing stock) was computed for the pre- and post-SEC eras.

Results suggest that the dispersion of abnormal returns is smaller in the period following the SEC than prior to the SEC. All subsamples of issues—seasoned, unseasoned, NYSE, and non-NYSE—exhibit significantly smaller forecast errors in the post-1934 era. . . .

Between the pre- and post-SEC periods issue-specific risk falls by 45 percent in the baseline market portfolio. This suggests that part of the decline in dispersion is due to factors unrelated to the Securities Act. Whether these factors are sufficient to explain risk reduction in new-issues samples is unclear.

All new-issues samples exhibit a substantially larger degree of risk reduction than does the baseline sample. Specifically, the variance of the forecast errors for seasoned NYSE issues falls, on average, by 60 percent. For unseasoned NYSE issues the decline is approximately 56 percent. Similarly seasoned, non-NYSE issues exhibit an average decline in error variance equal to 75 percent, while for unseasoned, non-NYSE issues post-SEC forecast errors are 85 percent lower. Segments of the market where private information may have been most costly before the SEC exhibit the largest declines in return forecast errors following the 1933 Act. . . .

The introduction of mandatory disclosure under the SEC was a one-time event. Its timing coincided with a great many other economic events—the effects of which may only be imperfectly controlled. While the results of this research suggest that a change in investment returns followed the SEC Act of 1933, confounding factors abound. . . .

This article does not address the costs of SEC regulation. It does suggest that the gains from regulation were small for seasoned issues, and for many issues traded on the NYSE. In fact, the 1933 Act and subsequent regulation contributed to the growth of the over-the-counter market as issuers sought lower cost, unregulated markets.

Notes and Questions

1. Even if information produced by mandated disclosure is desired by investors, there is a question whether the benefits of regulation are worth its cost. In a somewhat dated but still suggestive study, Susan Phillips and Richard Zecher estimated the costs of periodic disclosure in 1975 as $213 million and of new issue disclosure as $193 million. Phillips and Zecher, *The SEC and the Public Interest* 51 (Cambridge, Mass.: MIT Press, 1981). These estimates are derived by extrapolation to the universe of registered firms from a set of twenty-two firms that were required to report filing costs to an SEC Advisory Committee on Corporate Disclosure for evaluation of corporate disclosure programs. The estimates are conservative because, by SEC instruction, the firms excluded any allocation of overhead expenses in their calculations. In addition, they do not include the agency's costs of administering the system. The latter costs are not great: the SEC's total budget in 1975 was approximately $50 million, and Phillips and Zecher state that no more than one-quarter of it was spent on administering periodic disclosure. They further report that the advisory committee's data indicate that voluntary disclosure costs are ten times larger than the mandatory disclosure costs. Even if some of the costs the committee attributes to voluntary disclosure are misclassified, the difference is still impressive. Phillips and Zecher suggest that this indicates that substantial incentives exist for firms to provide information to investors

apart from the SEC's requirements. Does this bolster the position that disclosure should not be mandatory?

2. George Benston studied the efficacy of disclosure under the 1934 Act, which regulates the information distributed by firms with outstanding issues trading in secondary markets. Benston, "Required Disclosure and the Stock Market: An Evaluation of the Securities Exchange Act of 1934," 63 *American Economic Review* 132 (1973). Sales data was the one financial variable whose disclosure was required by the Act that had not already been reported by all firms; comparing the stock returns of corporations that did and did not disclose sales figures before the Act with their returns thereafter, Benston finds no significant differences. He concludes that the disclosure provisions of the 1934 Act were of "no apparent value" to investors. For a critique of Benston's study, and his rejoinder, see Irwin Friend and Randolph Westerfield, "Required Disclosure and the Stock Market," 65 *American Economic Review* 467 (1975); George Benston, "Required Disclosure and the Stock Market: Rejoinder," 65 *American Economic Review* 473 (1975). If investors did not benefit from the disclosure requirements of the 1934 Act, can you think of any other group who might? Members of the New York Stock Exchange (e.g., might investors increase their trading volume because the legislation restored their "confidence" in capital markets, which had been diminished by the great stock market crash of 1929 and a series of trading scandals that led to its enactment)? See G. W. Schwert, "Public Regulation of National Securities Exchanges: A Test of the Capture Hypothesis," 8 *Bell Journal of Economics* 128 (1977).

3. What is the relationship between market efficiency and disclosure regulation? For an interesting discussion, see William H. Beaver, "The Nature of Mandated Disclosure," in *Report of the Advisory Committee on Corporate Disclosure to the SEC,* 95th Cong., 1st Sess. 618 (1977). Should improving market efficiency be a goal of securities regulation? For a critique of such a goal, see Lynn A. Stout, "The Unimportance of Being Efficient: An Economic Analysis of Stock Market Pricing and Securities Regulation," 87 *Michigan Law Review* 613 (1988). Stout's principal contention is that market efficiency is unimportant because stock prices affect the efficient allocation of resources only when corporations raise capital by issuing new equity, which few firms do. Do you think that this is the only way in which stock prices affect the allocation of resources? There is a substantial body of evidence that stock returns predict changes in corporate investment, although one study finds that, except for the new issues market, the incremental predictive value of stock returns is small when fundamentals (sales growth and cash flows) are held constant. See Randall Morck, Andrei Shleifer, and Robert Vishny, "The Stock Market and Investment: Is the Market a Sideshow?" in 2 *Brookings Papers on Economic Activity* 157 (Washington, D.C.: Brookings Institution, 1990). Stout further considers market efficiency inconsistent with the market integrity and investor protection goals of the federal securities laws. Can a market be "fair and honest" if it is inefficient?

4. Consider the impact of managerial discretion in accounting choices on the effectiveness of mandatory disclosure. Linda DeAngelo finds evidence that managements engaged in proxy fights use accounting choices to report more favorable earnings during the contested election: firms releasing earnings information during the contest reported significantly higher earnings than prior years and the average unexpected accruals of these firms is significantly positive (increases in accruals increase reported earnings; this variable is a measure of managers' exercise of accounting discretion). DeAngelo, "Managerial Competition, Information Costs, and Corporate Governance: The Use of Accounting Performance Measures in Proxy Contests," 10 *Journal of Accounting and Economics* 3 (1988). There is, however, no contemporaneous increase in operating cash flows. This suggests that the firms' real profitability did not increase despite the increase in reported profitability.

The key question is whether management's manipulation of accounting performance figures affects proxy fight outcomes. The data are ambiguous: under some model specifications there is a significant correlation between reported earnings, unexpected accruals, and incumbents' electoral success but not under others. Should we expect to find any correlation? The efficient-market hypothesis suggests that shareholders are not fooled by changes in accounting earnings that are not indicators of changes in cash flows. DeAngelo contends that management would have been able to manipulate information in these contests because, in a majority of the cases, shareholders did not have "sufficient information to separate reported earnings into operating cash flows and accruals" since they did not receive complete financial statements before the election. In addition, they may not be able to anticipate the "extent of manager's available accounting discretion" as proxy fights are rare occurrences. DeAngelo, supra, at 27 n. 21. For an accounting choice study that finds that investors are not fooled by higher reported earnings that are due to the chosen accounting rule rather than real cash flow changes, see Hai Hong, Robert Kaplan, and Gershon Mandelker, "Pooling vs. Purchase: The Effects of Accounting for Mergers on Stock Prices," 53 *Accounting Review* 31 (1978). Ross Watts and Jerold Zimmerman's accounting text provides an excellent survey of studies on the stock price effects of accounting rules. Watts and Zimmerman, *Positive Accounting Theory* (Englewood Cliffs, N.J.: Prentice-Hall, 1986).

5. One use of modern finance theory in securities litigation involves plaintiff's proof of reliance on defendants' misstatements or omissions in civil cases. In anonymous markets like modern securities markets, reliance, a traditional element in common law fraud, is difficult to establish in the conventional sense because there is no privity between parties (investors in these markets do not know who is on the other side of their transactions). This problem has currently been resolved by the Supreme Court's acceptance of what is known as the "fraud-on-the-market" theory for proof of reliance. This doctrine establishes a rebuttable presumption of reliance, that investors rely on the "integrity of the market price," which is expected to reflect all publicly available information about the security (including any misrepresentations by

defendants). Has the Court wholeheartedly endorsed the efficient-market hypothesis, discussed in the Ross, Westerfield, and Jaffe selection in Chapter I? Consider, in this regard, the following footnote in the opinion:

> We need not determine by adjudication what economists and social scientists have debated through the use of sophisticated statistical analysis and the application of economic theory. For purposes of accepting the presumption of reliance in this case, we need only believe that market professionals generally consider most publicly announced material statements about companies, thereby affecting stock market prices.

Basic v. Levinson, 485 U.S. 224, 246–47 n. 24 (1988). Should the fraud-on-the-market presumption apply only to securities traded in thick capital markets like the New York Stock Exchange? Lower courts are divided on this question. For analyses that such a dichotomy is inappropriate, see Jonathan R. Macey and Geoffrey P. Miller, "Good Finance, Bad Economics: An Analysis of the Fraud on the Market Theory," 42 *Stanford Law Review* 1059 (1990); Jonathan R. Macey, Geoffrey P. Miller, Mark L. Mitchell, and Jeffry M. Netter, "Lessons from Financial Economics: Materiality, Reliance, and Extending the Reach of Basic v. Levinson," 77 *Virginia Law Review* 1017 (1991). Note that economists distinguish between internal and external market efficiency: That is, a market may be informationally efficient ("external" efficiency) but illiquid (the transaction costs of trading in the market are high; this is an "internal" efficiency measure). See Richard R. West, "On the Difference Between Internal and External Market Efficiency," 3 *Financial Analysts Journal* 30 (Nov.–Dec. 1975). European capital markets, for example, are very thin in comparison to U.S. markets, but empirical research finds that they are equally efficient in processing information. See Bruno Solnik, *International Investments,* 2d ed. (New York: Addison-Wesley, 1991).

6. As Easterbrook and Fischel note, the jurisdictional basis of state regulation of securities differs from corporate law: it depends on the site of the securities transaction and not the issuer's domicile. Another difference is that state securities laws are mandatory and not enabling statutes. Moreover, in contrast to the federal disclosure regime, many states engage in merit regulation: Registration is conditioned on a security's meeting a standard of investment worthiness. Why would states enact enabling corporation statutes and paternalistic securities regimes?

Common criteria used to evaluate a security's merit include the fairness of the offering price, underwriter fees and insider compensation, and the distribution of voting rights. Do the latter criteria suggest that merit regulation is a means to regulate corporate governance by the back door (that is, by a state securities commissioner's discretion rather than a corporation code), and hence that it could offset or blunt state competition for charters discussed in Chapter III? Would you expect states active in the corporate chartering market to be pioneers in adopting such regulation? See Jonathan R. Macey and Geoffrey P. Miller, "Origin of the Blue Sky Laws," 70 *Texas Law Review* 347 (1991). The reach by states into the governance of foreign corporations via securities regulation is, however, limited. This is because most states exempt from their regulation securities listed on a national exchange, as well as secondary market trading. In addition, major commercial states, such as New York, do not have merit regulation. Why do you think this is so? (Hint: Where are the major financial institutions that underwrite new securities located?) Finally, firms can avoid a state's merit regulation by not offering shares in that state. The importance of merit regulation in relation to the issues raised in Chapter III concerning the benefits of state competition depends, then, on whether the state has a substantial pool of potential investors. Would it be cause for concern if venture capital funds (see the Sahlman selection in Chapter IV) enable firms to reach investors residing in merit states without having to meet the state's merit criteria? How would you compare the investment screening function of venture capital fund managers with state securities commissioners?

The efficacy of merit regulation has been continually debated, as it entails a trade-off of the additional benefit accruing to investors in stemming promoter fraud from the use of merit instead of disclosure regulation against the increased cost of capital to new firms and consequent opportunity losses of businesses that cannot finance operations. For a recent review of the arguments, see the report of the Ad Hoc Subcommittee on Merit Regulation of the State Regulation of Securities Committee of the American Bar Association, "Report on State Merit Regulation of Securities Offerings," 41 *Business Lawyer* 785 (1986). Although there have been no studies of the cost of merit regulation, several studies have sought to evaluate the benefit by comparing securities registered in one state with securities either denied registration or withdrawn from that state and sold in another state. A common finding is that the return to nonmerit state securities is higher in the short term (one day, one month, one year, depending on the study) but lower in the long term (one year, three years) than the return to the merit state securities, although at least one study has found no significant difference. Where the risk of the two sets of securities is measured and the returns differ significantly, the nonmerit state securities are riskier. These findings are consistent with the extensive finance literature studying the stock price reaction of initial public offerings (IPO): IPOs on average exhibit positive abnormal returns over a short period of time after issuance but the returns steadily decline over the long term. For a comprehensive review of the merit regulation and IPO studies, see David J. Brophy and Joseph A. Verga, "The Influence of Merit Regulation on the Return Performance of Initial Public Offerings," University of Michigan School of Business Administration Working Paper No. 91-19 (1991). What does the finding that issues registered in merit regulation states are less risky than those registered in other states indicate about the efficacy of merit regulation? Does the assessment depend on whether investors living in states with merit regulation are more risk averse than investors in nonmerit states? Why should investors be prohibited from investing in securities deemed by a state official to be excessively risky? Does the regulatory rationale depend on a belief that state officials are better able to evaluate the information firms disclose concerning investment risk than investors?

B

Insider Trading Regulation

Insider Trading, Rule 10b-5, Disclosure, and Corporate Privacy

KENNETH E. SCOTT

[It is possible] to extract three different conceptions of [Rule 10b-5] and its objectives. The first and most common view is that the rule is principally intended to serve the ends of fairness and equity—to prevent, in the words of the commission, "the inherent unfairness involved where a party takes advantage of [inside] information knowing it is unavailable to those with whom he is dealing." This view we will denominate the Fair Play or "fair game" concept of the rule. It implies that the damaged party, who should be made whole by a remedy of rescission, is the one with whom the insider traded and of whom the insider took unfair advantage.

A second view of the rule is that it facilitates the flow of information to the market, so that it may better perform its functions of security evaluation and capital allocation. . . . Though not unrelated to the first concept, the focus here is on the entire market rather than the particular trading partner, and the implication is that damages were incurred by all investors who traded in the market in the opposite direction from the insiders during the period of nondisclosure. This rationale for the rule we will refer to as the Informed Market.

The third view of the function of the rule is that it affords protection to the property rights of the firm in inside information, which was described by the commission in its pathbreaking *Cady, Roberts* decision as "information intended to be available only for a corporate purpose and not for the personal benefit of anyone." This function is quite clear in a case like *Texas Gulf Sulphur,* where unusual trading by the drill-site geologist and others with knowledge of the assay analysis of the drill core would be likely to contribute to rumors of a strike and raise the costs of acquisition of the surrounding land by the company. This Business Property view, however, implies that the injured party was the company, and that damages would be better measured by the increase in land acquisition costs than by stock market price movements. . . .

Reprinted by permission from 9 *Journal of Legal Studies* 801. © 1980 by The University of Chicago.

Fair Play

The fair game approach to Rule 10b-5, which seems to have some kinship with
the layman's attitude that the stock market is just another form of gambling,
focuses on the individual parties to a particular trade and asks whether one
has an "unfair advantage" over the other in some respect. If pushed far
enough, of course, it will always be found that the parties are not on a parity
in all regards; there will be disparities in knowledge or intelligence or experience or capital or whatever. Among all these advantages and disadvantages,
which are "unfair" and why? . . .

In terms of our gambling analogy, secondary trading is apparently seen as a
zero-sum game in which insiders are playing with "percentage dice" and therefore winning abnormally often. . . . But is the conclusion from the gambling
analogy a valid one? . . . From this [modern finance (portfolio) theory—ED.]
perspective, the limited knowledge or ability of the individual investor is largely
irrelevant. He is "protected" by the price established by the market mechanism, not by his personal bargaining power or position. . . .

That is not to deny that those with superior (nonpublic) information can
reap higher returns. In a sense, they are "selling" their information and cor-
recting the market price in the process. Are other investors, then, not receiving the expected rates of return? Insider trading is hardly an unknown or
unanticipated phenomenon; the returns expected by investors would not include any gains unique to insiders. To return to the gambling analogy, if I
know you are using percentage dice, I won't play without an appropriate
adjustment of the odds; the game is, after all, voluntary.

Further, the excess returns received by insiders do not necessarily represent some sort of "unfair" or windfall gain for them. If a certain corporate
position carries with it the prospect of being able to obtain on occasion some
insider trading profits, the value of that prospect constitutes part of the total
compensation attached to that position and, like any other fringe benefit,
affects the level of direct salary payments. The opportunities for significant
insider trading profits are no doubt rather infrequent and unpredictable, however, so that management is in effect receiving long-shot lottery tickets as part
of its pay, and that may well be an inefficient form of compensation. The value
of such lottery tickets to their recipients may be substantially less than their
expected cost to the owners of the firm, but that is a different issue.

Manne has tried to make a stronger argument against the *TGS* rule[1] and in
favor of insider trading. His main thesis is that "profits from insider trading
constitute the only effective compensation scheme for entrepreneurial services in large corporations." The reason that insider trading profits are the
only effective means of compensation appears to be that they are measured,
or at least bounded, by the market's evaluation of the worth of the entrepreneur's innovation or discovery. Neither lawyers nor economists seem to have

---

[1]The rule of SEC v. Texas Gulf Sulphur, 401 F.2d 833 (2d Cir.), cert. denied sub nom. Coates v.
SEC, 394 U.S. 976 (1968) ("TGS"), is that insiders must disclose their inside information or
abstain from trading [EDITOR'S NOTE].

found this contention especially persuasive, judging by the subsequent literature. There are many forms of performance-related or stock-price-based compensation schemes, which do not depend on the recipient's capital resources
or skills in selling information to determine the amount realized. Nor is it
apparent that insider trading bears much relation, either in theory or in the
reported decisions, to entrepreneurial innovation. . . .

The Informed Market

In this conception of the rule, it is seen as intended to enhance the flow of
information to the market, to contribute to the accurate pricing of securities
and the efficient allocation of capital resources. . . . [L]et us examine further
the proposition that the *TGS* rule does serve to reduce, at least to a small
degree, delays in disclosure of available corporate information. When the
information is positive (giving rise to an increase in stock price), the proposition is plausible; if insiders cannot profit from a trading delay, they otherwise
have ample incentives to release promptly the information (after any requisite
corporate action, such as the land acquisition in *TGS,* is completed). Good
news benefits stockholders, which usually include the insiders, and correlates
with increases in management compensation. But if the news is bad, the
immediate incentives for insiders now point in the other direction, and therefore it is to be expected that one should be quite sure of the facts before
making a release, which should be framed to avoid overreactions by ill-
informed investors. In this situation, the *TGS* rule does not help, since insiders can delay or avoid the negative disclosure simply by not trading. Indeed,
the rule makes the situation somewhat worse, for by cutting off insider selling
it also cuts off an activity that is itself a source of information to the marketplace and removes an incentive for full disclosure promptly upon completion
of trading. . . .

Business Property

Many of the Rule 10b-5 decisions stress the importance of the defendant's
having acquired the information in some fiduciary capacity, knowing that it
was intended for a corporate purpose and not for personal use and benefit. In
this view, the wrong committed is essentially that of theft or conversion. The
information belongs to the firm, but an employee appropriates it for his own
use and gain. . . .

The application of this concept to a case like *TGS* is quite straightforward.
By buying on their knowledge of the strike, the trading insiders fed the rumor
mill to the likely detriment of their employer. What Manne saw as entrepreneurial compensation comes out in this instance as an agent acting in his own
interest and contrary to the interest of his principal—what Jensen and
Meckling refer to as an agency cost. The application is also clear in a case like

Chiarella, where the Second Circuit found it awkward to explain why the acquiring company could make market purchases with advance knowledge of the tender offer while the layout man could not: The trading printer tended to alert the target company and to drive up the price of its stock, to his own ultimate gain but to the detriment of the acquirer, from whom he appropriated the information of the impending bid. . . .

In both cases the firm had invested resources to make socially useful discoveries—new mineral deposits and companies that could be made more productive and profitable. Applying Rule 10b-5 in such situations to debar insider trading merely serves to protect the firm's property rights in the discovery it has made. Presumably express provisions in employment and other written contracts, where they exist, could accomplish much the same objective, but the incorporation of a legal rule may well be the more efficient method. . . .

It is by no means easy to measure the trade-off between costs and benefits in the various situations to which Rule 10b-5 applies or can be applied. My own impression is that application to protect investments in socially valuable discoveries is justifiable, but beyond that the case becomes increasingly dubious. Certainly the process of broadening the rule through simple extensions of the logic of the Fair Play and Informed Market rationales would take us into realms where the costs far exceed any discernible benefits.

The Regulation of Insider Trading

DENNIS W. CARLTON AND DANIEL R. FISCHEL

Insider Trading and the Coase Theorem

Critics of insider trading draw a sharp distinction between the proper legal response to insider trading and to other forms of managerial compensation. Salaries, bonuses, stock options, office size, vacation leave, secretarial support, and other terms of employment are all, it is generally assumed, properly left to private negotiation. Nobody would argue seriously that these terms and conditions of employment should be set by government regulation. . . .

Precisely the opposite presumptions have been applied with respect to insider trading. Most believe that existing government regulation is necessary and should be extended; virtually no one has considered the possibility, let alone has argued, that private negotiations between a firm and its employees can most efficiently determine whether insiders should be allowed to profit by trading on inside information.

Does whatever difference that exists between profits from trading in shares and other forms of compensation warrant such different legal responses? . . . If it is bad, firms that allow insider trading will be at a competitive disadvantage compared with firms that curtail insider trading.

Coase's famous insight is quite relevant in this regard. Whether insider trading is beneficial depends on whether the property right in information is more valuable to the firm's managers or to the firm's investors. In either case, the parties can engage in a value-maximizing exchange by allocating the property right in information to its highest-valuing user. If the critics of insider trading are correct, therefore, both the firm's investors and the firm's insiders could profit by banning insider trading, thereby allocating the property right in information to the firm's investors. . . .

The preceding discussion assumes, of course, that transaction costs will not interfere with the optimal allocation of property rights. While the costs of negotiating contracts banning insider trading in the employer-employee situation appear to be low, some have argued that the costs of enforcing such contracts are high. Firms must encourage managers to own shares, the argument runs, to induce them to act in shareholders' best interests. Once having permitted share ownership by managers, the firm, because it cannot separate proper from improper trades, cannot adequately enforce a rule against insider trading. These high enforcement costs render firms unable to prohibit insider trading, even if doing so would benefit all parties. The only practical method to ban insider trading, the argument concludes, is public enforcement with large penalties. . . .

The argument overstates the problems with private enforcement because it assumes that managers must own shares to be induced to act in shareholders' best interests. There is no reason, however, why this must be the case. The strategy, presumably, is to ensure that managers have a stake in the venture so that they will profit from good performance and lose from bad performance. But this incentive effect in no way depends on the actual ownership of shares. Firms are perfectly free to base compensation on share performance and thus create incentives for managers to increase the value of their firms even if managers own no shares. If managers are forbidden from owning shares, the problem of separating proper from improper trades disappears. Alternatively, firms could allow managers to own shares, but not trade them. That firms generally do not eliminate insider trading by employing these alternative methods for controlling share ownership while linking managers' fortunes with those of the firm again suggests that the explanation for the absence of such prohibitions is that they are inefficient, not that they are unenforceable. . . .

Information Effects

If insiders trade, the share price will move closer to what it would have been had the information been disclosed. How close will depend on the amount of "noise" surrounding the trade. The greater the ability of market participants to identify insider trading, the more information such trading will convey. . . .

Several reasons explain why communicating information through insider trading may be of value to the firm. Through insider trading, a firm can convey information it could not feasibly announce publicly because an announcement would destroy the value of the information, would be too expensive, not believable, or—owing to the uncertainty of the information—would subject the firm to massive damage liability if it turned out ex post to be incorrect. . . .

Efficiency Effects

Contracts that provide for periodic renegotiations ex post based on (imperfectly) observed effort and output are alternatives to contracts that ex ante tie compensation to output. Such renegotiations are constrained by the difficulty of monitoring the effort and measuring the output of individual managers, and the bargaining process itself is costly. To reduce these costs, firms seek to minimize the number of renegotiations. But reduction in the number of renegotiations itself creates a cost. If renegotiations occur too infrequently, they are less likely to exert the proper incentives at any given time. . . .

Insider trading may present a solution to this cost-of-renegotiation dilemma. The unique advantage of insider trading is that it allows a manager to alter his compensation package in light of new knowledge, thereby avoiding continual renegotiation. The manager, in effect, "renegotiates" each time he trades. This in turn increases the manager's incentive to acquire and develop valuable information in the first place (as well as to invest in firm-specific human capital). If a manager observes a possible valuable investment for the firm—such as a potential value-increasing merger or a possible new technology—he will be more inclined to pursue this opportunity if he is rewarded upon success. Insider trading is one such reward. The alternative is to tell others of the opportunity, explain that it can be realized with extra effort, and hope to be compensated by some form of ex post settlement. The insider trading alternative reduces the uncertainty and cost of renegotiation and thus increases the incentives of managers to produce valuable information. Moreover, because managers themselves determine the frequency of "renegotiations," they can tailor their compensation scheme to their particular attitudes toward risk.

A related advantage of insider trading is that it provides firms with valuable information concerning prospective managers. It is difficult for firms to identify those prospective managers who will work hard and not be overly risk averse in their choice of investment projects. Basing compensation in part on insider trading is one method for sorting superior from inferior managers. Because insider trading rewards those managers who create valuable information and are willing to take risks, managers who most prefer such compensation schemes may be those who are the least risk averse and the most capable. Thus, with insider trading, self-selection mini-mizes the costs of screening potential managers, the monitoring costs created by risk-averse managers, and the opportunity costs resulting from subopti-mal investment decisions.

Critics of insider trading correctly point out that compensation schemes that allow insider trading could backfire, particularly if short selling is permit-ted, since managers would have incentives to reduce the value of the firm. We think that the force of this argument has been exaggerated. . . . Insiders are worried about the value of their human capital. If a project succeeds, the insider's value as a manager is increased. But if the project fails, even if the investment was optimal ex ante, the manager will suffer a loss in the value of his human capital because he may be blamed for the failure (the usual monitor-ing problem). To avoid this loss, managers will tend to accept investment projects that reduce volatility of cash flows even if they do not maximize the value of the firm. By permitting managers to sell short and thereby profit from investment projects that are optimal ex ante, even if they do not turn out well ex post, insider trading may induce managers to take on projects with a high expected return even if they are riskier. The ability to profit by selling, there-fore, as well as the ability to profit by buying, may reduce divergence of interests between managers and shareholders by causing managers to behave in a less risk-averse manner. . . .

Fairness Arguments

We have left for last the most common argument against insider trading—that it is unfair or immoral. The prevalence of this intuition is so powerful that many commentators have argued that insider trading should be prohibited even if it is efficient. What is commonly left unsaid is how and why insider trading is unfair.

Kenneth Scott has pointed out that if the existence of insider trading is known, as it surely is, outsiders will not be disadvantaged because the price they pay will reflect the risk of insider trading. This is a useful insight and in some sense is a complete response to the claim that investors are exploited by insider trading. But the argument does not address the desirability of insider trading. If traders knew that a firm burned half of its assets, the price would fall and subsequent investors in the firm would have the same expected re-turns as any other asset of comparable risk. But the fact that investors would not be fooled does not mean that burning assets is a beneficial practice. On the contrary, firms that followed this strategy to a substantial degree, like firms that adopted inefficient compensation schemes, could not survive over time.

A more powerful response to the argument that insiders profit at the expense of outsiders is that if insider trading is a desirable compensation scheme, it benefits insiders and outsiders alike. Nobody would argue seriously that salaries, options, bonuses, and other compensation devices allow insiders to profit at the expense of outsiders because these sums otherwise would have gone to shareholders. Compensating managers in this fashion increases the size of the pie, and thus outsiders as well as insiders profit from the incentives managers are given to increase the value of the firm. Insider trading does not come "at the expense of" outsiders for precisely the same reason. . . .

Summary and Conclusion

Both the common law and state law place few, if any, restraints on insider trading. . . . No evidence suggests that firms generally have attempted to prohibit insider trading or, after 1934, attempted to plug the large gaps in the federal bans against insider trading. . . . [F]ederal regulation of insider trading would be justifiable only if the federal regulations could enforce contracts against insider trading at a lower cost than private firms or states. The evidence does not conclusively support (or refute) this conclusion.

Even if federal regulation is justified on the basis of low enforcement costs, firms should have the opportunity to opt out of the regulation in the absence of any showing of third party effects. Firms are the best judges of how to structure the terms of their employment contracts.

Notes and Questions

1. Carlton and Fischel propose permitting firms to opt out of the insider trading rules. Are the arguments presented in Chapter III in support of the enabling posture of state corporation codes apposite for federal regulation? What would the Coase Theorem imply regarding the merits of insider trading regulation?

2. Consider the incentive compensation arguments against insider trading regulation from the perspective of portfolio theory (Chapter I). Is insider trading an efficient compensation scheme if managers are less well diversified than shareholders (because the bulk of their wealth is in their human capital, which is tied up in the firm)?

3. Is absence of corporate contracts banning insider trading prior to the 1934 Act a complete answer to the question whether the practice is efficient? What other considerations could effect the writing of such contracts? Might detection be so difficult that very severe penalties, which private corporations cannot impose, are necessary to deter misconduct? See Frank H. Easterbrook, "Insider Trading as an Agency Problem," in J. Pratt and R. Zeckhauser, eds., *Principals and Agents: The Structure of Business* 81 (Boston: Harvard Business School, 1985). Consider Michael Dooley's finding, in a

study of all reported insider trading cases from 1966 to 1979, that the legal system is "ineffective" at regulating insider trading. Dooley, "Enforcement of Insider Trading Restrictions," 66 *Virginia Law Review* 1 (1980). Would you expect to find the same results in a study of insider trading after Congress's enactment of increased sanctions for insider trading in 1984 and 1988? See H. Nejat Seyhun, "The Effectiveness of Insider-Trading Sanctions," 35 *Journal of Law and Economics* 149 (1992).

4. Who benefits from the ban on insider trading? Note that the only investors who lose out to insiders trading on good (bad) news are those who have decided to sell (buy). Who is more likely to obtain inside information first, the average individual investor or a market professional (analysts, arbitrageurs, investment advisers, and brokers)? For the thesis that prohibiting managers from trading simply passes the informational advantage on to market professionals, see David Haddock and Jonathan Macey, "Regulation on Demand: A Private Interest Model with an Application to Insider Trading Regulation," 30 *Journal of Law and Economics* 311 (1987).

5. Carlton and Fischel question how widespread insider trading regulation is. Is it relevant that insider trading has only newly been prohibited in other countries, such as Japan and those of the European Community? Will the globalization of securities markets produce uniform insider trading regulation? Would you expect there to be international competition in a "race to the bottom" or "top" in this context?

6. Can the market efficiency argument of Carlton and Fischel against insider trading regulation be reconciled with the informed market rationale, discussed by Scott, in support of regulation? Why should third parties dealing with corporations, such as property owners in Timmons, Ontario, have less protection concerning misstatements or omissions than shareholders? For a good discussion of the issue of differential rules on corporate silence, comparing the disclosure requirements of securities and contract law, see Saul Levmore, "Securities and Secrets: Insider Trading and the Law of Contracts," 68 *Virginia Law Review* 117 (1982). Can shareholders benefit from a rule preventing corporations from misleading third parties in negotiations? See Marcel Kahan, "Games, Lies, and Securities Fraud," 67 *New York University Law Review* 750 (1992).

Lisa Muelbroek studied the stock price effects of insider trading detected by the SEC and subsequently cited in civil cases during the 1980s. She finds that the trades are associated with immediate significant price movements, generating abnormal returns of the same sign as the abnormal returns on the day the inside information is released, which supports the Manne and the Carlton and Fischel position that insider trading aids market efficiency. Muelbroek, "An Empirical Analysis of Illegal Insider Trading," 47 *Journal of Finance* 1661 (1992). Recall the discussion of the evidence against strong-form market efficiency in the Ross, Westerfield, and Jaffe selection in Chapter I, that several studies have found that a trader could realize an abnormal profit if she traded when insiders did.

7. The materiality standard under Rule 10b-5 is whether there is a substantial likelihood that investors would have considered the information important. If an event study of a corporation's press release identifies a significant abnormal return, should this be sufficient evidence to prove that insiders trading before the release had traded on material information? How could you use such a study to calculate the defendants' damages? Calculation of damages in securities cases can be a thorny issue from a theoretical perspective. For the view that courts get it right in practice, see Frank H. Easterbrook and Daniel R. Fischel, "Optimal Damages in Securities Cases," 52 *University of Chicago Law Review* 611 (1985).

Event studies are, in fact, used in securities litigation both to establish materiality and to measure damages for affirmative misrepresentations or material omissions. For an analysis of the SEC's use of event-study methodology in insider trading cases, see Mark Mitchell and Jeffry Netter, "The Role of the Efficient Market Hypothesis in Insider Trading and Other Securities Fraud Cases at the Securities and Exchange Commission," Center for Research in Securities Prices Working Paper, University of Chicago Graduate School of Business and Terry School of Business, University of Georgia (1992). When would you expect defendants to use the event-study methodology?

8. The principal issue in insider trading cases in recent years has involved trading by nonconventional insiders, such as market professionals, who are not affiliated with the issuer and hence have only an attenuated relation with the business property rationale, emphasized by Scott, from the perspective of a fiduciary duty to trading shareholders. Are the rules against corporate insiders' trades sufficiently clear-cut so as to reduce uncertainty over litigation outcomes? Consider also that the critical information during the 1980s involved corporate takeovers, where the source of material information is the bidder and not the issuer. For the SEC's response, see Chapter VI, part A, note 16.

9. Stock exchange specialists and market makers in over-the-counter stocks are paid for providing liquidity (that is, for permitting immediate execution of investors' transactions) by what is referred to as the bid-ask spread, which is the difference between the buying premium—the price at which they will buy shares—and the selling concession—the price at which they will sell. If these market professionals are concerned that they may be trading against insiders, could they protect themselves by increasing the spread? Would this be a persuasive rationale for prohibiting insider trading? Could insiders profit by becoming market makers themselves and offering a lower spread?